# ENVISIONING LANDSCAPES, MAKING WORLDS

T0179045

The past decade has witnessed a remarkable resurgence in the intellectual interplay between geography and the humanities in both academic and public circles. The metaphors and concepts of geography now permeate literature, philosophy, and the arts. Concepts such as space, place, landscape, mapping and territory have become pervasive as conceptual frameworks and core metaphors in recent publications by humanities scholars and well-known writers.

*Envisioning Landscapes, Making Worlds* contains over 25 contributions from leading scholars who have engaged this vital intellectual project from various perspectives, both inside and outside of the field of geography. The book is divided into four sections representing different modes of examining the depth and complexity of human meaning invested in maps, attached to landscapes, and embedded in the spaces and places of modern life. The topics covered range widely and include interpretations of space, place, and landscape in literature and the visual arts, philosophical reflections on geographical knowledge, cultural imagination in scientific exploration and travel accounts, and expanded geographical understanding through digital and participatory methodologies. The clashing and blending of cultures caused by globalization and the new technologies that profoundly alter human environmental experience suggest new geographical narratives and representations that are explored here by a multidisciplinary group of authors.

This book is essential reading for students, scholars, and interested general readers seeking to understand the new synergies and creative interplay emerging from this broad intellectual engagement with meaning and geographic experience.

**Stephen Daniels** is Professor of Cultural Geography at the University of Nottingham, UK.

**Dydia DeLyser** is Associate Professor of Geography at Louisiana State University, USA.

**J. Nicholas Entrikin** is Vice President and Associate Provost for Internationalization at the University of Notre Dame, USA.

**Douglas Richardson** is Executive Director of the Association of American Geographers, USA.

"This book provides powerful evidence of geography's intellectual and moral affiliations with the humanities. It boasts an impressive cast of contributors, with elegant and compelling essays that show why creativity, imagination and reflection matter to geographers, and why the insights of geography matter to the humanities as never before."

Professor Felix Driver, *Royal Holloway, University of London, UK.*

"For geography, this book vigorously promotes the significance of the powers of spatial and visual representation in evoking landscapes and places. For the humanities, it elegantly maps the variety of ways in which geographical concepts are helping respond to the so-called crisis of representation by grounding texts, performances, and visual art in landscapes and places."

Professor John Agnew, *UCLA, USA.*

# ENVISIONING LANDSCAPES, MAKING WORLDS

## Geography and the humanities

*Edited by*
*Stephen Daniels, Dydia DeLyser,*
*J. Nicholas Entrikin and Douglas Richardson*

Routledge
Taylor & Francis Group

LONDON AND NEW YORK

First published 2011
by Routledge
4 Park Square, Milton Park, Abingdon, Oxon OX14 4RN
605 Third Avenue, New York, NY 10017

*Routledge is an imprint of the Taylor & Francis Group, an informa business*

*British Library Cataloguing in Publication Data*
A catalogue record for this book is available from the British Library

*Library of Congress Cataloging in Publication Data*
  Envisioning landscapes, making worlds/Association
  of American Geographers.
    p. cm.
  Includes bibliographical references and index.
  1. Human geography. 2. Landscape assessment. I. Association
  of American Geographers.
  GF41.E556 2010
  304.2′3–dc22
                                                    2010028082

ISBN: 978-0-415-58977-2 (hbk)
ISBN: 978-0-415-58978-9 (pbk)
ISBN: 978-0-203-83928-7 (ebk)

Typeset in Bembo
by RefineCatch Limited, Bungay, Suffolk

# CONTENTS

# LIST OF ILLUSTRATIONS

## Figures

## Colour Plates (*Between pp. 160-161*)

# ACKNOWLEDGMENTS

The editors would like to thank the National Endowment for the Humanities and the Virginia Foundation for the Humanities for providing financial support for this book project and related activities. The editors also thank Ed Ayers, formerly of the University of Virginia (and now President of the University of Richmond), who co-hosted the AAG's 2007 Geography & Humanities Symposium at the University of Virginia. We are also grateful for the support of our editors Andrew Mould and Faye Leerink of Routledge and also to Candida Mannozzi of the Association of American Geographers for guiding the book through the process of compilation and proofing, assisted by colleagues Miranda Lecea, Megan Overbey, and Marcela Zeballos. Our special thanks go to artist Chase Langford for generously granting permission to use his artwork "Half Moon Bay" for the cover of this book.

# CONTRIBUTORS

**Stephen Cairns** has a Chair in Architectural and Urban Design at ESALA (Edinburgh School of Architecture and Landscape Architecture). His publications include *Drifting: Architecture and Migrancy* (Routledge 2003) and *The SAGE Handbook of Architectural Theory* (2010), and he co-curated the Reciprocity exhibit at the 2009 International Architecture Biennale Rotterdam (see *Open City: Designing Coexistence*, SUN, 2009).

**Edward S. Casey** is Distinguished Professor at SUNY Stony Brook and immediate past President of the American Philosophical Association, Eastern Division. In his work, he has taken inspiration from phenomenology, poststructuralism, and psycho-analysis. Since the early 1990s, he has focused on the role of place in people's lives. His primary books on place are *Getting Back into Place* (1993; 2nd edn, 2009) and *The Fate of Place* (1997). Casey extended his work on place in two linked volumes, *Representing Place in Landscape Painting and Maps* (2002) and *Earth Mapping: Artists Reshaping Landscape* (2005).

**Jim Cocola** is an Assistant Professor of Literature, Film, and Media in the Department of Humanities and Arts at Worcester Polytechnic Institute, in Worcester, Massachusetts. He has published articles, essays, and reviews in *Discourse, the minnesota review, n+1*, and *SEL: Studies in English Literature 1500–1900*. His work on Neruda draws and expands upon material extracted from his dissertation, "Topopoiesis: Contemporary American Poetries and the Imaginative Making of Place" (University of Virginia, 2009), completed under the auspices of the Georgia O'Keeffe Museum Research Center in Santa Fe, New Mexico.

**Tim Cresswell** is Professor of Human Geography at Royal Holloway, University of London. His research interests are broadly in issues surrounding the conceptualization of place and mobility, spatialities of ordering and geographical theory more generally. He is the author of *On the Move: Mobility in the Modern Western World* (Routledge, 2006), *Place: A Short Introduction* (Blackwell, 2004), *The Tramp in America* (Reaktion, 2001), and *In Place/Out of Place: Geography, Ideology and Transgression* (Minnesota, 1996). He has co-edited four volumes, most recently, *Gendered Mobilities* (Ashgate, 2008).

**Stephen Daniels** is Professor of Cultural Geography at the University of Nottingham. He is also Director of the Arts & Humanities Research Council (AHRC), where he is currently involved in research and project work on geographical education and citizenship in eighteenth-century England, and cultural geography of nineteenth-century arboretums. Daniels has co-edited several books, among them *Exploring Human Geography: A Reader* (Oxford University Press), and is the author of *Fields of Vision (Human Geography)* (Polity Press). He has contributed to many scholarly journals such as the *British Journal for the History of Science*, and *Cultural Geographies*. Daniels' other research interests include river history and representation, the cultural geography of gardens, and the art of Paul Sandby (1731–1809).

**Diana K. Davis** is an associate professor of History at the University of California at Davis where she teaches environmental history. She received her Doctorate of Veterinary Medicine in 1994 from Tufts University and her PhD in Geography in 2001 from UC Berkeley. She has conducted research with Afghan and Moroccan nomads and worked extensively in the British and French archives. Her first book is *Resurrecting the Granary of Rome: Environmental History and French Colonial Expansion in North Africa* and she has published articles in many scholarly journals. She has been named a Guggenheim and Ryskamp (ACLS) Fellow for her new research on imperialism and environmental history in the Middle East.

**Veronica della Dora** is Lecturer in Geographies of Knowledge at the School of Geographical Sciences, University of Bristol. She is the author of *Imagining Mount Athos: Visions of a Holy Place from Homer to World War II* (University of Virginia Press, 2011) and co-editor with Denis Cosgrove of *High Places: Cultural Geographies of Mountains, Ice, and Science* (IB Tauris, 2008). Her research interests and publications span cultural and historical geography, history of cartography, Byzantine and post-Byzantine studies, and science studies, with a particular focus on the Eastern Mediterranean region.

**Dydia DeLyser** is Associate Professor of Geography at the Louisiana State University, and North American editor of *Cultural Geographies*. Her work on landscape and social memory, as well as on qualitative research and writing, has appeared in journals and book chapters. Her book, *Ramona Memories: Tourism and the Shaping of southern California* was published in 2005 (University of Minnesota Press); a second book, *The Handbook of Qualitative Geography*, edited by DeLyser along with Stuart Aitken, Steve Herbert, Mike Crang, and Linda McDowell was published in 2009 (Sage).

**Jessica Dubow** is a lecturer in Cultural Geography at the University of Sheffield. She is author of *Settling the Self: Colonial Space, Colonial Identity and the South African Landscape*. Her journal publications include, *Critical Inquiry, New German Critique, Art History, Journal of Visual Culture* and *Interventions: The International Journal of Postcolonial Studies*. Her current research focuses on the relation between spatial mobility and exilic thought in Judaic philosophy and a Jewish-European intellectual tradition.

**J. Nicholas Entrikin** is Vice President for Internationalization and Professor of Sociology at the University of Notre Dame. He is also Professor Emeritus of Geography at UCLA. He has been a Guggenheim Fellow, a Visiting Director of Research with the Centre National de la Recherche Scientifique (CNRS), and a Faculty Fellow at the Yale Center for Cultural Sociology. He is the author of *The Betweenness of Place* (1991), co-editor with John Agnew of *The Marshall Plan Today: Model and Metaphor* (2004), and the editor of *Regions* (2008). His articles and book chapters have appeared in both English- and French-language publications.

**Franco Farinelli** is the President of the Corso di Laurea Magistrale in Geography and Territorial Processes, University of Bologna, where he is a Professor in the Department of Communication Studies. He has published widely on issues of geographic epistemology and the history of geographical thought. He has held visiting positions at the University of Geneva, UCLA, University of California, Berkeley, and the University of Paris IV, (Sorbonne). He is Vice-President of the Association of Italian Geographers.

**Matthew Gandy** is Professor of Geography at University College London and Director of the UCL Urban Laboratory. He has published widely on cultural, urban and environmental themes. His book *Concrete and Clay: Reworking Nature in New York City* (MIT Press, 2002) was co-winner of the 2003 Spiro Kostof award for the book "that has made the greatest contribution to our understanding of urbanism and its relationship with architecture." He is currently writing a book on cultural histories of urban infrastructure. He has held visiting positions at Columbia University, Humboldt University, Newcastle University and UCLA.

**Derek Gregory** is Distinguished University Scholar and Professor of Geography at the University of British Columbia at Vancouver. The author of *The Colonial Present: Afghanistan, Palestine and Iraq* (2004) and the lead editor of the fifth edition of *The Dictionary of Human Geography* (2009), his recent research focuses on late modern war. He has completed a major study of its cultural and political geographies (*War Cultures*), and is presently working on a geography of aerial bombing, *Killing Space*.

**Michael Heffernan** is Professor of Historical Geography at the University of Nottingham in the UK. He has previously taught and researched at Loughborough University in the UK, the University of Heidelberg in Germany (where he was an

Alexander von Humboldt Research Fellow), and as a Visiting Professor at UCLA. He is interested in the history of geography and cartography in Europe and North America from the eighteenth to the twentieth centuries. He edited the *Journal of Historical Geography* from 1996 to 2005. His most recent book is *The European Geographical Imagination* (2007).

**Sheila Hones** is a Professor in the Department of Area Studies (North American Division) of the Graduate School of Arts & Sciences, The University of Tokyo, Japan. Her main area of interest is the interaction of texts and geography, particularly in relation to American fiction and to geographies of academic practice in American Studies. www.http://sheilahones.wordpress.com

**Jane M. Jacobs** has a Chair in Cultural Geography at University of Edinburgh. Her research interests fall into two broad, and sometimes related, areas: postcolonial geographies and geographies of architecture. Her publications include: *Edge of Empire: Postcolonialism and the City* (Routledge, 1996), *Uncanny Australia: Sacredness and Identity in a Postcolonial Nation* (Melbourne University Press, 1998), and *Cities of Difference* (Guilford, 1998).

**Keith D. Lilley** is Reader in Historical Geography at Queen's University Belfast. Using particularly mapping and cartography as interpretative frames, his research focuses on the materiality and imagining of space, place, and landscape during the later Middle Ages. His publications include *Urban Life in the Middle Ages, 1000– 1450* (London, 2002), and *City and Cosmos – the Medieval World in Urban Form* (London, 2009). He has also directed a series of funded research projects exploring medieval mappings, each of which has led to the creation of innovative web-based digital resources, such as "Mapping Medieval Chester" (see www.medievalchester. ac.uk) and "Linguistic Geographies" (see www.goughmap.org).

**David Livingstone** is Professor of Geography and Intellectual History at Queen's University Belfast and a Fellow of the British Academy. He is the author of several books including *Nathaniel Southgate Shaler and the Culture of American* Science (1987), *Darwin's Forgotten Defenders* (1987), *The Geographical Tradition* (1992), *Putting Science in its Place* (2003), and *Adam's Ancestors: Race, Religion and Politics of Human Origins* (2008). He is currently working on two books, *Locating Darwinism* and *The Empire of Climate*.

**David Lowenthal**, emeritus Professor of geography and honorary research fellow at University College, London, formerly secretary of the American Geographical Society, has taught at a score of universities in America, Europe, Australia, and the Caribbean, and has been a Fulbright, a Guggenheim, a Leverhulme, and a Landes Fellow. Among his books are *West Indian Societies, Geographies of the Mind, The Past Is a Foreign Country, Landscape Meanings and Values, The Politics of the Past, The Heritage Crusade and the Spoils of History, George Perkins Marsh, Prophet of Conservation*, and

*Paysage du temps sur le paysage.* He is currently working on problems of small islands and on the troubled relations between the sciences and the humanities, fore-grounded in geography and in the insights of Denis Cosgrove.

**Fraser MacDonald** is a senior lecturer in human geography at the University of Melbourne, Australia. His research interests include visual culture, geopolitics, and the histories of social and scientific knowledge. He is currently writing a history of the world's first nuclear missile.

**Gunnar Olsson** is a Professor Emeritus at Uppsala University with previous appointments first at the University of Michigan (1966–77) then at the Nordic Institute for Studies in Urban and Regional Planning (1977–97). Among his publications are *Birds in Egg/Eggs in Bird* (Pion, 1980); *Lines of Power/Limits of Language* (University of Minnesota Press, 1991); and *Abysmal:A Critique of Cartographic Reason* (University of Chicago Press, 2007).

**Kenneth R. Olwig** has an MA in Scandinavian Studies (essentially Nordic philology) and PhD in Geography from The University of Minnesota, where his dissertation advisor was Yi-Fu Tuan and he worked with David Lowenthal. He has lived and worked most of his life in Scandinavia where he is presently a Professor of landscape theory and planning in the Landscape Architecture, Planning and Heritage Department of The Swedish University of Agricultural Sciences. Recent publications include: *Landscape, Nature and the Body Politic* (Madison, University of Wisconsin Press: 2002), and *Nordic Landscapes: Region and Belonging on the Northern Edge of Europe* (Minneapolis, University of Minnesota Press: 2008), edited with Michael Jones.

**Anthony Pagden** is Distinguished Professor of Political Science and History at UCLA. He has been University Reader in Intellectual History and Fellow of King's College, Cambridge, and the Harry C. Black Professor of History at the Johns Hopkins University. He has written widely on Europe's contacts with the non-European world, and his most recent publications are *Peoples and Empires* (2001) and *Worlds at War:The 2500Year Struggle between East and West* (2008). He is currently working on a history of cosmopolitanism.

**Mike Pearson** studied archaeology in University College, Cardiff (1968–71). He was a member of R.A.T. Theatre (1972–3) and an artistic director of Cardiff Laboratory Theatre (1973–80) and Brith Gof (1981–97). He continues to make performance as a solo artist and in collaboration with artist/designer Mike Brookes as Pearson/Brookes (1997–present). He is co-author with Michael Shanks of *Theatre/Archaeology* (2001) and author of *In Comes I: Performance, Memory and Landscape* (2006); and *Site-specific Performance* (2010). The monograph *Mickery Theatre: An Imperfect Archaeology* (2011) is forthcoming. He is currently Professor of Performance Studies, Aberystwyth University.

**Douglas Richardson** is Executive Director of the Association of American Geographers (AAG). He previously founded and was President of the firm GeoResearch, Inc., which invented, developed, and patented the first real-time interactive GPS/GIS technology, leading to major advances in the ways geographical information is collected, mapped, integrated, and used within geography and in society at large. He has worked closely with American Indian tribes for many years on cultural and ecological issues, directs the NEH-funded Historical GIS Clearinghouse, and is co-editor of the recent book, *Geohumanities: Art, History, Text at the Edge of Place* (Routledge, 2011).

**Gillian Rose** is Professor of Cultural Geography at the Open University, and has also taught social, cultural and feminist geographies at the Universities of London and Edinburgh. She is the author of *Feminism and Geography* (1993) and *Visual Methodologies* (2001, 2007, 2011), as well as many papers and essays on images and their spaces. Her book on family photography, *Doing Family Photography: The Domestic, the Public and the Politics of Sentiment*, was published by Ashgate Press in 2010.

**Julie Sanders** is Professor of English Literature and Drama at the University of Nottingham, UK. She is the author of books and articles on Shakespeare, Jonson and early modern drama as well as adaptation studies. Her monograph *The Cultural Geography of Early Modern Drama 1620–50* will be published by Cambridge University Press in 2011.

**Susan Schulten** is a history Professor at the University of Denver, and the author of *The Geographical Imagination in America, 1880–1950* (University of Chicago Press, 2001). She is currently writing a history of thematic mapping in the United States, and was a contributing author to *Maps: Finding Our Place in the World* (2007). Her recent articles include "The Cartography of Slavery and the Authority of Statistics," *Civil War History* 56 (2010), and "Emma Willard and the Graphic Foundations of American History," *The Journal of Historical Geography* 33 (2007). In 2010 she received a fellowship from the John Simon Guggenheim Foundation.

**Joan M. Schwartz** teaches History of Photography at Queen's University, Kingston, and is co-editor of *Picturing Place: Photography and the Geographical Imagination* (2003) and *Archives, Records, and Power*, two double issues of *Archival Science* (2002). From 1977 to 2003, she was a specialist in photography acquisition and research at the National Archives of Canada. Her current research and writing combine interests in archival theory, photographic history, and historical geography.

**Ignaz Strebel** is with the Centre for Research on Architecture, Society and the Built Environment at ETH Zürich (ETH Wohnforum – ETH CASE). He spent a number of years doing project work in geography and architecture at the Universities of Glasgow and Edinburgh (www.ace.ed.ac.uk/highrise/). His research touches upon science and technology studies of workplace activities, decision-making and social complexity, and related issues of urban change.

**Yi-Fu Tuan** is the J.K. Wright and Vilas Professor Emeritus of Geography at the University of Wisconsin–Madison. His many academic honors include the Cullum Medal of the American Geographical Society, the Lauréat d'Honneur of the International Geographical Union, and the Charles Homer Haskins Lectureship of the American Council of Learned Societies. He is the author of more than two dozen critically acclaimed books.

**Kathryn Yusoff** is a Lecturer in geography at the University of Exeter and Director of the MA in Climate Change. Kathryn's primary research interest is in the political aesthetics of environments within the context of climate change (past and present). She is particularly interested in how we understand dynamic earth processes and environmental change through aesthetic experience, and how these experiences configure our political relations in human and non-human worlds.

# FOREWORD

## Converging worlds: geography and the humanities

*Douglas Richardson*

While the discipline of geography has traditionally embraced and contributed to the humanities, the recent resurgence of intellectual interplay between geography and the humanities in both academic and public circles has been nothing short of remarkable. Ideas, terminology, and concepts such as space, place, scale, landscape, geography, and mapping now permeate both academic and popular cultures as conceptual frameworks, methodologies, and core metaphors.[1] The explosive growth of new geographical technologies also has brought elements of geography into direct interaction with the humanities and the creative arts, as well as within society more broadly.[2]

What the many diverse scholars in this book share is a growing interest in exploring the relationships and perspectives of geography within exciting new contexts, whether pragmatic, aesthetic, or philosophical. By doing so, many humanists and geographers – working together and separately – have helped initiate changes in geography itself, stretching its traditional boundaries and applications in creative new directions. This book examines this reciprocal process and suggests new pathways for potential future collaborations. In bringing together artists, authors, geographers, humanities scholars and others, we believe ideas will flow that will have significant impact for years to come.

The contributors to *Envisioning Landscapes, Making Worlds: Geography and the Humanities* address core geographical and human questions such as the depth and complexity of human meaning invested in maps; the human construction of geographical and philosophical worlds through place making and landscape creation; struggles between cultures and peoples thrown into conflict by globalization; opportunities and problems engendered by the pervasive increase of all kinds of geographical information; and the ways that new technologies continue to profoundly alter our experiences of place, landscape, history, and nature. The authors engage a world of uncertainty and continuous change by productively transgressing the bounds of traditional categories to open up new directions in scholarship, literature, and the arts.

The range of books currently available on the intersections between geography and the humanities, and the fledgling interdisciplinary research being carried out by researchers and academics in these fields, is still very limited. New publications in this arena therefore become immediately important. Given the intellectual range and profile of the contributors to this volume, I believe that *Envisioning Landscapes, Making Worlds* may well form a foundational contribution in these emerging fields, a book that has the potential to become influential and heavily cited, and to generate heated debate.

This volume also fills a gap as a much-needed resource for university-level humanities and geography courses, with particular relevance to courses covering historical, cultural, and environmental geography, art history, literary analysis and history, environmental studies, landscape design, and social theory and thought. It is my hope that this book and its engaging debates might also appeal to a broader audience of educated readers beyond the arts and humanities, who are interested in the latest interdisciplinary developments in these fields. The maps, illustrations, art and photographic reproductions will be an added attraction for non-academic readers, as well as rendering the interplay between the various disciplines in the volume more clearly.

The book is one of two distinctive publications[3] with roots in a seminal Geography and the Humanities Symposium organized by the Association of American Geographers in 2007 at the University of Virginia. That highly productive and inter-active gathering brought together and forged new relationships among scholars and artists from three groups: geographers who have engaged humanities topics in creative ways; humanities scholars who have integrated concepts from geography's traditional core strengths or its significant recent innovations in their work; and popular artists and writers whose literary or artistic works investigate geographical concepts or perspectives, or perhaps even use new geographical technologies in unintended and thought-provoking ways. While *Envisioning Landscapes* had its inception in the AAG Symposium, the contributions to the book have been carefully solicited broadly from scholars from around the world and shaped by the editors to create a strong and well-rounded coverage of the traditions of the humanities in geography, and of geography's new interactions with and within the humanities.

Finally, and on a more personal note, this book is dedicated to the memory and contributions of Denis Cosgrove. *Envisioning Landscapes, Making Worlds* would not have been possible absent his vision and intellectual legacy. Denis and I were the book's initial co-editors during its very early formative stages and his insight and conceptual framing of the volume during that early period still reverberates throughout the book. His untimely death in 2008 was an incalculable loss to both geography and the humanities, and personally to each of the book's current editors, all of whom knew facets of him well, from different but intertwined pathways of our lives. The memory of his grace, intellect, and friendship has guided the trajectory of this book and sustained its creation. Denis Cosgrove's extraordinary essay, "Geography Within the Humanities," which follows immediately as the prologue to this book, was one of his very last pieces of scholarly writing.

Denis Cosgrove believed deeply in the necessity of this book project, stating that, "I think this is a very important, potentially course-altering project for American geography," which "has the opportunity to document and push forward a significant rapprochement between our discipline, the humanities and the creative arts."[4] I hope that we have been able to realize, at least in some small part, his vision for this book.

## Notes

1  D. Richardson, "Geography and the Humanities," *AAG Newsletter* 41(3), 2006, 2, 4.
2  D. Richardson and P. Solis, "Confronted by Insurmountable Opportunities: Geography in Society at the AAG's Centennial," *The Professional Geographer* 56(1), 2004, 4–11.
3  The second book, *Geohumanities: Art, History, Text at the Edge of Place* (Routledge, 2011), explores the more experimental and experiential engagements by humanities disciplines themselves as they seek to understand and incorporate geographical methods and concepts of space and place into their own work.
4  D. Cosgrove, personal correspondence with D. Richardson, September 14, 2007.

# PROLOGUE

## Geography within the humanities[1]

*Denis Cosgrove*

"The Humanities" denotes both an approach to knowledge and a specific set of scholarly disciplines. In this brief presentation I explore Geography's relationship to each of these aspects of the Humanities in the past and today, and comment upon the current status of humanities geography in the context of theoretical, technological and institutional change.

As an *approach* to knowledge, the Humanities are characterized by hermeneutic, interpretative methods. They work through cycles of commentary and criticism rather than the establishment of theory and law, although rules of evidence and logical argument are vital to their practice. Their characteristic mode of communication is the lecture, the monograph or essay, rather than the demonstration, research paper or report. Individual authorship, and the use of the discursive footnote or endnote rather than the author citation system for sourcing, reflect "conversation" rather than progressive and cumulative advance of knowledge. The humanities approach thus foregrounds the active role of the author in the construction of knowledge and understanding.

As *disciplines*, the Humanities study the evidence of distinctively human actions and works, for example History, Philosophy, Philology (language, literature, linguistics), Theology and the study of Art. By convention, History is "queen" of the Humanities; human achievement by definition is judged in retrospect. Philosophy is central for determining enduring (even universal) criteria for judgment. Rhetoric is foundational to the pedagogic and communicative goals of the Humanities. Although focused on the products of human creativity, and thus closely connected to the arts, the Humanities are concerned with interpretation and criticism rather than creativity as such. There is no historically consistent definition of the Humanities and each of today's humanities disciplines is the product of continuous evolution.

The *goal* of humanities' study is still best encapsulated by the Greek aphorism: "Know thyself," and the idea that we best come to that knowledge through the reflective study of exemplary human achievement. Learning is lifelong and becomes a signature of "authority" in the broadest sense of legitimacy to adjudicate and exercise power. Historically, therefore, study of the Humanities has consistently been aligned with the exercise of authority and power over places and people: thus they have formed the bedrock of tuition for Renaissance princes, military officers, administrators and diplomats. Although challenged, the humanities disciplines remain at the core of the Western university. The Humanities have an inescapable element of elitism, rendering them subject to criticism from leveling forms of democratic thinking.

Within the Humanities, *Geography* was long regarded as the "eye" of history, providing the necessary knowledge of the physical earth and its spaces without which the human record cannot be understood: an insight easily obscured by the seductions of grand theory and some forms of historicism, which nevertheless current work on geographies of science and knowledge more generally has gone a long way in restoring. For all its significance, however, this formulation relegates Geography to a subordinate role, providing context for historical events, unless we are able to demonstrate decisively the active role that space and place play in those events. Geographers' intense study of spatiality in recent years has moved us toward this goal without resurrecting the specter of determinism, and this is one reason for the current interest in geographical scholarship evident across the Humanities. Focus on the human subject, and specifically on *"imaginative geographies,"* allows us to "map" the self into the world, a second dimension of geography as a humanity. My own studies of the Venice region during the sixteenth century demonstrated the mutual and reciprocal shaping of human selves and material landscape through a complex discourse of science and engineering, planning and planting, mapping and painting. Fundamental to the making of that landscape was a philosophy and vision of a perfectible world, drawn from reading and reflection on Classical texts. This suggests a third aspect of geography as a humanity: the human modification of the earth provides in landscape one of the richest records of human achievement that we possess.

Formal connections between Geography and the Humanities have varied historically. They have been strong in periods of cultural inquisitiveness, of "discovery" and cultural expansion when imagination encounters the resistance of material reality: Herodotus and Strabo recorded the expanded geographical knowledge of imperial Greece and Rome respectively; Renaissance geographers and map-makers Sebastian Munster, Abraham Ortelius, Gerardus Mercator, Andre Thevet and John Dee among so many others registered and interpreted European navigation and "discovery"; modern geography began, according to David Stoddart, when James Cook entered the Pacific and Geography came into the nineteenth-century university at the high point of European imperial confidence. Connections are correspondingly weaker when universalism and instrumental or managerialist concerns hold sway. Geography's contributions were necessarily constrained in the theocratic worlds of seventeenth-century Europe, and geographical knowledge was reduced to the status of gazeteer in the mid-eighteenth-century Enlightenment. As

Susan Schulten has shown,[2] geography here in the United States lost prestige and influence as American education at all levels became increasingly instrumental over the course of the last century. The formal response was to emphasize geography's "hard" scientific, technical and research potential rather than its connections to humanities learning.

Modern university Geography relies strongly and properly on both natural science and social science epistemologies and methods. But we need to be conscious and cautious of the rhetoric of "science" within Geography which, alongside the focus on research, policy and "critique" as opposed to pedagogy, can subordinate our broader humanities tradition. Our connections with the Humanities have been maintained principally through historical geography. But historical geography has mutated in significant ways over recent decades. While the self-styled "humanistic" geography in the 1970s and 1980s owed less to conventional humanities concerns than to a revived interest in subjectivity derived from behavioral theory and phenomenological method, it re-opened Human Geography to questions of perception, imagination and interpretation long associated with historical geography and humanities scholarship more generally. Geographers' embrace of cultural, feminist and other post-structuralist theories has introduced an acute critical awareness into the very notion of "human achievement," challenging many of the comfortable verities associated with conventional humanities scholarship.

Geographers today share a widespread skepticism toward universalistic claims of "humanity," and thus toward the focus on individual agency (both in subject matter and authorship) conventionally embraced in the Humanities. The "cultural turn" finds geographers working with materials and methods conventionally associated with the Humanities, for example the interpretation of texts and images, but taking these concerns into the world beyond the library and the study in more active and engaged ways. This has strengthened the connection with the arts in practice (well reflected in the section devoted to "Cultural Geographies in Practice" in *Cultural Geographies*, the journal that best reflects the contemporary alignment of Geography with the humanities).

But I think there is a note of caution to be sounded here. The 'new' cultural geography of the late twentieth century sought to collapse the boundaries between Social Science and the Humanities. Methodologically and theoretically, I think this has had remarkable success: the complexities of representation are fully acknowledged and interpretative, hermeneutic methods widely adopted in social science geography today, even within economic geography. By the same token, historical geographers draw heavily on the insights of social and cultural theory. But in their goals, I believe that Social Science and the Humanities still differ. The individualistic, reflective and pedagogical concerns of the Humanities remain distinct from the collectivist, interventionist and scientific research concerns of Social Science. This divergence is reflected within Geography; it is apparent in the choice of materials for study, the statement of scholarly goals, the framing of argument and the use of language more than in explicit choices of theory and method. And it remains an important part of the richness of our discipline.

Today, in the context of an anti-intellectual public culture and an "audit" academic culture, the Humanities face powerful challenges, and humanities Geography, always a minority concern in the modern discipline, shares those challenges. Too often, however, the threat to humanities geography comes from within the discipline more than from outside. While geographical concepts such as space, place, landscape and mapping preoccupy scholars in the traditional humanities (why else would a recent edited collection of quite conventional scholarly essays on Strabo be titled *Strabo's Cultural Geography?*),[3] too many within the discipline remain seduced by the idea that new techniques or contributions to science and policy will demonstrate our worth to others within and beyond the academy. Humanities geography can and does make good use of evolving theory and new technology, but it will meet its challenges and contribute to our discipline's reputation and significance by remaining true to its central goals of scholarship, and education, and its commitment to a breadth of learning, critical reflection, and clear and imaginative communication. *Know thyself.*

## Notes

1 Plenary paper delivered at the Geography and the Humanities symposium organized by the Association of American Geographers at the University of Virginia in 2007.
2 S. Schulten, *The Geographical Imagination in America, 1880–1950*, Chicago University Press, 2001.
3 *Strabo's Cultural Geography:* ed. by Daniela Dueck, Hugh Lindsay and Sarah Pothecary, Cambridge University Press, 2005.

# INTRODUCTION

## Envisioning landscapes, making worlds

*Stephen Daniels, Dydia DeLyser, J. Nicholas Entrikin and Douglas Richardson*

### Geographical knowledge and imagination

The title of this book describes two reciprocal aspects of human geography: envisioning landscapes – imaginative, reflective and representational; making worlds, ethical, performative and material. These aspects dovetail and interact, as individual chapters show in case studies of geographical ideas, images, media, sites and situations. They capture the rich diversity of geographical engagements with the humanities.

Contributors explore the power of the geographical imagination, its range, substance and complication, as expressed in maps, photographs, paintings, films, novels, poems, performances, monuments, buildings, traveler's tales and geography texts. They show how places represented in pictures and writings and those constructed on the ground are shaped by interacting imperatives, including morality and aesthetics, mentality and materiality, perception and practicality. Maps, primary media of the geographical imagination, are understood as artifacts as well as images, physically made no less than the material worlds they envision; moreover the meaning of maps may be in actively realizing, or reshaping, those worlds on the ground as well as representing them on paper, with ethical implications for relations between people and place, land and life. So too landscapes, another main focus of geographical inquiry, are understood as experienced as well as envisaged, as lived in as well as looked at, as produced by natural processes as well as by cultural processes. Landscapes on the ground, these studies reveal, are more than a projection of subjective ideas and images, shaping inert nature, but are themselves a medium, shaping the way we look at the world.

Geography is a significantly visual discipline and form of knowledge, and as a consequence this book focuses on a repertoire of ways of seeing, methods of depiction, and wide-ranging fields of vision. Modes of factual recording and image making, such as documentary survey and topographical mapping, are shown to be no less transformational, no less visionary, than works of fiction or fantasy, indeed shown to share conventions of representation and fabrication. Geographical knowledge is a matter of description and depiction, its form often a mixed medium of image and text, designed for telling as well as showing, plotting time as well as space, including making and remaking the terrain of cultural memory. So chapters in this volume explore modes of narrative in the production of geographical meanings, including the development of expressly spatial and visual stories, of prospect and retrospect, cast in stone, built in brick and mortar, as well as printed on the page. Their authors examine the capacities and complexities of geographical knowledge in specific times and places, its limits as well as potential, and its variation from optic to sonic, haptic and affective experience. They rework a tradition of geography scholarship which connects academic knowledge to a wider world of geographical experience and imagination, an actual world of mental reverie as well as material reality, of conflict and uncertainty as well as creativity and possibility.

## Discipline and scholarly domain

In examining the making and meaning of places, in a variety of media, this book shows why geography matters, and how it matters to the humanities in particular, to that domain of inquiry concerned with interpreting and evaluating people's place in the world.

Geography as an academic discipline and the humanities as a scholarly domain have faced similar challenges in the twenty-first-century university. Each has had to address questions of relevance in an era of increased specialization and accountability in research universities. For the humanities this question has come from divergent sources ranging from scholarly reactions against theory to financial or socially instrumental assessments of their public value. Questions of relevance directed at geography are generally of a different sort, based in relatively naïve conceptions about what geographers in fact do. To many, geography is what is there in the world to be straightforwardly experienced, explored and described, a naïve realism to be found in places within and beyond the academy. For some academics outside of the discipline, geography's core concepts of maps, places, landscapes, territories and environment remain useful, everyday concepts of little theoretical import or cultural complication; utilitarian devices for describing what is there for all to see.

In other quarters, of humanities research as well as public understanding, geography as a form of knowledge enjoys a high profile, notably in the fascination with maps and travel writing, but too often this interest seems disconnected from geography as an academic discipline. The connection problems can unwittingly be created by the way geographers position themselves to the outside world. Human geographers have often looked well beyond the boundaries of the discipline for resources of theory and method to address questions of space and place, rather than looking into and developing longer, perhaps less fashionable, traditions of geographical scholarship, in a way which can compromise a clear and effective contribution to a cross-border, multi-disciplinary humanities conversation. All disciplines of course engage in cross-border traffic, deploying and translating concepts, and the tendency to customize intellectual imports for purely domestic consumption is not purely the province of human geography. However, the recent claim to reduce, or even nullify, the place of the representational in human geography is a coinage which has little or no currency beyond the discipline. But significantly, within the discipline, not only would such efforts sever connections with a humanities domain which is centrally concerned with the history and theory of representation (even as it is mindful of the limitations of some of its forms), but they would also reinstate a narrow realism in human geography, a form of subjective and objective knowledge shorn of cultural perspective and focusing on the practical imperatives of the present.

## The humanities: changing conventions

The humanities have always designated a changing, multi-disciplinary domain, and an increasingly permeable one, that reflects a wider world of cultural change within and beyond places of learning. With the traditional pantheon of history, philosophy, languages and literature the humanities now includes more modern subjects of media studies, critical theory, and various forms of practice from music, dance and design to creative writing, exhibition curation, information technology and cultural policy. The expansion of culture as a field of inquiry, to include popular, material and commercial forms and practices, ways of life as well as media of representation, has entered into significant exchanges with the social sciences, and multi-domain subjects like archaeology, anthropology, area studies and geography, especially on questions of imagery, identity, ideology and authority, and their material conditions and effects. A grounded, culture and society model of humanities scholarship has widened the world of contextual significance for traditional subjects – for the study of canonical works in history, literature and art – as well as expanding the repertoire to include lesser-known, more vernacular works, like diaries and family photos, concerned with recording and commemorating the everyday world.

Traditions of individual scholarship in the humanities have been supplemented by newer conventions of collective and collaborative research, self-reflection by social action, including public engagement, particularly in publicly funded institutions where the humanities have to demonstrate a use value beyond older, liberal arts ideals of cultivating a critically aware, educated citizenship. The humanities are no longer the preserve of humanists. Unitary and universal notions of humanity, including the integrity and authority of the individual human subject, are displaced in research emphasizing, for better or worse, the relativism of knowledge, the decisiveness of cultural difference and dispute, and the social construction of reality, including the cultural framing of nature. This culturalist position has itself come under challenge by a renewed naturalistic perspective on culture, on evolutionary explanations of art and other creative forms of human consciousness, on the agency of the non-human or beyond-human world, both through advanced, so-called cyborg forms of technological-human assemblage, and more atavistic forms of animal and environmental determination. Researching the worldliness of texts and images, including their material substance, thus involves recognizing the integrity and complexity of culture as a bearer of meaning and effective presence – not a thin super-structural surface that requires more basic socio-economic support and explanation, but an integral part of a wider world of physical processes and social relations.

## New geographical sensibilities in the humanities

While history has traditionally occupied the core of the humanities, geography has also in periods and places assumed a prominent position, often in harness with history, particularly when the domain is being reformed along more liberal lines. As a subject of both scholarship and practice, geography played a key role in envisioning mankind's place in the world during the period of European empire building, from the sixteenth to the nineteenth century, as both a cultural endorsement of empire, its commerce and politics, and a form of social criticism and utopian imagining. An example can be found in the developing of ecumenical, sometimes cosmopolitan versions of citizenship, ethically attentive to both the colonization of other cultures and to the exploitation of the natural world, to human rights, including rights of access to land and learning about it. Such liberal traditions of geography have continued to enrich humanities education and citizenship, particularly those concerned to connect learning about local landscapes, in all their physical and cultural complexity, with the place of those landscapes in a wider world, and to communicate this accessibly. This tradition has been cultivated beyond established academic centers, by professional writers as well as lay enthusiasts, and it is worth remarking that it is geography's integrative vision of human-centered learning with field science and information technology (including traditional forms of topographical mapping) that has always appealed to this constituency.

Within the humanities, and across its borderland with the social sciences, there has, over the past twenty years, been a pronounced turn to a range of geographical questions, with a lexicon of terms like place, space, mapping, landscape, locality, globalism, environment and region gaining a new currency, sometimes independently of geography as a discipline, more recently in conversation and collaboration. This new geographical sensibility has a number of sources, including a post-colonial consciousness of zones of cultural dominion and resistance, the recasting of grand social narratives of development as explicitly environmental ones, the de-centering of the human subject and the rise of ecological criticism, the re-framing of subjectivity and cultural identity in terms of position and perspective, and of knowledge (including self-knowledge) in terms of location and movement.

This multi-disciplinary geographical turn, and geography's disciplinary contribution to it, has been developed in a number of monographs, edited collections and journals. It has resulted in major funded thematic programs such as the UK Arts and Humanities Research Council's Landscape and Environment programme as well as forums sponsored by learned societies and academic associations. Arising from a symposium at the University of Virginia, organized by the Association of American Geographers, this volume includes authors from philosophy, history, literature, political science and performance studies, and is designed to enhance the ecumenical spirit of the conversation between geography and the humanities, while showing how geography as an institutional discipline, and itself a broad church, can enlarge and enrich the humanities as a forum of and for theory and practice.

## Four themes

This book is organized in terms of four themes: mapping, reflecting, representing and performing. While these modes of knowing the world overlap, and their differences are often a matter of degree not kind, they frame specific geographical practices. These thematic frames display the power of ideas and images, including specific theories and iconographies, in what may seem at first sight merely matters of fact and expediency or mainly material imperatives. The material world retains its power as a proving ground for such ideas, articulating their effects and implications.

Mapping as a term of cultural description in the arts and humanities has moved beyond the practice of cartography to a broader, metaphorical sense of interpreting and creating images and texts and of making sense of a fast modernizing or post-modernizing of the world. This academic reorientation raises fundamental geographical questions about knowing and experiencing the world as cartographic scholars consider maps in seemingly novel ways, focusing not so much on their factual accuracy but more on their politics and their beauty, scholarship which has captured the public imagination in books and exhibitions. Contributors to this section consider the projective scope of mapping as a graphic form of storing and communicating knowledge, focusing on techniques of medieval and modern cartography, including their claims to factual precision, to position cartography within the wider world of its deployment from nation building to battle planning. They are concerned too about the place of mapping as a form of ontology, as more than a surface for the display of other forms of knowledge, but as a fundamental form of meaning creation, manifesting deeper and wider earthly powers, to make the world present and absent, erasing as well as inscribing space, making some places and dismantling or destroying others.

Part of geography's philosophical purpose has always been as a form of practical wisdom, concerned with changing the world as well as interpreting it, and for many cultural geographers the discipline is connected to forms of intervention and activism from urban planning to community participation. The chapters in this section adopt a more reflective, but no less engaged, perspective on the past as well as the present to probe the moral and cultural complexity of places and landscapes, and to reveal the dense web of meanings and social relations which lie behind that which seems in plain view. In their essays on the relation of places and events and the bounding of each – the encounters with others through travel and communication that hide political dominion under the veneer of cosmopolitanism, the social choreography of human bodies moving through space–time, the cultural significance of children's geographical texts, the local environment in the creation and reception of scientific knowledge, and the experience of environmental inequality and the distribution of goodness – the authors all begin with seemingly uncomplicated geographical descriptions. Such simplicity quickly gives way, however, to the critical probing into the endless variety and complexity of humanly constructed worlds.

Humanities scholars have played a provocative role in identifying the so-called "crisis of representation," claiming that even the most straightforwardly factual descriptions and depictions of the world are not mimetic correspondences with material reality, accurate or inaccurate, but projections of human values, hopes and fears, sometimes coercive ones. Given that core geographical concepts and methods have long been regarded as realistically descriptive, excavating geography's field of representation in many visual and written forms within and beyond the discipline has proved a revealing form of inquiry. In focusing on a range of media – including prospect poetry, landscape painting, pollen cores, travel writing and commemorative statuary – the chapters in this section are centrally concerned with questions of time, with forms of narrative which plot places as well as events; these are both local topographical tales, and larger national and global stories. Moreover authors give radical attention to the "re" prefix of representation, the possibilities and predicaments of making present, reclaiming the long ago and far away, in the restoration of material landscapes and the writing of travel texts.

The currency of performance as a concept and method of humanities research has been part of a broad move beyond written and visual forms of representation, or rather beyond the limits of too screen-like or desk-bound an interpretation of images and texts, to an appreciation of their relations with various embodied, multi-sensory practices, of sound, smell and touch, and the expressly physical engagement with the material worlds of their making and meaning. A performative perspective has in turn emphasized the desk and screen as material spaces, with their own acts and ritual spaces of reading and looking. Chapters in this section consider the performative spaces of books within and beyond their bindings, practices of reading which make novels events as well as objects, travel writings as a test of recreating the physical, experiential impact of encountering extreme environments, including the agency of the physical world. Other chapters explore the ritualistic role of a range of objects, including religious icons, ice cores, apartment windows and family photographs, examining how they shape codes of conduct and generate imaginative and affective worlds that extend well beyond the sites in which they are placed. In reaching out from, or to, perspectives from other disciplines these chapters also return us to a long-established performative sensibility in the geographical imagination, in mapping, reflecting and representing, Theatrum Orbis Terrarum, the Theater of the World.

## Selected References

Cosgrove, D., *Geography and Vision: Seeing, Imagining and Representing the World*, London and New York: I.B. Tauris, 2008.

Cosgrove, D. (ed.), *Mappings*, London: Reaktion, 1999.

Cosgrove, D. and Daniels, S. (eds), *The Iconography and Landscape*, Cambridge: Cambridge University Press, 1988.

Curtis, N. (ed.), *The Pictorial Turn*, London: Routledge, 2010.

Daniels, S., *Fields of Vision: Landscape Imagery and National Identity in England and the United States*, Cambridge: Polity, 1993.

Daniels, S., Pearson, M. and Roms, H. (eds), *Field Works,* special issue of *Performance Research* vol. 15, Number 4, 2010.

Driver, F., *Geography Militant: Cultures of Exploration and Empire*, Oxford: Blackwell, 2001.

Entrikin, J.N., *The Betweenness of Place: Towards a Geography of Modernity*, Basingstoke: Macmillan, 1991.

Gregory, D., *Geographical Imaginations*, Oxford: Blackwell, 1994.

Livingstone, D., *Putting Science in its Place: Geographies of Scientific Knowledge*, Chicago: Chicago University Press, 2003.

Lowenthal, D. and Bowden, M.J. (eds), with assistance of M.A. Lamberty, *Geographies of the Mind: Essays in Historical Geosophy in Honor of John Kirkland Wright*, New York: Oxford University Press, 1976.

Mitchell, W.J.T. (ed.), *Landscape and Power*, Chicago: Chicago University Press, 2002.

Ogborn, M., *Global Lives: Britain and the World, 1550–1800*, Cambridge: Cambridge University Press, 2008.

Said, E.W., *Culture and Imperialism*, London: Chatto and Windus, 1993.

Tuan, Y.-F., *Morality and Imagination: Paradoxes of Progress*, Madison: University of Wisconsin Press, 1989.

Williams, R., *The Country and the City*, London: Chatto and Windus, 1973.

Withers, C.W.J., *Placing the Enlightenment: Thinking Geographically about the Age of Reason*, Chicago and London: University of Chicago Press, 2007.

Wright, J.K., *Human Nature in Geography: Fourteen Papers 1925–1965*, Cambridge, MA: Harvard University Press, 1966.

Wylie, J., *Landscape*, London: Routledge, 2007.

**PART I**

# Mapping

# 1

# WHY AMERICA IS CALLED AMERICA

*Franco Farinelli*

## The map and the coin

Agathemerus writes that Anaximander, a pupil of Thales in the sixth century BC, "was the first one who had the audacity to draw the Ecumene on a small table" made of terracotta or bronze.[1] For this reason, Anaximander was judged most impious: he dared to represent the earth from above in a way that only Gods could do. So the story goes. But as a matter of fact things are more complicated than this. For the Greeks, nature was not a combined group of things but an ongoing perpetual process; nature was movement. From this perspective, Anaximander's project appeared scandalous because it aimed to immolate Earth as a function of knowledge (of the dominion) of Earth itself. It is only because, with Anaximander, the Earth becomes a corpse that the rigor (the rigidity) of death becomes the equivalent of the rigor of science; *rigor mortis* allows us to measure only what was once alive but is no more. Anaximander might not have been the first one to talk about such a reduction in the Western world but he is certainly the first one to celebrate it in geometrical forms, and make it so pervasive that it comes to stand at the center of a complex *Weltanschauung*.[2]

It is such a deliberate reduction of reality into a corpse-like geometrical scheme that explains Anaximander's tragic supremacy, his role of precursor that Western tradition strangely has assigned to him, despite the fact that maps existed long before his time. Michel Serres[3] writes that the difference between the Babylonian tables dating back to the third millennium BC and Anaximander's table corresponds to the difference between local and global. The tables have at their center the city and the river Euphrates; Anaximander's table, on the other hand, provides a model of the world as a whole. These reflections, however, concern merely the descriptive plane. More significant in this respect are a series of bronze and silver circular tetradrachms coined in Ephesus during the fourth century BC. On the reverse these show a map of the city's hilly surroundings traversed by the tributaries of the river Meander: on the recto a Persian satrap, holding an arch and the baton of command, is hurrying on; on the opposite side a series of valleys stretch amidst woody hills depicted with plastic precision. The absence of anything that would suggest human intervention, such as roads and cities or any other kind of nomenclature, is noteworthy. Precisely because of such absences it can be concluded that this coin was meant for local circulation limited to Ionia.[4] However, one could object that until recently the Brazilian cruzeiro and the Taiwan dollar also carried on their reverse a map showing rivers and mountains, reminiscent of the silent maps of our childhood. The images impressed on these coins and the stories they continue to convey show that we still believe in maps just as we still believe in coins.

## The map is the money

What is common to modern maps and coins is not merely a question of more or less limited circulation, rather it is their symbolic regime; this is functional to the exchange value rather than the use value of things. Whether or not the exchange value contains as much as an "atom" of use value, as Marx holds,[5] or whether exchange value and use value are, on the contrary, driven by the same logic[6] does not interest us here. What is important is the fact that the role of geographical maps at this point is not concerned with the recognition and localization of things on earth but rather with their possibility to be transformed into commodities, as the case of maps and coins devoid of any names plainly shows. This idea is clearly illustrated at the end of the nineteenth century by the author of the translation of that part of earth which we today call India into a geodetic map: on a "sound, square, geographical" map "based on systematic measurement" (i.e. triangulation), every point "on a boundary-line, every peak in a mountain system, every landmark of any importance in the country-side, has a value whose correctness can be proved just as easily in a London office as in the open field. And this value is not only incontrovertible, but absolutely distinctive, because every point on the whole world's surface has its own special position in terms of latitude and longitude, with which no other point can interfere."[7]

The analogy between cartographic representation and the market, which is its realization, can be clarified by rephrasing Holdich's assertion using the language of Marxism: natural forms become value forms, commodities show a phenomenal form different from natural forms only with respect to the exchange value these have with other commodities. Things on a map share the same destiny: despite their differences, all things co-exist one next to the other; at the same time, they also submit to the same regime which assimilates them all. Commodities have a value only because all other commodities have a value in relation to the same equivalent, that is to say an equivalent commodity excluded from the realm of commodities represented on the map.[8] The map is in turn the agent that produces a general form of value. Such a general equivalent is space, intended in the Ptolemaic sense of the standard linear interval between two geometrical points,[9] in relation to which each use value, that is to say each place, is destined to disappear. In the middle of the sixteenth century the introduction of graphic scale into maps marks the beginning of the systematic use of space as the phenomenal form for the value of goods,[10] in other words, as a universal commodity in relation to single types of commodities. In this way the map with its properties becomes the model of territory and produces the general form that stands for modern territorial value.

In other words, space and money become the same thing; cartographic symbol and money function in exactly the same way: the former in relation to the map, the latter in relation to the market. The consequence of their mutual, incessant tension toward the equivalence of all things has led to the poor generalizations that today mark our relations with the world. A few years ago, Immanuel Wallerstein wondered whether India really existed.[11] He reached the conclusion that India existed only as "an invention of the modern world-system," in the same way as any other nation-state. Much earlier, as early as the end of the nineteenth century, the major expert of the land-systems of British India explained that India did not exist at all because no country in the world is called with this name: "within the borders of the area that is designated with this name on the maps, there are a series of provinces populated by different races often speaking different languages."[12] What today escapes English social science was perfectly clear to the British Empire's civil servants and beforehand to Metternich, who, two centuries earlier, at the Congress of Vienna, triumphantly declared that reality is its geographical, or more precisely, cartographic expression.

## The cartographic project

The above statement remains valid even when it is re-phrased in more radical terms: reality is the product of its treatment, and the latter is in turn the product of the geographical – better cartographic – expression. On the reverse of the above-mentioned tetradrachms, space does not feature yet. In its place we find the portrait of a territory delimited by money circulation, the image of the natural form of things, of their use value. More importantly the coin signifies natural data and simultaneously the potential equivalence of each element with the others and their possibility of being interchanged. This is the reason why names or representations of roads that already existed are missing:[13] these are not significant. Such a mutual function has a two-fold preliminary characteristic. First, it consists of a simultaneous representation secured by the image itself. This is a fundamental characteristic of commodities that enables them to reveal their values in the same equivalent, that is to say, according to a general expression of values.[14] Secondly, it coincides with the form of the vehicle of the representation itself, that is to say, the money, which is also the main agent in the market. Alfred Sohn-Rethel's reflections are useful for understanding the functional identity of map and money in the form of coins; both of these, he remarks, are Ionic inventions that can be dated to the seventh century BC. It could be objected that the invention of money occurs at the beginning of the seventh century,[15] while the first map appears a century after. This, however, means that the map can be defined, using the words of Marx, as an "extended reproduction" of money, a general sign that testifies to the existence of the market, which in this period is on the point of being born.[16] At the same time, the coin also works as an unavoidable interface between what, according to Sohn-Rethel, is already directly connected: on the one hand, the money, that is to say, the real abstraction in the exchange process, and on the other, thought in abstract form. It follows that when, for example, Kant breaks up the object in abstract substance and phenomenon, he makes a distinction between the exchange value and use value of the goods.

According to Sohn-Rethel, concepts in the intellect exist in consciousness but do not originate from the latter. It is precisely money, the expression of the abstract exchange of things, that establishes the connection between social reality and conceptual ideals of formal abstractions. These are precisely the same abstractions of modern science represented by Galileo's mathematical knowledge of nature, whose data have no direct relation to perceptions.[17] This knowledge contributed to the destruction of the unity of mind and hand typical of artisanal production and paved the way to capitalistic production[18] through the imposition of laws based on exchange-abstraction and money: "laws of uniformity, divisibility, of a particular kind of movement and quantification." The introduction of mathematics in the productive process – in construction and in particular in military architecture concomitant with the development of firearms – further favored the development of capitalism.[19] Considering all this, we can reasonably conclude that the first and main agent of the introduction of mathematics was the map rather than money. Sieges, obsidional techniques, and the use of openings for firearms required precisely drawn fortifications[20] and the development of the geometrical image for the modern city. The isometric drawing used to represent the front of buildings – their use value – as we are accustomed to see them gives way to the geometrical drawing produced by the impassable and dehumanized vertical viewpoint and expression of their exchange value; at the same time the quality of manufactured goods gives way to issues of dimensions and measures, a mere matter of quantities.[21] If money changes the nature of the relationship among human beings, it is the map that changes the way we think of things and, prior to that, the way we name things.

## "The age of the world image"

There is a history of the invention and construction of the Orient.[22] There are descriptions of the way the Western world is seen today by its enemies.[23] Despite all the volumes that have appeared in the West, a real history of the invention of the concept of the West and how this appears in the eyes of those who are part of it remains unwritten. Foucault would say that there is still no complete genealogy of the Western world. This, I wish to argue, is the reason why there are still problems in recognizing the real value of geographical knowledge: its central role in molding all ideal models.

The West is geography and modernity is the latest invention of western geography. Heidegger would have been more precise had he defined modernity as the era of geographical representation rather than "the era of the image of the world."[24] In the same way as space - the *form* of modernity[25] - modernity itself originates from a grin; that same implicit and deferred grin that gives a name to America. It is known that a name is a petrified laugh,[26] a laugh caught in the process of becoming something else, deferred, crystallized and finally made permanent. The American laugh, the name America, marks the end of the Colombian tragi-comedy, the end of the adventure of an explorer who did not realize what he had discovered, unaware of his real function to reduce the world to a map.[27] Garry Wills is wrong when he accuses Columbus of having sailed on his adventure with "the map of the *old* world in his mind," a map whose image nothing would have shaken or erased.[28] He is wrong because it is not so much a question of what is old and what is new but rather it is of the relation between the map and the world, of the subordination of the world to the map, a relation that was unknown to the ancients and which starts with the modern period. The difference between Vespucci and Columbus is the degree of consciousness that characterizes and shapes their actions.

Columbus is the protagonist in a cartographic project that is controlled by someone else; more precisely he is the unaware agent of Paolo dal Pozzo Toscanelli's project. Vespucci's role also is decided in the end by a cartographic act, but this occurs after everything had already happened. In both cases cartography prevails over geography, cartographic ethos imposes itself on individual will. However, there are differences between the two explorers: the first does not understand almost anything about what is happening to him; the second understands almost everything. Such a difference is shaped by their different relationships with carto-graphic representation: Columbus receives a complete cartographic representation and produces the prognosis that it incorporates; Vespucci, the first cartographer and the one responsible for the *Padron real* in the Spanish Court,[29] constructs and produces, he does not undergo. The famous letter the Florentine sends from Cape Verde on the 4th of June 1501 to the Medici is very significant. He complains about the Alvarez Cabral expedition because it includes neither mathematicians nor cosmographers and, as a consequence, one can discuss the nature and the form of the Brazilian coast only "*discontortamente*," tortuously.[30] This adverb uses a contrast, or better a negation, to highlight the new values of rectilinearity and orthogonality, that is to say, the new spatial values required and imposed by the modern carto-graphic image of the world to the world itself. In this respect the fortune of Vespucci is the ruin of Columbus, and their intertwined stories are in fact one single story: the story of the modern relation between reality and linguistic, cartographic sign. It is a story that right now, at the end of the era of the New World, should be recon-sidered in order to attempt to shed some light on the nature of the very New World that is ahead of us. Such a reconsideration can be mapped out through the following two questions: why was it an arbitrary act to call America by this name? And if this is the case, what does this name really mean?

## The name "America"

Émile Benveniste identified a contradiction in the way Ferdinand de Saussure defines the linguistic sign and the central importance he attributes to it. Saussure states that language is form and therefore the linguistic sign does not bring together a thing and a name but rather a concept and an acoustic image. He adds, however, that the sign "is *unmotivated*, that is to say that it is arbitrary in relation to meaning, that it has no natural relation with reality."[31] In this way, Benveniste says, Saussure explicitly excludes reality from the definition of sign and therefore constructs a fundamental contradiction: if language is form and not substance, linguistics should be a science of forms, centered on an understanding of the sign and not at all concerned with reality and things. Benveniste interprets such a huge anomaly as a more or less conscious homage to the historical, relativistic, and comparative ethos typical of the late nineteenth century. Such an attitude begins with the realization of the infinite diversity of cultural phenomena to conclude that nothing was necessary; it recognized "universal dissimilarity" to end up professing "universal contingency." According to Benveniste, on the other hand, the sphere pertaining to the arbitrariness of language should be narrowed: this does not concern the relation between signifier (that is, the word) and signified, as Saussure thought, but the mere fact that a certain sign is related to a certain aspect of reality. However, the relation between signifier and signified is necessary as both are harmoniously impressed in the mind of the speaker and, by so doing, evoke each other: the "mind does not contain empty forms, concepts without names."[32] Benveniste, apparently more than Saussure, believed that the mind functioned like a map.

Let us leave this issue for the moment and return to Vespucci, or better to Vespucci's luck. Let us consider the well-known *amapamondo*, the "flat figure" (as opposed to the "spherical body") that Waldseemüller adds to the 1507 *Cosmographiae Introductio*, which bears for the first time the name of America. This, like the mind, contains forms that bear names and therefore convey concepts. The names that appear on the newly discovered continent are not many: there are names for the Antilles; a handful of toponymies mainly around the Brazilian and Florida coasts, point by point; and in the inland, together with the expression "TERRA INCOGNITA" repeated twice, north and south of the Gulf of Mexico, there are only two other names: "PARIAS" around the Tropic where New Spain is and "AMERICA" in the heart of Brazil. Local toponymies are written in small letters, all other names are in capital letters. Additionally, the last one appears to be written with letters bigger than the others, similar to the letters of the name "SPAGNOLLA" (the present Haiti), a name followed by the predicate "INSULA" written in smaller letters. As Varela has noted,[33] the popularization of the name America happened thanks to this map rather than the text that accompanied it. Whichever might be the case, besides definite descriptions of the still unknown New World's natural landscape, Waldseemüller's map contains only two kinds of names: those which refer to points and lines (i.e., to localities, rivers, capes and mountain chains) and other names which designate whole areas. In other words, the process of nomination on the map follows a geometric logic organized around the articulation of a Euclidian syntax constructed on the relationship between points, lines and surface areas. There are two consequences: firstly, there are no names on the map, except proper names – when I say names here I mean those which refer (and can only refer) to one and only one of the three elements, according to a bi-univocal relation. Secondly, when names refer to surface areas, there is a relationship between the body of the printed letters and the extension of the area itself. This explains why the name "AMERICA," the one written in the largest print and devoid of all punctual values because it refers to a wide area, becomes the name for the whole island and later for the whole continent *once it is printed on the map*. And the *Introductio* significantly confirms that this was precisely the intention of its author.

## Deconstructing the deconstruction

Now we can return to semiology. As Saussure has stated,[34] "the method that uses words to define things, is a bad method." So, let us attempt to reverse this process; let us start from the map to attempt to define words. The definition of America is an exemplary case of arbitrariness (intended in an historical sense, and not in the semiological sense according to Saussure) of the linguistic sign. This is not to say, as Benveniste would put it, that this word has "no natural relation" with the reality it refers to. Rather I would like to note that the example of America shows Benveniste and Saussure in agreement, demonstrating the common cartographic roots of their thought, the very essence of their common position. On the one hand, the map is the only representation in which sign, meaning, and reality enter into a relation with one another, causing the contradiction that Benveniste notes in Saussure. On the other hand, in the attempt to remove such contradiction, Saussure reaffirms the necessity of the relation between signifier and signified that actually occurs only in the geographical image. This does not contain empty forms and anonymous concepts. For this reason, according to Saussure and Benveniste and following a tradition that dates back at least to Hobbes and Locke and stretches on to Kant, it can be considered a paradigm for the mind. If we agree that the map stands for a unique and originary sign, this idea is also expressed by Saussure, when he writes that: "it is not thought that creates sign, rather it is sign that guides thought in a primordial manner (as a matter of fact sign creates thought and in its turn provokes thought to create signs that are not so dissimilar from those that it has received)."[35] These reflections can be useful in reflecting on the *Galleria delle Carte Vaticane* (Gallery of Vatican Maps) painted after 1580. At the top of some maps is the word "Ditio" (*Ditio Veronensis, Ditio Bononiensis* and so on: from the Latin word *dire*, which means "to say"). This is proof of the exhaustive character of a definition that only ostensibly refers to the linguistic expression, but in fact is the product of a pure and simple tabular extension. This is so much so that the borders of the thing represented, that is to say, the borders of the representation itself, have become unimportant; such borders are those of the representation itself because the thing represented has become its image and vice versa. With modernity the limits of the world become the limits of the geographical table, which is the thing that establishes or expresses what exists on earth. This started perhaps with the *Carta dell'Oceano* by Paolo dal Pozzo Toscanelli, whose copy Columbus carried with him and which, along with the map of the New World by Waldseemüller, open modernity. According to this map, the New World is not simply the fourth part of earth that had just been discovered but rather it is the whole world, including that part of the world that was already known to the ancients and which in this way acquires new meaning, existence and essence.

This has decisive consequences for the analysis of cartographic logic. In international geographical debates, reflection on the nature of maps still remains where Brian Harley[36] had left it some years ago, that is to say, at the point at which the map acquires a textual (i.e., literary) character, which is possible to understand by way of deconstructive techniques, on the belief of the supremacy of models derived from reflections about natural language over all other models. But we have another possibility, the possibility of finding another explanation, according to which cartographic language is more important than verbal language and explains, as I have attempted to show here, the aporias of the latter, in the belief that exactly this kind of consciousness is the only and authentic reason for which Vespucci was able to understand better than others what was really happening.

## Notes

1  K. Muller (ed.), *Geographi Graeci Minores*, Paris: Firmin Didot, 1882, p. 471.
2  F. Farinelli, "Salomé," in *I segni del mondo. Immagine cartografica e discorso geografico in età moderna*, Firenze: La Nuova Italia, 1992, pp. 8–9; F. Farinelli, "Did Anaximander Ever Say (or Write) any Words? The Nature of Cartographical Reason," *Ethics, Place and Environment* 1, 1998, 135–44; F. Farinelli, *Geografia. Un'introduzione ai modelli del mondo*, Torino: Einaudi, 2003, p. 81.
3  M. Serres, *Le origini della geometria*, Milano: Feltrinelli, 1994, pp. 61–2.
4  A.E.M. Johnston, "The Earliest Preserved Greek Map: A New Ionian Coin Type," *The Journal of Hellenic Studies* LXXXVII, 1967, 86–94.
5  C. Marx, *Il Capitale*, I, Roma: Editori Riuniti, 1967, p. 70.
6  J. Baudrillard, *Pour une critique de l'économie politique du signe*, Paris: Gallimard, 1972, p. 160.
7  T. Holdich, "The Use of Practical Geography Illustrated by Recent Frontier Operations," *The Geographical Journal* XIII, 1899, 473.
8  Marx, *Il Capitale*, pp. 88, 93, 98–9.
9  Farinelli, *Geografia*, pp. 12–13.
10 P.D.A. Harvey, "The Spread of Mapping to Scale in Europe, 1500–1550," in C.C. Marzoli, G.C. Pellegrini, and G. Ferro (eds), *Imago et Mensura Mundi, Atti del IX Congresso Internazionale di Cartografia*, Roma: Istituto dell'Enciclopedia Italiana, 1985, pp. 473–77.
11 I. Wallerstein, *La scienza sociale: come sbarazzarsene*, Milano: Il Saggiatore, 1995, pp. 141–5.
12 B.H. Baden-Powell, *The Land Systems of British India*, I, Oxford: Clarendon Press, 1892, p. 5.
13 Johnston, "The Earliest Preserved Greek Map," p. 92.
14 Marx, *Il Capitale*, 1967, pp. 98–9.
15 A. Sohn-Rethel, *Lavoro intellettuale e lavoro manuale. Per la teoria della sintesi sociale*, Milano: Feltrinelli, 1977, p. 9.
16 Farinelli, *Geografia*, p. 163.

17  A. Sohn-Rethel, *Il denaro. L'apriori in contanti*, Roma: Editori Riuniti, 1991, pp. 46–7, 15, 9, 17–23.
18  Sohn-Rethel, *Lavoro*, pp. 90–106.
19  Sohn-Rethel, *Il denaro*, pp. 29, 73.
20  F. De Dainville, *Le language des géographes*, Paris: Picard et Cie, 1964, p. 218.
21  F. Farinelli, "Dallo spazio bianco allo spazio astratto: La logica cartografica," in T. Maldonado (ed.), *Paesaggio: immagine e realtà*, Milano: Electa, 1981, pp. 200–4.
22  E. Said, *Orientalismo. L'immagine europea dell'Oriente*, Torino: Bollati Boringhieri, 1991.
23  I. Buruma and A. Margalit, *Occidentalismo*, Torino: Einaudi, 2004.
24  F. Farinelli, "Certezza del rappresentare," *Urbanistica* 97, 1989, 7–16.
25  F. Farinelli, "Squaring the Circle, or the Nature of Political Identity," in F. Farinelli, G. Olsson, and D. Reichert (eds), *Limits of Representation*, München: Accedo, 1994, pp. 11–28.
26  M. Horkheimer and T. Adorno, *Dialettica dell'Illuminismo*, Torino: Einaudi, 1966, p. 62.
27  Farinelli, *Geografia*, pp. 18–19.
28  G. Wills, "Foreword," in L. Formisano (ed.), *Letters from a New World: Amerigo Vespucci's Discovery of America*, New York: Marsilio, 1992, p. ix.
29  A. Managhi, *Amerigo Vespucci. Studio critico, con speciale riguardo ad una nuova valutazione delle fonti e con documenti inediti tratti dal Codice Vaglienti (Riccardiano 1910)*, Roma: Fratelli Treves, 1924, pp. 14–23.
30  Managhi, *Amerigo Vespucci*, vol. II, pp. 35.
31  F. De Saussure, *Cours de linguistique générale*, Paris: Payot, 1972, pp. 98, 101.
32  É. Benveniste, "Natura del segno linguistico," in *Problemi di linguistica generale*, Milano: Il Saggiatore, 1990, pp. 60–4.
33  C. Varela Bueno, *Amerigo Vespucci. Un nome per il Nuovo Mondo*, Milano: Fenice 2000, 1994, p. 116.
34  F. De Saussure, *Cours de linguistique générale*, p. 31.
35  F. De Saussure, *Écrits de linguistique générale*, Paris: Gallimard, 2002, p. 46.
36  B. Harley, "Deconstructing the Map," in T. Barnes and J. Duncan (eds), *Writing Worlds: Discourse, Text and Metaphor in the Representation of Landscape*, London: Routledge, 1992, pp. 231–47.

# 2

# ABOVE THE DEAD CITIES*

*Derek Gregory*

> When I think of myself then [in Hamburg in 1943], hurrying home with
> the pram, I also think of all the things unknown to me. German officials
> who had placed our names on deportation lists; Royal Air Force officers
> studying aerial photographs of our city; bombers revving on runways in
> the flat fields of eastern England. All of them were about to impinge on
> my life.
>
> *Marione Ingram, "Operation Gomorrah" (2006)*

## Introduction

When 28-year-old Heinrich Böll saw his "first undestroyed city" at the end of the
Second World War, he broke out in a cold sweat. It was Heidelberg. Böll was a
native of Köln, which had been virtually destroyed by the Allied bombing campaign,
and he was haunted by the suspicion that Heidelberg had been spared the same fate
as other major German towns and cities for purely aesthetic reasons. In post-war
Germany "dead cities" were normal cities, so much so that W.G. Sebald, who was
born just one year before the war ended, did not attribute the ruins to the bombing
and shelling at all. Almost every week on newsreels "we saw the mountains of
rubble in places like Berlin and Hamburg," he wrote, yet for the longest time he
"did not associate [them] with the destruction wrought in the closing years of the
war" – he knew "nothing of it" – but "considered them a natural condition of all
larger cities."[1]

The British and American air war against Nazi Germany from 1940 to 1945
was brutal by any measure: necessarily so according to its protagonists, needlessly
so according to its critics. By the end of the war 131 German towns and cities
had been bombed, and 80 per cent of those with populations of more than 100,000
had been either seriously damaged or devastated. And yet in a series of lectures

delivered in Zürich 50-odd years later, Sebald drew attention to what he saw as the extraordinary absence of the Allied bombing campaign from the public memory of post-war Germany:

> The destruction, on a scale without precedent, entered the annals of the nation as it set about rebuilding itself, only in the form of vague generalizations. It seems to have left scarcely a trace of pain behind in the collective consciousness, it has been largely obliterated from the retrospective understanding of those affected, and it never played any appreciable part in the discussion of the internal constitution of our country. As Alexander Kluge later confirmed, it never became an experience capable of public decipherment.[2]

Sebald's purpose was to explore the possibility of its public decipherment through what he called "a natural history of destruction," which subsequently became the title for an expanded English translation that included the original lectures.

## The natural history of destruction

Sebald unearthed the phrase from an essay proposed by British scientist Solly Zuckerman. Trained in zoology and anatomy, Zuckerman had joined the UK Ministry of Home Security's Research and Experiments Department early in the war to study the effects of blast on the human body. His expertise soon widened to include a systematic study of the statistics and logistics of bombing as part of the fledgling science of operations research, and he was appointed Scientific Director of the RAF's Bombing Analysis Unit. He was a fierce critic of Bomber Command's area bombing strategy, and determined to complement his theoretical and analytical objections by observations in the field. "Once the noise of exploding bombs had died away, and the sense of fear that went with it," he wrote, "I always wanted to get as quickly as possible to the places that suffered."[3] Nothing prepared him for what he saw when, in December 1944, he arrived in Aachen, which had been subjected to a pulverizing series of raids over the previous 18 months. The devastation was "greater in extent than anything I had ever seen," he wrote, and on his return to Britain he promptly persuaded Cyril Connolly, editor of the literary periodical *Horizon*, to commission an essay from him to be called "The Natural History of Destruction."[4] He returned to Germany in the spring, and, following the Allied advance, arrived in Köln in early April. This city had been the target of the first "Thousand Bomber raid" at the end of May 1942 and, like so many others, had been bombed repeatedly thereafter. His first view of the devastated city, and particularly of the area around the cathedral, made it impossible for him to complete his essay for Connolly: he said it cried out for more eloquence than he could muster.[5] Sebald notes Zuckerman was so "overwhelmed by what he had seen" that he was unable to convey the enormity of the destruction. Years later, when Sebald asked him about it, all Zuckerman could remember was a surreal still life, "the image of the blackened cathedral rising from the stony desert around it, and the memory of a severed finger that he had found on a heap of rubble." It is immediately after this passage that Sebald asks: "How ought such a natural history of destruction to begin?"[6]

This raises two important questions. The first, naturally enough, is how Zuckerman understood the phrase. Since his report was never written, we can't know for sure; but, given Zuckerman's training, it is not surprising that his interventions over the direction of the bombing campaign should have had recourse to biological-physiological metaphors which conjured up a natural history of sorts. Zuckerman intended these metaphors to convey the effects of bombing not on the human body, however, but on the body politic. He made it clear that he was interested in "the functional inferences" that could be drawn "from aerial photographs of devastated towns," "translating areas of physical destruction into a functional assessment," and when he attempted to persuade his opponents of the need to target transportation nodes he said he "constantly resorted to biological analogies" like arteries, circulation, and paralysis to show that the first priority ought to be "to disrupt a system."[7] Similar metaphors could be found in the RAF's *War Manual* in 1935, and Richard Overy notes that "biological metaphors were commonly used in describing targets" while "paradoxically ignoring the many thousands of real bodies that bombing would destroy."[8] There was nothing paradoxical about it, of course: it was a studied exercise in abstraction. Sebald's enumeration of possible prefaces to a "natural history of destruction" identifies other strategies that work to the same end: "a summary of the technical, organizational and political pre-requisites for carrying out large-scale air-raids;" "a scientific account of the previously unknown phenomenon of the firestorms;" "a pathographical record of typical modes of death." But in each case, significantly, these are followed by a question mark.[9]

Sebald's rhetorical hesitation is significant, I think, because what he understood by a "natural history of destruction" was something different. This is the second question, needless to say, and the most common answer to it has attracted the fiercest criticism. Many commentators have focused on a series of images that Sebald deploys in his description of the air raids on Hamburg between 24 July and 2 August 1943, "Operation Gomorrah," which killed 45,000 people in a single week. Sebald writes of "the whole airspace [as] a sea of flames," a firestorm on the ground "of hurricane force" whose flames "rolled like a tidal wave" through the streets, and smoke rising high in the air to form "a vast, anvil-shaped cumulonimbus cloud." He then describes "horribly disfigured corpses," flames still flickering around them, "doubled up in pools of their own melted fat," and "clumps of flesh and bone" and bodies reduced so completely to ash that the remains of whole families "could be carried away in a single laundry basket."[10] Sebald's critics object that the opening images reduce the air war to a natural disaster "for which no ordinary person was responsible but from which everyone eventually suffered," that a "natural history" of destruction conceived in such terms "assimilates a human-induced and -produced cataclysm into an event of nature" so that it "ontologizes and neutralizes a human product, an historical event."[11] Others conclude that

Sebald's anatomy of grotesquely deformed bodies privileges "the hyper-physical effects of this kind of destruction" and shows that he has no interest in excavating the cultural landscape of terror, pain, and suffering: that, in effect, he multiplies Zuckerman's abstracted image of the cathedral and the finger.[12]

I think these are unfair criticisms, not least (but not only) because they ignore the testimony of those who survived. The extraordinary firestorms released by the raids *were* acutely physical in their causes and in their effects, and survivors repeatedly used the same images to describe them: a "sea of flames," "a hurricane," and even "a volcanic eruption."[13] But I think these objections are also misplaced because they fail to take seriously Zuckerman's inability to convey what he had seen: and it is on this that Sebald sets his sights.

## Language in/of ruins

Zuckerman's attempt to render the devastation of Köln in "natural" or "physiological" terms (which is what I suggested *his* sense of natural history required) was overwhelmed by an inability to make the ruined landscape meaningful. This failure of ordinary language is central to Sebald's account. The survivors, even more than Zuckerman, confronted "a world that could no longer be presented in comprehensible terms," and like them (though also, of course, unlike them) Sebald is struck by the incapacity of "ordinary language" to convey the extraordinary; it was simply inadequate to the task of rendering "the reality of total destruction."[14] More than this, however, there is an intimate connection between the destruction of a city and the ruin of language, which novelist Peter Ho Davies conveys through an arresting image of bomb-damaged Liverpool:

> Esther stares out at the ruins around her … A single gutted house still stands at the end of one flattened terrace like an exclamation mark, and she suddenly sees the streets as sentences in a vast book, sentences that have had their nouns and verbs scored through, rubbed out, until they no longer make any sense.[15]

Simon Ward suggests that "ruin" in Sebald's *oeuvre* more generally marks a site of "broken narration." These ruins are dispersed and they mark traumas that rupture language and leave visible, often photographic, traces that evade or confound linguistic expression.[16] This is the very ground of a natural history of destruction, which in its turn implies that most of the critical responses to such a project have also failed to take Sebald seriously. Indeed, Eduardo Mendieta objects that the reference to a "natural history" in the title of the posthumous English translation of the Zürich lectures is misleading and asserts that it is "not one that Sebald would have chosen."[17] Yet, as I have shown, Sebald clearly regarded the possibility of a natural history of destruction as a crucial question and had in fact explored the idea in relation to the air war in an earlier essay, where he noted Zuckerman's abandoned project and also discussed a radically different concept of natural history from the one vilified by his critics.[18] For the concept of natural history – and it is a concept – derives from Adorno and Benjamin, and marks both the difficulty that Zuckerman faced – the resistance of a ruined, reified world to interpretation – and the ground of Sebald's own inquiry: the site at which memory falters.

"Natural history" conceived in these terms is located at the dialectical intersection of Nature and History or, as Max Pensky has it, of "physical matter and the production of meaning." It brings into view a reified, obdurately physical world – for Adorno and Benjamin the commodity landscape of capitalist modernity, for Sebald the moonscape of modern war – that has been hollowed out and emptied of human meaning. These landscapes thus appear to be "artificially natural."[19] Pensky's is an intricate discussion and it is impossible to convey its subtleties here; but the crux of the matter is captured in a remarkable image in A.L. Kennedy's novel *Day*, where Allied aircrew are being flown back to Britain after the liberation of their prisoner-of-war camps:

> [T]hey flew low and level above the bombed thing that was Germany, above their work. As if the cities had been eaten, as if something unnatural had fed on them until they were gashes and shells and staring spaces, as if it was still down there like a plague in the dust.[20]

There is a hideous literalness to this image, and Sebald describes a "striking change in the natural order of the cities" in the weeks after the air raids: a "sudden and alarming increase in the parasitical creatures thriving on the unburied bodies," a "multiplication of species that are usually suppressed in every possible way," in short the burgeoning populations of rats and flies, the "repulsive fauna of the rubble" – as if the cities were being ravaged all over again.[21] These are extraordinary passages, in which Sebald documents sensation without feeling, morphology without meaning, and life reasserting itself without language. In doing so, his descriptions of a mutely physical geography evoke an altogether different plane that is also conveyed through Kennedy's image: the existential difficulty of recognizing the ruined landscape as the product of human action. Such a "charnel house of rotting interiorities," as Lukàcs described the fetishized landscape of capitalism, cannot be recovered through memory: Pensky insists that natural history is directed *against* the claim of memory as recuperation or recollection and that it works instead to recover "only concrete, singular and utterly empirical facts and bodies, each 'transient', which is to say incapable of being incorporated into a meaning-giving conception of historical continuity and historical experience."[22] And against those who propose a counter-natural or "moral" history of destruction to Sebald's, it is necessary to insist that "transience" here *is* a moral term, a mark of what Pensky calls "the forgetting of the bodily suffering that constitutes the materiality of historical time." In short, there is, in the production of this reified, ruined world, "a functional equivalence between 'that which suffers' and 'that which cannot (must not) be remembered.'"[23]

These are significant elaborations, but Pensky also shows that natural history operates through a particular "way of seeing" or a scopic regime.[24] This speaks directly to Sebald's project too: not only to the optical anxiety to which he draws attention – "we are always looking and looking away at the same time" – but also to the visual register that enframes his own account. Sebald's use of photographs in his work has attracted considerable critical commentary but Carolin Duttlinger has argued that their incorporation in *Luftkrieg und Literatur* (the same images reappear in the English translation) departs from the photographic strategies which inform his literary texts. She is concerned that the totalizing aerial views of destroyed cities (an unsourced photograph of Frankfurt and a photograph of Halberstadt borrowed from Kluge) are not subjected to interrogation. They invite the viewer "to adopt a detached stance" by staging "an abstract geometrical survey which gives the viewer a sense of mastery in the face of chaos," she contends, and in doing so "starkly parallel the perspective of the Allied planes during the attacks."[25] This optical detachment animates Sebald's narrative of the bombing of Hamburg, which is punctuated by a series of aerial perspectives, opening with what Todd Presner calls "a high-angle establishing shot," from the viewpoint of the bombers, and then following "a kind of cinematic logic" that swoops down to the ground only to return to the air.[26] Duttlinger glimpses a critical potentiality in this movement, but it is at best a fleeting one: "What starts out as a position of mastery, an ordered overview, can suddenly tip over into a state of vertiginous disorientation at the sight of destruction."[27]

But I think there is another critical potentiality to be seen in Sebald's account. For these shifts between the air and the ground are mirrored in a stream of measurements: ten thousand tons of bombs, a target area of twenty square kilometers, flames rising two thousand meters, fire advancing at over one hundred and fifty kilometers an hour, smoke rising to a height of eight thousand meters. The result is, as Presner shows, a modernist montage that multiplies different perspectives; it furnishes what Sebald himself saw as "a synoptic and artificial view," which, so Julia Hell argues, "positions us between the illusion of immediate visual access and the consciousness that our 'seeing' is highly mediated."[28] That (critical) awareness of mediation, which is what Duttlinger believes Sebald to marginalize, is dependent on the "establishing shot" and the return to the enframing of the city-as-target but also, as I want to show, on a matrix of measurements, a lethal calculus that abstracted, ensnared and transformed living cities into dead ones.

## Mapping the kill-chain

This was a progressively developing chain – what is today described as the "kill-chain" – that extended from the identification of targets to their destruction. It included the collection and analysis of aerial reconnaissance photographs and the collation of target books and target folders; from January 1943 it included the assignment of numerical and graphical key point ratings and key point factors to establish a hierarchy of targets, and by the end of the year it also included the production of zone maps of target cities, based on the work of geographer R.E. Dickinson, showing population and building densities that were essential for calibrating and setting firestorms (Plate I).[29] The chain also included the production of stylized target maps, which were limited to outline shapes in gray, purple, black and white – since the RAF conducted area bombing at night farther detail would have been superfluous – together with the position of anti-aircraft batteries, Luftwaffe airfields and decoy fires; a series of concentric circles radiated out from the target at one-mile intervals (Plate II). Novelist Len Deighton describes target maps as "sombre things: inflammable forest and built-up areas defined as grey blocks and shaded angular shapes. The only white marks were the thin rivers and blobs of lake. The roads were purple veins so that the whole thing was like a badly bruised torso."[30] From January 1943, ground-scanning H2S radar allowed these maps to be supplemented by crude real-time images of the outline of the target on the aircraft's Plan Position Indicator screen. Finally, there was the intricate choreography of the raid itself, which from December 1942 was orchestrated by a "Master Bomber," a Pathfinder circling above the target to direct the bombers through a shifting grid of flares and red and green markers dropped to outline the target area and, to correct for creep-back, to re-center the force over the aiming point(s). The aerial photographs taken during each raid fed back to start the next cycle, and when *Life* published Bob Landry's photographs of the Commander-in-Chief of Bomber Command, Arthur Harris, poring over the views of destruction, the headline read: "The brain behind the death of Berlin looks at his work from afar."[31] As that caption implies, the kill-chain was thus a concatenation of aerial views produced through a process of calculation that was also a process of abstraction.

Sebald's stream of measurements mimics this process, although he does not refer to it. But drawing out the chain, even in this incomplete form, makes it clear that the mediations involved in enframing the city as a target were by no means secure. None of the images was stable, including the maps that were nominally fixed; far from being "immutable mobiles," as Bruno Latour would have it, they were all subject to constant revision, annotation, and interpretation at each of the points through which the chain extended.[32] And their tacit promise to produce the effects they named – the reduction of a city to a target and thence to rubble – was always conditional. The bomber did not "always get through," as Stanley Baldwin had predicted in 1932, and mortality rates in the air were extremely high because there was an elaborate German counter-imaginary, a parallel system that tracked and enframed the bomber as the target. The two systems moved in counterpoint, and each rationalized its own kill-chain by subdividing its production and regulating its practices through standard operating procedures. This entailed not only an abstraction of the target *but also an abstraction of the process through which the target was produced*, which was made to appear inevitable – target as telos – and its destruction the terminus of a more or less "natural" history. It is in this sense, I think, that Sebald "regards the events of war precisely as a kind of condition that captures participants in its logic whatever their intentions."[33] In other words, the chain enframes the target *and entrains its operators*. The execution of an air raid was animated by a volatile mix of emotions – anger and fear, rivalry and comradeship, excitement and exultation among them – but these were filtered to leave what Alexander Kluge, in his montage of the Allied bombing of Halberstadt, called the *Angriffsmethode*: the pure method of the strike.

In the flight and the bombing, in the gradual purification from the trouble-some ballast of reality, such as personal motivation, moral condemnation of what is to be bombed (moral-bombing), in the calculated know-how, the looking which is replaced by radar control, etc., there is a formalism. It is not aeroplanes ... that are flying here; instead, a conceptual system is flying, a structure of ideas clad in tin.[34]

This effect was produced not only for those in the chain, enveloped in the "conceptual system" and the practices through which it was performed – the manuals and the maps, the drills, and the procedures – but also for a watching public. In July 1941 the Crown Film Unit released Harry Watt's *Target for Tonight*, a dramatized documentary of the bombing of a military-industrial target in Germany, which had been made with the cooperation of Bomber Command using RAF personnel instead of actors. At the beginning of the film the objective is described as "a peach of a target" and its plucking, mediated by a series of aerial photographs and maps meticulously followed by the camera, becomes purely axiomatic and perfectly natural: so much so that Graham Greene writing in the *Spectator* praised the film – and by implication the process that it represented – because everyone involved in the operation had carried out "their difficult and dangerous job in daily routine just like shop or office workers" so that "what we see is no more than a technical exercise."[35]

If the visualizations that produced the target had performative force, however, it is not only the sight of destruction on the ground that has the power to call the aerial mastery of this "technical exercise" into question. That is a common critical response, which, as Duttlinger suggests, depends on memory-work that deliberately abandons detachment: hence Kluge's distinction between "the strategy from above" and "the strategy from below." But in his illuminating discussion of the crisis of representation and modern war Bernd Hüppauf argues that "an iconography based on an opposition between the human face and inhuman technology oversimplifies complex structures," and this opposition reappears in the usual separations between above and below, air and ground, bomber and bombed.[36] I understand the gesture of imaginatively crouching beneath the bombs and establishing an affinity with their victims, but I also believe that by the time we do so it is too late. Another critical response is necessary to precede, supplement, and reinforce this act of empathy and its mobilization of memory: one that has the power to reveal and de-naturalize the "conceptual system" through which the world is reduced to a target.[37] This is the task of a truly critical natural history of destruction capable of addressing the present and future as well as the past.

## "Doors into nowhere"

Such a project can take many forms, but here I continue to focus on the explicitly visual register through a remarkable series of more than 60 images by American artist and scholar elin o'Hara slavick. I have space for only one of them: *Dresden*

(1945) (Plate III).[38] Although the sources for slavick's work are the media of modern war – the aerial photographs, surveillance imagery, and maps I have been discussing – she works by hand rather than, say, video in the hope that her viewers will, like her, "take their time" with them and "work to understand them on a deeper and more complicated level than they might when seeing a photograph." She begins by dropping ink or watercolor onto wet paper, and uses "this common ground of abstract swirling or bleeding" to suggest "the manner in which bombs do not stay within their intended borders."[39] In doing so, she adopts an aerial view – the position of the bombers – in order to stage *and to subvert* the power of aerial mastery. The drawings are made beautiful "to seduce the viewer," she says, to draw them into the deadly embrace of the image only to have their pleasure disrupted when they take a closer look. "Like an Impressionist or Pointillist painting," slavick explains, "I wish for the viewer to be captured by the colors and lost in the patterns and then to have their optical pleasure interrupted by the very real dots or bombs that make up the painting."[40] Her strategy is thus one of deliberate abstraction, and she is not alone in this. Other contemporary artists have drawn attention to the violence of abstraction that is focal to bombing from the air: Martin Dammann's *Über Deutschland* series, for example, or Raquel Maulwurf's *Trümmerfelder* project and her brilliant renderings of RAF photographs of air raids over German cities. But slavick remains uncomfortable at its implications, and she confesses that Sebald's criticism of the production of aesthetic effects from the ruins of an annihilated world "both challenges and paralyzes" her:

> What then is an artist to do? Should I put these drawings away? Should I display images of shrivelled and burnt corpses, photographs of the guilty military generals, pictures of ruins next to the drawings?[41]

And yet here is historian Howard Zinn, who served as a bombardier with the USAAF during the Second World War:

> As I look at [slavick's] drawings, I become painfully aware of how ignorant I was, when I dropped those bombs on France and on cities in Germany, Hungary, Czechoslovakia, of the effects of these bombings on human beings. Not because she shows us bloody corpses ... She does not do that. But her drawings, in ways that I cannot comprehend, compel me to envision such scenes.[42]

That compulsion arises, I suggest, because slavick makes visible a temporality that is contained within the logic of targeting, and even invites its desperate, because agonizingly impossible, reversal. In layering the ghosts of maps and air photographs over the bomb bursts on the ground, and composing beneath and around them a spectral, almost subliminal cellular imagery that, in slavick's own words, "conjures up the buried dead" ("replicated stains in the background, connected tissue in the foreground, concentric targets like microscopic views of damaged cells"), these drawings produce precisely that dizzying, vertiginous glissade that Duttlinger wants to topple the assumption of aerial mastery: but they do so by setting it in motion *from within the aerial view itself.* The bomb-aimer asks the pilot to hold the aircraft steady and as the bomb-doors open so the viewer is precipitated into the dying city. A.L. Kennedy achieves a similar effect in reverse:

> Walk anywhere and you'll catch yourself calculating out from where the first cookie [blockbuster bomb] would fall and blast the buildings open, let the incendiaries in to lodge and play.
>
> And so you see targets beside targets: nothing but targets and ghost craters looping up from the earth, shock waves of dust and smoke ringing, crossing. *You feel the aerial photograph staring down at you where you stand, waiting to wipe you away.*[43]

That extraordinary last sentence breaches the separation between above and below and captures the percussive force of targeting that is also shown in slavick's drawings. For these cities had been reduced to rubble – they were already "dead cities" – *before any bombs were dropped.* This is not only because the violence of representation ("the target") is a necessary condition for the violence on the ground, which ought to be obvious, but also because the one *precipitates* the other: as Kluge puts it in a fictionalized interview with a USAAF Brigadier-General, "The town was erased *as soon as the plans were made.*"[44] Similarly, in his commentary on *Target for Tonight*, Jez Stewart explains that "the logic of the film is that, from the moment the intelligence photographs land safely at Bomber Command, the fate of Freihausen is sealed."[45]

The bomber stream was the advancing edge of a process of abstraction that represented bombing as a domain of pure objects (aircraft and bombs); in some degree, those objects could be personalized, even domesticated – the names and artwork on the bombers, the messages on the bombs – but that humanizing conceit was not extended to the objects of the targeting process. The visualizations within the kill-chain converted cities into numbers and coordinates, shapes and images, so that eventually, as Jörg Friedrich notes, the bombers simply "dropped their load into this abstraction."[46] As a navigator in one bomber crew wrote in a letter to his wife in the summer of 1943: "Were it more personal, I should be more regretting I suppose. But I sit up there with my charts and my pencils and I don't see a thing. I never look out."[47] A natural history of destruction conceived in the terms I have been describing would force us to look out – to see our "not-seeing." A.L. Kennedy captures this, and so much else, when Alfred Day, a tail-gunner in a Lancaster, looks back at the bombing war:

> My, but wasn't it all just a big, free university – the university of war – with HE and armour piercing and incendiaries, just for a lark. And so much to find out: the far edges of people and the bloody big doors into nowhere that you don't want to know about.[48]

Marione Ingram didn't know about them either: but we surely owe it to all the Days and the Ingrams to find out.

## Notes

★ This chapter draws on a much more extended discussion published as "'Doors into Nowhere': Dead Cities and the Natural History of Destruction," in P. Meusburger, M.J. Heffernan, and E. Wunder (eds.), *Cultural Memories*, which will be published in 2011. The title of this version is taken from A.C. Grayling, *Among the Dead Cities: The History and Moral Legacy of the WWII Bombing of Civilians in Germany and Japan*, New York, NY: Walker and Company, 2006, who derived it from an Allied report on possible sites for the trial of Nazi leaders for war crimes; Nuremberg was chosen not only because it had been the scene for spectacular Nazi rallies but also because its location "among the dead cities of Germany" would provide a vivid illustration of Allied retribution, p. 12.

1 H. Böll, "Which Cologne?" *Missing Persons and Other Essays*, Evanston, IL: Northwestern University Press: 1994, pp. 22–7, 25; W.G. Sebald, *Vertigo*, trans. M. Hulse, New York, NY: New Directions, 2001, p. 187; first published in German in 1990.

2 W.G. Sebald, *On the Natural History of Destruction*, trans. A. Bell, New York, NY: Vintage, 2004, p. 4; the Zürich lectures form the first part of this translation, pp. 1–104.

3 S. Zuckerman, *From Apes to Warlords*, London: Hamish Hamilton, 1978, pp. 218, 324. For lucid discussions of the strategy of "area bombing" by night preferred by RAF Bomber Command and the "precision bombing" by day preferred by the US Eighth Air Force – and of the battles that raged between the two and the practical difficulties in making clear distinctions between them – see T.D. Biddle, *Rhetoric and Reality in Air Warfare: The Evolution of British and American Ideas about Strategic Bombing, 1914–1915*, Princeton, NJ: Princeton University Press; R.G. Davis, *Bombing the European Axis Powers: A Historical Digest of the Combined Bomber Offensive 1939–1945*, Alabama: Air University Press, 2006; R. Hansen, *Fire and Fury: The Allied Bombing of Germany 1942–1945,* Doubleday Canada, 2008. In what follows I focus on Bomber Command for purposes of clarity; it was responsible for the lion's share of the destruction of German cities.

4 Zuckerman, *From Apes to Warlords*, pp. 309, 322, 324.

5 Zuckerman, *From Apes to Warlords*, p. 322.

6 Sebald, *Natural History*, pp. 31–2, 33.

7 Zuckerman, *From Apes to Warlords*, pp. 218, 240–2, 289–90.

8 R.J. Overy, "Allied Bombing and the Destruction of German Cities," in R. Chickering, S. Förster and B. Greimer (eds), *A World at Total War: Global Conflict and the Politics of Destruction, 1939–1945*, Cambridge: Cambridge University Press, 2005, pp. 277–96, 284.

9 Sebald, *Natural History*, p. 33.

10 Sebald, *Natural History*, pp. 26–8.

11 D. Crew, "Sleeping with the Enemy," *Central European History* 40, 2007, 117–32, 132; E. Mendieta, "The Literature of Urbicide: Friedrich, Nossack, Sebald and Vonnegut," *Theory & Event* 10:2, 2007, note 14.

12 D. Barnouw, *The War in the Empty Air: Victims, Perpetrators and Postwar Germans*, Bloomington, IN: Indiana University Press, 2005, p. 115.

13 K. Lowe, *Inferno: The Devastation of Hamburg*, London and New York, NY: Viking/ Penguin, 2007, p. 213.

14 This explains his skepticism about eyewitness reports. "The apparently unimpaired ability – shown in most of the eyewitness reports – of everyday language to go on functioning as usual raises doubts of the authenticity of the experiences they record": Lowe, *Inferno*, p. 25.

15 P.H. Davies, *The Welsh Girl*, New York, NY: Houghton Mifflin, 2007, p. 282.

16 S. Ward, "Responsible Ruins? W.G. Sebald and the Responsibility of the German Writer," *Forum for Modern Language Studies* 42, 2006, pp. 183–98, 190.

17 Mendieta, "The Literature of Urbicide," § 14.

18 W.G. Sebald, "Zwischen Geschichte und Naturgeschichte," *Orbis Litterarum* 37, 1982, 345–66, 365–6 note 54. Sebald's translator confirms that the English-language title was his own:

The title was Max's idea [Sebald was known to his friends and colleagues as Max]. I would never have made such a sweeping change of title on my own initiative. In the early stages of the translation project, Max was still referring to it as "Air War and Literature" but he soon decided that it would not cover all the material in the book .... His rationale for the wording is in fact present in his reference to the account of bombed-out Cologne that Solly Zuckerman planned to write, but never did, for *Horizon*.

A. Bell, pers. comm., 27 March 2009.

19  M. Pensky, "Natural History: The Life and Afterlife of a Concept in Adorno," *Critical Horizons* 5, 2004, 227–57: 232.
20  A.L. Kennedy, *Day*, London: Jonathan Cape, 2007, p. 271.
21  Kennedy, *Day*, pp. 34–5.
22  Pensky, "Natural History," pp. 233–4.
23  Pensky, "Natural History," p. 243.
24  Pensky, "Natural History," p. 232; see also S. Buck-Morss, *The Dialectics of Seeing: Walter Benjamin and the Arcades Project*, Cambridge, MA: MIT Press, 1989.
25  C. Duttlinger, "A Lineage of Destruction? Rethinking Photography in '*Luftkrieg und Literatur*'," in A. Fuchs and J.J. Long (eds), *W.G. Sebald and the Writing of History*, Wurzburg: Königshausen and Neumann, 2007, pp. 163–77, 166, 172.
26  T.S. Presner, "'What a Synoptic and Artificial View Reveals': Extreme History and the Modernism of W.G. Sebald's Realism," *Criticism* 46, 2004, 341–60, 354–5.
27  Duttlinger, "A Lineage of Destruction?'," p. 177.
28  Presner, "'What a Synoptic and Artificial View Reveals'," p. 26; J. Hell, "The Angel's Enigmatic Eyes, or the Gothic Beauty of Catastrophic History in W.G. Sebald's 'Air War and Literature'," *Criticism* 46, 2004, 361–92, 370.
29  U. Hohn, "The Bomber's Baedeker: Target Book for Strategic Bombing in the Economic Warfare against German Towns, 1943–1945," *Geojournal* 34, 1994, 213–30.
30  Len Deighton, *Bomber: Events relating to the last flight of an RAF Bomber over Germany on the night of June 31st 1943* (London: Grafton, 1978; Harper, 2009) p. 383.
31  *Life*, 28 February 1944, p. 40.
32  In the British case a minimum mapping would include the (Allied) Central Interpretation Unit at Medmenham, near Marlow (responsible for the analysis of aerial photographs), the Ministry of Economic Warfare and the Air Ministry in London (which identified potential targets), the Air Ministry's Air Intelligence section AI 3 (c) at Hughenden Manor ("Hillside"), near High Wycombe (responsible for producing descriptions of targets for operational planners, and target maps, illustrations and files for briefing officers and aircrew), Bomber Command HQ at High Wycombe, Six to Eight Bomber Command Groups and their bases, and individual flight crews. The chain is a different version of the registers through which Latour tracks the appearance of the Amazon rainforest on the pages of a scientific journal in Paris, and in this sense it too marks the passage of a parallel "natural history": B. Latour, "The 'Pedofil' of Boa Vista: A Photo-philosophical Montage," *Common Knowledge* 4 (1), 1995, 144–87.
33  T. Osborne, "Literature in Ruins," *History of the Human Sciences* 18, 2005, 109–18, 111.
34  A. Kluge, "Der Luftangriff auf Halberstadt am 8. April 1945," in *Neue Geschichten: Hefte 1–8*, Frankfurt-am-Main: Suhrkamp, 1978, 33–110, 65–6, 76.
35  K.R.M. Short, "Bomber Command's 'Target for Tonight' (1941)," *Historical Journal of Film, Radio and Television* (17) 1997, 181–218, 195.
36  B. Hüppauf, "Experiences of Western Warfare and the Crisis of Representation," *New German Critique* 59, 1993, 41–76, 46.
37  R. Chow, *The Age of the World Target*, Durham, NC: Duke University Press, 2006.
38  elin o'Hara slavick, *Bomb after Bomb: A Violent Cartography*, Milan: Charta, 2007; see also her website "Protesting Cartography: Places the United States has Bombed," at http://www.unc.edu/~eoslavic/projects/bombsites/index.html

39 elin o'Hara slavick, "Protesting Cartography: Places the United States has Bombed," *Cultural Politics* 2, 2006, 45–54, 247, 249; slavick, *Bomb*, 97–8.
40 slavick, *Protesting*, p. 249.
41 slavick, *Bomb*, p. 98.
42 H. Zinn, "Foreword," in slavick, *Bomb*, p. 9.
43 Kennedy, *Day*, pp. 202–3 (my emphasis).
44 Kluge, "Der Luftangriff auf Halberstadt," p. 80 (my emphasis); Sebald mistakes this interview for fact, but draws a similar conclusion from Kluge's montage. "So much intelligence, capital and labour went into the planning of destruction that, under the pressure of all the accumulated potential, it had to happen in the end": Sebald, *Natural History*, p. 65.
45 Jez Stewart, "Target for Tonight," BFI Screenonline at http://www.screenonline.org.uk/film/id/577991/index.html.
46 Friedrich, op. cit., p. 25.
47 P. Bishop, *Bomber Boys*, London: Harper, 2008, p. 155.
48 Kennedy, *Day*, p. 271.

# 3

# DIGITAL CARTOGRAPHIES AND MEDIEVAL GEOGRAPHIES

*Keith D. Lilley*

Combining geography, cartography, art history, textual criticism and architecture, *The Iconography of Landscape* is one of the more influential collections of papers for geographers with humanities leanings, especially historical-cultural geographers. Published in 1988, the volume's editors argued for the inherent "mutability" of landscape in the context of a then increasingly invasive and conspicuous digital world, and in their introductory essay concluded how:

> ...landscape seems less like a palimpsest whose "real" or "authentic" meanings can somehow be recovered with the correct techniques, theories or ideologies, than a flickering text displayed on the word-processor screen whose meaning can be created, extended, altered, elaborated and finally obliterated by the merest touch of a button.[1]

Since these words were written, two decades ago, computing and information technologies have pervaded scholarly thought and activity far beyond "flickering text displayed on the word-processor screen" – and yet the consequences of these encroaching digital technologies, as anticipated by Daniels and Cosgrove, have received comparatively little comment from humanities geographers. This is in contrast to other humanities-based disciplines: whether in art history, textual criticism or architecture, scholarly engagements with computing and computers have coalesced into a new intellectual enterprise labeled "digital humanities."

## Digital humanities and humanities geography

Over the past two decades humanities research has capitalized on information communication technologies, providing a basis for multi-media web-distributed versions of particular texts or images, for example, and for disseminating electronic publications and digital editions of textual, visual, oral and cinematic sources.[2] As well as broadening access to humanities research in different fields, specialized computing software has assisted in other aspects. Applications typically include data archiving and retrieval, and computational analyses of "texts" broadly defined, ranging through landscapes and objects, to music and drama, with various means of visualization through computer-generated graphics. As well as this growing volume of digital output from different areas of arts and humanities research, "digital humanities" is emerging as a distinctive field in its own right.[3] Yet for all the intellectual investment and fervour in digital humanities research in recent years, comparable input by humanities geographers has been minimal.[4]

Where geographers might be expected to make a strong contribution to digital humanities is with Geographical Information Systems (GIS). GIS is a multi-media platform for all kinds of spatial information, whether qualitative or quantitative in nature, and highly compatible with various humanities computing outputs and applications, for example in internet dissemination, data storage and retrieval, visualization and computational analysis.[5] While its specialized methods and technical software might have previously acted as something of a barrier in certain humanities subjects more familiar with other digital techniques, such as text encoding and mark-up, current open-access GIS software together with increasingly flexible and more user-friendly tools and interfaces have made GIS both more attractive and widely accessible. However, GIS is a relatively recent phenomenon in digital humanities research, and rather than geographers it is largely archaeologists who are raising its profile, and to some extent also art and architectural historians.[6]

Archaeologists were quick to realize the potential of GIS in visualizing past environments to see how, within various spatial and temporal contexts, landscapes, monuments and buildings were used and experienced.[7] This *qualitative* approach to GIS has literally opened our eyes (and sometimes our ears) to the power of using GIS to understand better the values and practices of past cultures. In human geography, in contrast, GIS applications have tended to follow a social-science model of research, geared toward mapping and analyzing *quantitative* spatial datasets, such as census returns or mortality rates, rather than using the kinds of qualitative sources that are generally the norm in arts and humanities research.[8] This is especially apparent in studies of contemporary urban, social and economic geography, where GIS has had particular impact; yet in historical-cultural geography, too, where some overlap with cognate humanities subjects might be expected, a broadly social-science approach and use of quantitative data and techniques have also predominated in what has become known as "Historical GIS" (HGIS), where qualitative data are rarely used.[9] This prevailing social-science bias in geographers' use of GIS might thus in part account for geography's absence in digital humanities, while its perceived "science" basis perhaps also explains why humanities geographers seem to avoid using GIS, for even its vocabulary and nomenclature, of "analysis" and "data," is seemingly at odds with the qualitative approaches of the humanities and its interpretative "discourses." A further and related factor might be that GIS remains largely un-theorized.

Applications of GIS and their resultant forms of "digital mapping" have largely escaped critical, theoretical study, by either geographers or other GIS-users such as archaeologists. John Pickles is one of the few geographers to have taken the philosophical ideas of social and cultural commentators, such as Paul Virilio and Bruno Latour, to provide an *exposé* of geographers' "uncritical valorization" of GIS, noting how GIS-based studies are typically empirically focused with more attention given to modeling potential, methodological development, and GIS applications rather than contemporary theoretical debates.[10] In his reappraisal of GIS, Pickles draws on the "critical" philosophy of cartography developed and deployed by J.B. Harley in the 1980s and 1990s.[11] Harley's approach to maps and mapping is familiar to many historical-cultural geographers – it made an appearance in Cosgrove and Daniels' *Iconography of Landscape*, for example – yet among users of GIS it is often overlooked, even in Historical GIS, despite Harley's work having much to offer as a basis for theorizing digital maps and mapping.[12] This again points to the conceptual and methodological divide in geography between users of GIS on the one hand and humanities geographers on the other. The task that geographers have, then, is to bridge this gap, and for humanities geographers to embrace more the use and application of spatial technologies in our work. Doing so might not only raise the external profile of humanities geographers in burgeoning areas of digital humanities research, but also offer us new theoretical and empirical insights into areas of our own particular interests, in understanding past worlds and cultures for example.

## Medieval mappings and "cartographic consciousness"

As a geographer, Harley began his academic career working on medieval rural settlement but like so many historical geographers in the past few decades he moved forwards in time to focus more especially on the post-medieval world.[13] Geographers' neglect of the medieval period and its geographies and cartographies has had the unfortunate effect of foreshortening our temporal horizons.[14] In some areas of historical geography and cartographic history, maps and map-making of the Middle Ages are described still in teleological terms as a step in the development of "modern," "rational," "scientific" and "accurate" cartography, as if medieval maps were simply a staging post on the long road toward "cartographic consciousness."[15] This is not at all how historians of medieval geography and cartography see things, however, as recent critical studies of theirs have shown.[16] However, where historical geography might have something new to offer this emerging history of medieval geography is the application of GIS in reinterpreting medieval maps and map-making, particularly in pursuing Harley's interest in the *agency* of maps and the *agents* of map-making, for in medieval contexts neither authorship nor provenance is usually known, even in celebrated cases such as *mappaemundi*.[17] What GIS can provide is an analytical approach to medieval maps that allows us to begin to explore how maps were made during the Middle Ages, the roles that maps played, and the extent to which there was in medieval Europe, too (despite the persistent misconception), a "cartographic consciousness" among its inhabitants.

The Gough Map of Great Britain is one of the few medieval maps specifically produced to show a whole country, yet little is known of its provenance. Based upon its palaeographical and historical details the map is usually dated to *c.* 1360 while of its authorship nothing is recorded, though it is typically seen to belong to an English royal context.[18] Now in the ownership of the Bodleian Library in Oxford, the unique manuscript map was digitally scanned in 2005.[19] This provided an opportunity to begin to analyze the cartographic and geographical content of the Gough Map, to see what these might reveal of its origins and purpose. The methods used for this are described in detail elsewhere, and involved importing the digital map into ArcGIS and then digitizing its various features, including routes, settlements, rivers, and coastline.[20] Through this process of digitization the map "raster" is thus "vectorized," meaning that a qualitative source – in this case a manuscript – is converted into digital, quantifiable spatial data. It is these spatial data that then provide a basis for exploring how the Gough Map was made, by whom and why.

Once the map is in the GIS it becomes possible to begin to analyze statistically the information that it shows, such as its settlement pattern. In all, the Gough Map locates more than 600 settlements through a system of "vignettes" covering England, Wales and Scotland. Visually, the location of each on the manuscript seems to match locations where modern maps would place them, suggesting that the creators of the Gough Map were seeking to produce a map that was to scale and of a practical purpose, for travel and administration for example. Because each place on the manuscript is given a map position through the digitization process, it is relatively straightforward (through using latitude-longitude or National Grid coordinates for the same locations) to see how places on the Gough Map relate to modern conceptions of geographical space. The key to this is to use regression procedures and to compare locations in "map space" and "grid space." There are different ways of doing this, and also of mapping out the results of the statistical analyses, to explore the map's relative compression or exaggeration of distances between the places shown. This gives a guide to variations in the map's "distortion."[21] The important proviso here is that this is not to reinforce ideas that medieval maps were "inaccurate" compared to mapping today, but rather to expose the Gough Map's subjectivities to see which parts of Britain were better known by the map's maker(s), or more important to them, and then use this as a guide to suggest where they were geographically located. Using a Distortion Grid, the statistical outputs can be expressed cartographically, and this reveals that the places mapped in south-eastern England were more consistently positioned on the map than those in northern and western parts of Britain (Plate IV). With a distinct Anglo-centric orientation and configuration, the Gough Map's *relative* "distortion" maps out an imagined geography of medieval Britain as perceived by its unknown creator(s).[22]

Using GIS to digitize and analyze the Gough Map might well risk the charge of "presentism," and yet the map's cartographic and geographical content offer the only means to understand how this map was made. What it reveals on the part of its maker(s) is his or their desire for a map that was to scale, broadly, as far as the places it showed were concerned. Through regression analysis the map's variations in scale can be explored but overall, taking the map as a whole, its creators had somehow gained enough geographical information about the locations of places to depict their positions systematically and so make the Gough Map a practical document. From it estimates might be made for the length of time it would take to move from one place to another or to direct travelers and administrators around the realm in the most cost-effective way. As Harley showed, all maps are at once both utilitarian and symbolic, and the Gough Map is no different.[23] But what GIS in this case shows is that in Plantagenet England, probably in the court of King Edward III, there were the means and also the desire to map the land in the modern sense, to create a spatially realistic image of Britain. This perhaps should not come as a surprise, for Matthew Paris, a producer of maps in the mid thirteenth century, recognized the value and importance of scale in drawing and in the same period the Oxford scholar Roger Bacon devised a system of creating maps using a coordinate system.[24] However, with the Gough Map and the application of GIS we now have an indication that such ideals were being put into practice and that there was a cartographic consciousness which was decidedly "modern" in its outlook; seeking to map as accurately as possible, measuring and surveying the land for the purposes of knowing the realm.

Analyzing the Gough Map within a GIS shows its potential for looking "behind the map," as it were, at the agents who created it as well as its agency. There is also scope for using GIS to explore other aspects of medieval "cartographic consciousness" even where maps no longer exist. One example concerns the design and planning of towns in medieval Europe. Manuscript plans for the architectural design of some buildings are known and a detailed contemporary parchment plan of the new town of Talamone in Italy might have been drawn up deliberately for laying out its streets and plots on the ground, but otherwise, despite the widespread formation of urban landscapes in the Middle Ages, sometimes to complex geometrical designs, the evidence for layouts being first mapped out for planning purposes is quite sparse; unless the surviving physical layouts of the towns themselves are examined.[25] Here GIS provides a useful analytical framework, partly to help reconstruct the likely early layout of medieval "new towns," and show their physical forms as they were laid out at the time of a town's foundation, and also as a basis for analyzing the form of its urban landscape. Having used GIS to map medieval urban layouts (from qualitative sources such as historic maps and plans, archaeological field-surveys and documentary records), it is then straightforward for the geographer to overlay their forms to see how closely (or not) they correspond and whether they share any common design traits.[26] A case in point concerns the new towns founded on sites adjacent to castles in North Wales established in the late thirteenth century under English authority by the formidable Plantagenet king, Edward I.[27]

Historians have long known that Edward employed Master James of St. George, an architect, in the construction and design of his Welsh castles, but under whose responsibility the castles' adjacent towns were created has remained less clear.[28] Master James was present early during the construction of castles at Conwy and

Beaumaris, in 1282–4 and 1294–6 respectively, and the layouts of these two towns, like their adjacent castles, show stylistic similarities in design. Within the GIS it is apparent just how close the two town-designs are to one another (Plate V).[29] The street plan of Conwy matches up perfectly when overlaid on that of Beaumaris, suggesting not only that the same designer was responsible for both – presumably in this case Master James – but also that a plan had been drawn up, presumably on parchment, as part of the planning process. Such a practice would not be unusual for architects in the thirteenth and fourteenth centuries, but for Conwy and Beaumaris no parchment plan survives, just the layout that arose as a consequence. What this suggests is a "cartographic consciousness" on the part of a town's creator; with overtones of "modern" urban planning and design in the need to work out first a plan and then implement it on the ground. The survey work required for this task is in effect the reverse of the measuring needed to make the Gough Map. Either way, it shows how individuals in the Middle Ages were quite capable of thinking spatially and cartographically, accurately surveying and mapping, and applying techniques of ground-measurement to meet particular ends. Such a view of the medieval designer, architect, cartographer and surveyor is at odds with the image so often portrayed of them and yet it is a thoroughly modern digital cartography that provides a means to explore their geographical imaginations via "medieval mappings."[30]

## Conclusion

GIS is particularly effective in helping us to reinterpret historic maps and map-making, and Harley's "philosophy of cartography" provides a useful conceptual framework not only for raising questions about historical geographies and geography's histories but also for promoting a more critical use of GIS in humanities research. Thus, Harley's concern with both the agency of maps and the agents who were involved in making maps are as relevant and important in the production of new digital maps as they are for (re)interpreting past mappings, forcing those of us using GIS for "mapping the past" to think a little more deeply about the processes we are involved in and also the effects (and agency) our own digital maps may have. Harley's "philosophy of cartography" therefore provides a useful critique of GIS-based research that helps to align it with contemporary concerns of humanities geographers. At the same time, taking an empirical yet reflexive approach which interprets qualitative material digitally and spatially also makes the case to "digital humanities" scholars that geographers have a particularly distinctive contribution to offer, not least on the subject of past and present maps and map-making. This short chapter arises from an attempt to begin to forge such a connection.[31] It shows how digital cartographies have intellectual advantages for humanities geographers: offering a definably geographical input into ongoing debates among geography's historians, both within the discipline and also outside geography, thus furthering the study of medieval geographies and cartography, and strengthening scholarly relationships between those working across the humanities in related disciplines. At a time when figurative and metaphorical "mappings" are becoming particularly prominent in other humanities areas, such as literary criticism and philosophy, it is perhaps worth underlining the benefits of still thinking about maps and "map-making" in a more conventional and literal sense.[32] Here a critical approach to GIS, focusing on an historical period that is often now neglected by geographers, highlights how geographers in particular, with our spatial aptitude and awareness, have much to contribute to the developing field of "digital humanities."

## Notes

1  S. Daniels and D. Cosgrove, "Introduction: Iconography and Landscape," in Cosgrove and Daniels (eds), *The Iconography of Landscape: Essays on the Symbolic Representation, Design and Use of Past Environments*, Cambridge: Cambridge University Press, 1988, p. 8.

2  For an indication of the kinds of areas covered by digital humanities see: http://www.arts-humanities.net/, accessed July 22, 2009.

3  e.g., through a range of new scholarly journals and periodicals such as *Humanities Computing, Digital Humanities Quarterly* and the *Digital Medievalist*, along with specialized academic conferences such as "Digital Resources in the Humanities and Arts"; and dedicated research centers, such as the Center for Digital Humanities in the University of South Carolina and the Centre for Computing in the Humanities (CCH) at King's College, London. The accepted importance of digital technologies in contemporary humanities research is further underlined, at least in a UK context, by the place that it now holds in the strategies and decision-making processes of the main funding council for British arts and humanities research, the Arts and Humanities Research Council (AHRC).

4  This is made clear, for example, by a recent set of essays collected under the title *A Companion to Digital Humanities*, which though broad in its coverage of arts and humanities disciplines, ranging through archaeology, to classics, to literary studies, music and the performing arts, neglects to include geography either as a humanities subject area or as a mode of critical inquiry and analysis. See S. Shreibman, R. Siemens and J. Unsworth (eds), *A Companion to Digital Humanities*, Oxford: Wiley-Blackwell, 2004. For a recent discussion see D.J. Bodenhamer, J. Corrigan and T.M. Harris (eds), *The Spatial Humanities. GIS and the Future of Humanities Scholarship*, Bloomington, IN: Indiana University Press, 2010.

5  On GIS generally see P.A. Longley, M.F. Goodchild, D.J. Maguire and D.W. Rhind (eds), *Geographical Information Systems: Principles, Techniques, Management and Application*, 2nd edn, abridged, Hoboken, NJ: Wiley, 2005.

6  See M. Jessop, "'The Inhibition of Geographical Information in Digital Humanities Scholarship," *Literary and Linguistic Computing* 23, 2008, 39–50; H. Eiteljorg, II, "Computing for Archaeologists," in Schreibman, Siemens and Unsworth (eds), *Companion to Digital Humanities*.

7  See D. Wheatley and M. Gillings, *Spatial Technology and Archaeology: The Archaeological Applications of GIS*, London: Taylor and Francis, 2002; J. Conolly and M. Lake, *Geographical Information Systems in Archaeology*, Cambridge: Cambridge University Press, 2006.

8  e.g., I.N. Gregory and P.S. Ell, "Analysing Socio-Spatial Change Using National Historical GISs: Population Change Before and After the Great Irish Famine," *Historical Methods* 38, 2005, 149–67; I.N. Gregory, "Different Places, Different Stories: Infant Mortality Decline in England and Wales, 1851–1911," *Annals of the Association of American Geographers* 98, 2008, 773–94.

9  See D. Martin, *Geographic Information Systems and their Socio-Economic Applications*, 2nd edn, London: Routledge, 1996; I.N. Gregory and P.S. Ell, *Historical GIS: Technologies, Methodologies and Scholarship*, Cambridge: Cambridge University Press, 2007, in which qualitative data and GIS get but a cursory mention at the end of the volume, pp. 195–9.

10  J. Pickles, *A History of Spaces: Cartographic Reason, Mapping and the Geo-coded World*, London: Routledge, 2004, pp. xvii–xviii. To this end Pickles sees in GIS a "democratization" of geographical and cartographic knowledge, especially when coupled to the

internet, as well as a "malleability" in GIS-based cartography, providing a viewer of a map – or mapped spatial information – the scope to interact with it, and even adapt it and add material of their own, pp. 145–75. See also J. Pickles, *Ground Truth: The Social Implication of Geographic Information Systems*, New York, NY: Guilford Press, 1995.

11  J. Pickles, *A History of Spaces*, pp. 47–9.

12  For a critique of Harley's "philosophy of cartography" see P. Laxton, "Preface," in J.B. Harley, *The New Nature of Maps: Essays in the History of Cartography*, Baltimore, MD: Johns Hopkins University Press, 2001, pp. ix, xv.

13  On Harley's career see M.H. Edney, "The Origins and Development of J.B. Harley's Cartographic Theories," *Cartographica Monograph* 54, Toronto: University of Toronto, 2005.

14  R. Jones, "What Time Human Geography?," *Progress in Human Geography* 28, 2004, 287–304.

15  Pickles is guilty of this, for example: see Pickles, *A History of Spaces*, pp. 75–106. In discussing "innovations in mapping practice," and their correspondence with "European Renaissance ways of seeing," he uses the phrase "map consciousness," pp. 96–9 at page 99. Pickles is not alone in promoting this view, however; e.g., see N. Thrower, *Maps and Civilization: Cartography in Culture and Society*, Chicago: University of Chicago Press, 1996. It is a modernist conceit that can be traced back through twentieth-century geographers' representations of medieval cartography, e.g., G.H.T. Kimble, *Geography in the Middle Ages*, London: Methuen, 1938; J.K. Wright, *The Geographical Lore of the Time of the Crusades: A Study in the History of Medieval Science and Tradition in Western Europe*, New York, NY: American Geographical Society, 1925; R.C. Beazley, *The Dawn of Modern Geography*, London: John Murray, 1897, Beazley, *The Dawn of Modern Geography*, 2, London: H. Frowde, 1901; Beazley, *The Dawn of Modern Geography*, 3, Oxford: Clarendon, 1906.

16  e.g., N. Lozovsky, *"The Earth is Our Book: Geographical Knowledge in the Latin West, ca. 400–1000*, Ann Arbor, MI: University of Michigan Press, 2000; N.R. Kline, *Maps of Medieval Thought: The Hereford Paradigm*, Cambridge: Boydell Press, 2001; A.H. Merrills, *History and Geography in Late Antiquity*, Cambrdge: Cambridge University Press, 2005; K. Lavezzo, *Angels on the Edge of the World: Geography, Literature, and English Community, 1000–1534*, Ithaca, NY: Cornell University Press, 2006; R.J.A. Talbert and R.W. Unger (eds), *Cartography in Antiquity and the Middle Ages: Fresh Perspectives, New Methods*, Leiden: Brill, 2008.

17  On tracing map authorship see, for example, P. Barber, "The Evesham World Map: A Late Medieval English View of God and the World," *Imago Mundi* 47, 1995, 13–33.

18  E.J.S. Parsons, *The Map of Great Britain circa A.D. 1360 Known as the Gough Map: An Introduction to the Facsimile*, Oxford: Bodleian Library, 1958; N. Millea, *The Gough Map: The Earliest Road Map of Great Britain*, Oxford: Bodleian Library, 2007. See also D. Birkholz, *The King's Two Maps: Cartography and Culture in Thirteenth-Century England*, New York, NY: Routledge, 2004. New insights into the Gough Map's date and origins are being revealed by the Linguistic Geographies research project, funded by the Arts and Humanities Research Council (AHRC): see www.goughmap.org for further details.

19  For details see K.D. Lilley and C.D. Lloyd (with contributions by B.M.S. Campbell), "Mapping the Realm: A New Look at the Gough Map of Britain (c. 1360)," *Imago Mundi* 61, 2009, 1–28.

20  See C.D. Lloyd and K.D. Lilley, in "Cartographic Veracity in Medieval Mapping: Analyzing Geographical Variation in the Gough Map of Great Britain," *Annals of the Association of American Geographers* 99, 2009, 27–48.

21  For a fuller discussion of this, see Lloyd and Lilley, in "Cartographic veracity."

22  Other regression procedures using the same data complicate this picture, however. Geographically Weighted Regression (GWR), for example, reveals localized variations; thus places along the coast of southern England seem to be more consistently positioned on the map than those located inland, suggesting that ease of distance measurement (on

the ground) was an important factor in creating the Gough Map. See Lilley and Lloyd, in "Mapping the Realm," pp. 17–18.

23 J.B. Harley, "Maps, Knowledge, Power," in Cosgrove and Daniels (eds), *Iconography of Landscape*, pp. 277–312. For a "symbolic" reading of the Gough Map see Birkholz, *King's Two Maps, passim.*

24 See Lilley and Lloyd, "Mapping the realm," pp. 17, 19.

25 On this see K.D. Lilley, *City and Cosmos: The Medieval World in Urban Form*, London: Reaktion, 2009, pp. 41–73.

26 See K.D. Lilley, C. Lloyd and S. Trick, *Mapping Medieval Townscapes: A Digital Atlas of the New Towns of Edward I*, York: Archaeology Data Service, University of York, 2005, http://ads.ahds.ac.uk/catalogue/specColl/atlas_ahrb_2005 (accessed July 22, 2009).

27 See K.D. Lilley, C.D. Lloyd and S. Trick, "Designs and Designers of Medieval 'New Towns' in Wales," *Antiquity* 81, 2007, 279–93.

28 For Edward I's Welsh castles and their historiography see A.J. Taylor, *The Welsh Castles of Edward I*, Hambledon, 1986.

29 The correspondence between the two designs is apparent in their street-layouts especially, but their exact match is evident only if the plan of Conwy is "flipped" (i.e., inverted) in the GIS and then when overlaid onto the plan of Beaumaris the coincidence of street-orientations and angles is indisputable. The two plans are kept at the same scale in this process and there is no manipulation of the designs to make them fit.

30 I wish to acknowledge here the generous funding I have received from the British Academy and the Arts and Humanities Research Council (AHRC) for past research projects that are referred to in this chapter. I am also grateful to my colleagues Chris Lloyd, Steven Trick and Bruce Campbell for their various practical and intellectual inputs into these projects. Additional influences and inputs into the ideas put forward in this chapter have come from Denis Cosgrove, Stephen Daniels and Ian Gregory, to whom my thanks are due, and also the contributors and participants of the Mapping Medieval Geographies conference held at the Center for Medieval and Renaissance Studies at UCLA in May 2009.

31 For example, through involvement in a collaborative project, *Mapping Medieval Chester* (see http://www.medievalchester.ac.uk), linking humanities computing, human geography, and literary history. See P. Vetch, C.A.M. Clarke and K.D. Lilley, "Between Text and Image: Digital Rendering of a Late Medieval City," in B. Nelson and M. Terras (eds), *Digitizing Medieval and Early Modern Material Culture*, Tempe, AZ: RSA/Medieval and Renaissance Texts and Studies in press.

32 On the various permutations of "mapping" see D. Cosgrove, "Introduction: Mapping Meaning," in D. Cosgrove (ed.), *Mappings*, London: Reaktion, 1999, pp. 1–23; N. Howe, *Writing the Map of Anglo-Saxon England: Essays in Cultural Geography*, London: Yale University Press, 2008, pp. 4–7. Matthew Edney has pointed out the conceptual and terminological importance of not conflating the practices of "mapping" and "map-making," M. Edney, *Origins and Development of J.B. Harley's Cartographic Theories*, 9.

# 4

# MAPPING THE TABOO

*Gunnar Olsson*

Mapping the taboo is in itself forbidden. And in my understanding the most forbidden of everything forbidden is that which refuses to be categorized, that which is neither this nor that, ungraspable forces which do not sit still but hop capriciously about. Aristotle consequently knew what he did, when he inserted between the two concepts of identity and difference a third position called "the excluded middle," a non-bridgeable gap which in the same figure unites and separates, liberates and imprisons – an unruly space located beyond the realm of conventional reason, a no man's land which the well-behaved must never enter, perhaps because in the midst of life we never know what will happen next. But Aristotle also argued that what one cannot do perfectly, one must do as well as one can. It follows that even though what I say about the sea-battle tomorrow is strictly speaking neither true nor false, not every comment about it is equally empty.

One reason why the forbidden remains forbidden is that it is protected by the taboo, a situation which is etymologically connected not merely with the terms "under prohibition" and "not allowed," but with "sacred" and "holy" as well. The taboo is therefore doubly tied first to the forbidden itself and then to the strongest form of the taken-for-granted, more precisely to those aspects of the unconscious which are important enough to be blessed by the gods themselves, by definition beyond reach; sirens singing COME, sirens blaring DANGER. But why should I devote my professional life to issues which are not worthy of being taboo? How could I possibly ignore the question of what it means to be human? How could I ever fail to wonder how and why we are made so obedient and so predictable?[1]

★

For these reasons of power and socialization I am once again reminded of *Enuma elish*, the Babylonian tale of how the god Marduk gained and retained his elevated position as the Lord of lords. The premise of this oldest creation epic extant is that in the beginning of the beginning nothing has yet been formed, because in the beginning of the beginning nothing has yet been named. All that there is are merely the spatial coordinates of above and below, cardinal positions waiting to be inundated by the fluids of masculine Apsu and feminine Tiamat, the former sweet, the latter bitter. And, as if to underline the spatiality of its own structure, the term *apsu* literally means "abyss" and "outermost limit," by linguistic coincidence connected also to "the great deep," "the primal chaos," "the bowels of earth," "the infernal pit;" a perfect example of proper name and definite description merged into one.

Eventually there is a tremendous power struggle and sweet Apsu is killed by Ea, the most outstanding of his offspring. On top of the corpse, that is, *across the abyss*, Ea then builds a splendid palace for himself and his wife Damkina. There, in the Chamber of Destinies, their son Marduk is conceived, the most awesome being ever to be:

> Impossible to understand, too difficult to perceive.
> Four were his eyes, four were his ears;
> When his lips moved, fire blazed forth.
> The four ears were enormous
> And likewise his eyes; they perceived everything.[2]

Marduk's weapons are numerous, but most decisive are the magic net in which he goes to capture the recalcitrant Tiamat and the four winds by which he eventually blows her up. When it is finally over, Marduk, great lord of the universe, "crossed the sky to survey the infinite distance; he stationed himself in the apsu, that apsu built by [Ea] over the old abyss which he now surveyed, measuring out and marking in."[3] No longer dressed in the warrior's coat of mail but in the uniform of a land surveyor, he then proceeds first to the construction of a celestial globe and finally to the creation of a primeval man, the prototype of you and me, a creature explicitly designed to serve as slaves of the ruler's vassals, three hundred stationed as watchers of Heaven, an equal number as guardians of the Earth. This is not an invention formed in the image of the Almighty, though, but a savaged concoction stirred together from the blood of the slaughtered Kingu, Tiamat's lover and commander in chief; mankind a dish of *Boudins à la Mésopotamie*; nothing like a perfect copy of the perfect original, merely a black sausage.

Throughout these events the abyss remains the power center *par excellence*, the broken clay tablets of *Enuma elish* the ultimate proof of the Babylonians' insights into the secret workings of human thought-and-action. And therein lies, in my mind, the real reason for keeping the abysmal gap between categories taboo, for it is in the ontological transformations of the excluded middle that the magicians of power are performing their tricks. Hence it is only by entering this forbidden space of imagination that the analyst can ever hope to understand how the absent can be made present, the present made absent.

Most vividly these connections are expressed by the figure of Janus, my own favorite among gods. What intrigues me with this pivotal symbol of gate-keeping is less that he is equipped with a body which makes him see in opposite directions at the same time, but more that he has a mind which allows him to merge seemingly contradictory categories into one meaningful whole. From his watchtower at the middle of the bridge he is consequently able to keep both sides of the abyss under constant surveillance, in the same glance catching a glimpse of those pasts that once were and of those futures that have yet to come.

Given the Greek fear of the void – itself well expressed by Aristotle's decision to introduce the concept of the excluded middle – it is not surprising that Janus was invented in Rome and not in Athens. In the lands surrounding the *Mare nostrum*, though, he was everywhere to be seen, for not only was his image stamped on practically every coin, but in religious prayers this janitor of janitors was also the first to be mentioned and in cultural rituals this son of January was equated with the beginning of all beginnings. Diana was his godly consort, a connection which explains why the doors of his temple stayed open in times of war and why they were shut in times of peace. Like ordinary lovers, gods need their privacy too.

Janus' main concerns were one with my own – creativity, power, socialization. Defiantly I therefore pray again:

> Oh Janus! Help me become a sinner. Let me understand how you break definitions and thereby create. Show me how you erase what others see as irresolvable paradoxes. Teach me the equation of that third lens inside your head whereby contradictory images are transformed into coherent wholes. Speak memory, speak! SPREACH, Janus, SPREACH! And Babel's walls come mumbling down.[4]

Accordingly, and throughout my scholarly and artistic life, I have been searching for a place inside Janus' head. From that zero-point of the excluded middle I have then tried to grapple with the taboos of limits, the sins of trespassing, the braiding of epistemology and ontology, the challenge of writing in such a way that the resulting text actually *is* what it is about. With the aim of understanding how Janus stayed sane while ordinary people in similar situations of double bind go crazy, I have therefore tried to place him on the operating table, cut his skull open, lay his brain bare, investigate how his mind is wired. Why and how, for instance, did the Romans elevate this categorical juggler to godly status, when we, their descendants, diagnose his counterparts as schizophrenic madmen? Why did they afford him a special place in their pantheon, while we isolate his likes in the soundproofed cells of the asylum?

Perhaps the reason is that without distinctions our thoughts-and-actions would have nothing to stick to, our lives nothing to share. Such vacuities are in fact the norm in the Realm of Psychosis, that literally unthinkable province where there are no initiation rites, no scars, no individuals, hence no society either. And this empti-ness may well explain why the deeply psychotic is so frightening, because the deeply psychotic lives outside the laws of thought, an inhabitant of the excluded middle, an alien beyond both identity and difference; a non-mappable world without fix-points, scales and projection screens.

Lest it be thought that my understanding of the void is too closely tied to the Abrahamitic world, I now recall a stunning visit to the city of Kandy, once capital of the Sinhalese kingdom which in 1815 was annexed by the British and included as part of colonial Ceylon.[5] There the high priest of the temple of Sri Dalada Maligawa – the shrine that among other relics houses the tooth of Buddha, historically *the* national symbol – granted me and my wife a rare audience. Not just any audience though, but a visit to the *Vedahitina Maligawa*, the holiest of the holy, a small room on the upper floor with an altar bestrewn with jasmine flowers and the sacred tooth enshrined in a casket of gold. Before entering this forbidden place, we were most carefully instructed how to behave, especially *not to step on the threshold*, the barrier that separates the commoners in an antechamber and the higher classes in a middle room, on the one side, from the inner sanctum with the king, his closest ministers and the *Diyawadana Nilame*,[6] on the other: a wonderful illustration of how the excluded middle can be materialized in a silver-encased janitor that must not be touched.

The mind boggles as it encounters the walls of Babel, Kreml and Berlin in yet another setting, the hierarchical structure of the three chambers of the temple highly reminiscent of the narthex, nave and sanctuary of the orthodox church, the Kandyan threshold effectively serving the same exclusionary functions as the Russian iconostasis. Most astonishing is nevertheless the story that before King Vimaladharmsuriya I entered the same room in 1592 as we did in December 2007, he kneeled and put his forehead on the polished threshold; the stamp of power in the place of power, the mark of Cain in a Buddhist context, a clear warning that anyone who sets foot on the threshold is trampling not on a material object but on power itself. This circumstance, rather than the Greek fear of the void, is in my analysis the real reason why the excluded middle is excluded. And for protecting his own holiness from possible usurpers, the LORD put a mark on the restless wanderer so that no one who found him would kill him. In the same breath a blessing and a curse, yet another indication that it is in the nature of absolute power to violate every rule of behavior, to do exactly as it pleases. The reason is, of course, that in a norm system where both *a* and *not-a* are valid at the same time, everything is permitted.

No wonder, therefore, that it is from a position in the excluded middle that the Almighty rules, his words-and-deeds predictably unpredictable, his palace surrounded by a non-penetrable defense system, his propaganda machine everywhere to be heard and nowhere to be evaded. Yet everything codified in the constitutional law of Moses' first stone tablet, in my heretic (hopefully not blasphemous) interpretation the most penetrating show of power and submission ever performed.

★

The first stone tablet is, in the same document, a blessing and a curse, hated whip and pleasurable carrot brought together in a socialization instrument that no one can escape. A rhetorical masterpiece firmly rooted in the concept of trust, a social glue which under the label *pistis* was foundational to Aristotle as well; the point is that without pistis there is no communication and that is regardless of whether the chosen language is that of money, poetry, logic, geometry or anything else. This in turn suggested to Aristotle that dialectics and rhetoric are the twin sisters of each other, just as to me it illustrates how the two activities of logic and geometry are forms of rhetoric which after long use have become so credible that they now count as categories of their own. It cannot be said more clearly: reasoning is a persuasive activity grounded in the tension between personal trust and social verification.

In a very general sense it is this question of how we find our way in the unknown that lies at the heart of European culture, perhaps of all cultures. In Erich Auerbach's influential analysis of mimesis it is located exactly in the taboo-ridden interface between the certainties of Odysseus' scar and the ambiguities of Abraham's fear, you and I dangling in the abyss in-between.[7] Two modes of understanding, two modes of being, two ways of living which over the centuries have been condensed, purified and eventually codified, one in Aristotle's Laws of Thought, the other in the biblical formulation of the commandments, the latter not merely the 10 that can be counted on the fingers,[8] but a staggering total of 613. The interpretations vary accordingly, even though it is generally agreed that the ten words of the Decalogue may be divided into two groups such that the first three or four govern the relations between God and man (the Constitutional Law) and the rest regulate the relations between man and man (the Civil and Criminal Law).

In times of crisis it is the words of the first tablet that tell the ruler how to rule, and that is regardless of whether the potentate happens to be a Machiavellian Prince, a dictatorial Führer, an elected Prime Minister, a concerned parent. It is hard to imagine a more power-filled statement, not the least because it is there that YHWH for the first time reveals his own name, an expression so closely related to the Hebrew word for "to be," *hvh/hjh*, that it is often translated as "The Being." As a way of farther stressing its importance, it was this invisible entity itself, not one of its usual emissaries, who in the prologue let his subjects know that *it was I who liberated you, I who let you out of the land of Egypt, I who cut your chains.* The implication is, of course, that since I have proven myself to be such an outstanding leader in the past, you are wise to trust me also in the future; accordingly, every incumbent assures the voters that they never had it so good, that they should read his lips and scrutinize his record. Although you should prepare yourself for blood, sweat and tears, at the end of day there will be milk and honey. Thus I decree, because I am who I am.

Immediately following this naked piece of rhetoric comes the first paragraph of the Constitutional Law, a proposition as stunning now as when it was first uttered: *I shall be your dictator!* Wherever this Almighty happens to be – and by definition he is at the same time everywhere and nowhere – he shall rule over everyone and everything, like the surveying Marduk measuring out and marking in, showing mercy to those who love him and killing those who hate him.

The unknown genius who was the first to coin the phrase that there must be no power before (or according to some translations, "beside") me was certainly wise enough to realize that whoever declares that he shall be my dictator leads a dangerous life. For that reason he erected around the apsu palace a two-tier defense system consisting of both a wall and a moat, the former constructed as a ban on the (mis)use of metaphor, the latter as a rule against the creative associations of metonymy. The purpose of the second paragraph is consequently to ensure that the weapons gathered in the rhetorical arsenal will not fall into enemy hands, rephrased as: any critique must be silenced before it is uttered. In that mood the jealous LORD now decrees that you shall for ever know your place, never commit the sins of trespassing, never question his authority. In particular *you shall not possess the means for making of me a graven image, picture, statue or any other caricature, never tie my name to a definite description.*

The recent debacle about the Danish Mohammed pictures in its proper light,[9] for the graven image has always been the master key to idolatry and thereby to the doors of competing ideologies and potential usurpers. In the present context it is especially noteworthy that the Hebrew term for "image" refers more to the dwelling place of the divine than to the pictorial representation of its invisible being.[10] It follows that if you tell me *where* you are, I shall tell you *what* you are. Yet, as soon as I attempt to make the invisible visible, I run the risk of falling into the trap of misplaced concreteness, of deifying the reified. However, by outlawing the *as-if*, the Untouchable guarantees that no news will ever issue from his subjects but only from himself.

It cannot be said more clearly: the second paragraph amounts to a devastating *auto-da-fé*, a combined prohibition against picture-making and story-telling, the two primary modes of translation, understanding and reasoned critique. Even so, the declaration that I shall be your dictator is so outrageous that no censor will ever be strong enough to get it generally accepted. Other socialization techniques must therefore be mobilized as well and that is indeed the purpose and function of the third paragraph. With that goal firmly in mind, the law-maker therefore once again reminds the congregation that it was he who took them out of the land of bondage, he who gave them the freedom which he himself is now determined to take back. Therefore, after all these ordeals, I hereby declare that you deserve a rest. However, this precious time you must not spend alone but always in the company of your likes. In the synagogue and the church, at the play group and the faculty meetings, the confirmations, funerals and family dinners – it is at these gatherings that my officials will teach you how to think-and-act. The Kantian thesis about the necessary unity of consciousness in another form, for you must always remember that *you are nothing but a cog in my machinery. I am the spiritualized embodiment of your unconsciously taken-for-granted, the pivot of the world.* And provided you honor your father and mother I shall grant you a long lease on the land that I give you.

Like the drip drip drip of the raindrops, when the summer shower's through, so a voice within me keeps repeating you, you, you. And so it is that I read the commandment to keep the Sabbath day holy as the most crucial paragraph of the Constitutional Law, the ultimate guarantee that the power structure of monotheism will survive. And so it also is that Aristotle's Laws of Thought and Moses' Laws of Submission may be read as alternative maps of power, two codifications with the double purpose of showing how in the same breath you can tell the truth and be believed when you do it. It is difficult to imagine two formulations of greater historical significance, layers of meaning deeply embedded in the taken-for-granted, a palimpsest of the already but not yet.

★

Every map is a palimpsest, a product of imagination, that uniquely human faculty which assigns to the semiotic animal the privilege of making the absent present and the present absent. Simsalabim and the vistas from elsewhere lie open in front of us, the image of a reality never seen before, a no-where miraculously changed into a now-here, a shade of blue turned into an ocean, a line into a road, a dot into a city. By all accounts a most remarkable version of the story "Let there be and there is," an outstanding case of rhetoric performed on the high wire.

No wonder, therefore, that in absolute regimes even the most innocuous map is treated as a state secret, for just as no magician wants his tricks to be revealed, so every ruler guards his palace and masks his face. And that in turn explains why the biblical redactors let the LORD say to Moses:

> "I will cause all my goodness to pass in front of you, and I will proclaim my name, the LORD, in your presence. I will have mercy on whom I will have mercy, and I will have compassion on whom I will have compassion. But," he said, "you cannot see my face, for no one may see me and live." Then the LORD said, "There is a place near me where you may stand on a rock. When my glory passes by, I will put you in a cleft in the rock and cover you with my hand until I have passed by. Then I will remove my hand and you will see my back; but my face must not be seen."[11]

What a remarkable passage, nothing less than an exhibition of power in its nakedness, an image viewed from an abysmal cleft, a name spoken in an utterance of self-reference. Even more remarkably, I here detect an allusion to the second paragraph of the Constitutional Law with its double ban on picture and story, the two modes of representation that lie at the heart of cartographic reason. No wonder that the surveyor of power leads such a dangerous life, for how can his analyses be trusted when the faceless phenomenon he wishes to capture is itself steeped in distrust. The liar's paradox in a different context, for you never know who in the early hours may be knocking on your door.

And therein lies the profound difference between the social ethics of the first and second stone tablet. For even though the concept of pistis permeates both documents, the form of trust which ties you and me together is mutual, while the trust between the ruler and his subjects is at best (or is it at worst?) one-sided; since the Absolute is by definition self-referential, his name (if a name it is) cannot be translated into a definite description. Indeed the very sign of the covenant that the LORD makes with Noah is a rainbow, a palette of fleeting colors in the clouds rather than a material object on the ground.

Even so, the doubters refuse to be silenced and that explains why Abraham took the LORD to task for not keeping his promises of many children and why Job sued him for slandering, a court case never to be forgotten. In between is the story of Jacob, one of the greatest crooks ever born, yet one of the richest rewarded.[12] Of Jacob much can be said, but nothing more important than the fact that in the chronicles it was he who was the first to claim that he had seen God's face and survived; in the eyes of the Almighty the blasphemy of blasphemies, to the present analyst a propaganda trick of astonishing proportions. It may in fact be instructive to approach the first third of the Hebrew Bible as the story about a power struggle so violent that the self-proclaimed LORD is eventually forced to withdraw. Thus, after the Book of Job he never speaks again, and, as if to continue the assault, the New Testament contains many references to the commandments of the second stone tablet but makes no explicit mention of the first: fascinating glimpses of the interface between the knowledge of power and the power of knowledge, between geography and the humanities.

★

Located in the interface between knowledge and power is the art of mapping; just as no map can be a perfect map, so every account of power and knowledge depends first on the chosen fix-points and then on the scales through which the points are translated into connecting lines. Even more important than these instruments, however, is the cartographer's projection screen or *mappa*, the taken-for-granted plane onto which the pictures and travel stories are cast and preserved. It is of course tempting to associate the fix-points with the first paragraph of the Constitutional Law, the scales with the second and the *mappa* with the third.

Fix-points first, for have I not already noted that in the Realm of Power nothing sits still, that its jealous ruler never sleeps in the same bed two nights in a row? Since the earliest accounts his palace has been variously located in the abyss between categories, in the untouchable threshold between this and that, in the face which must not be seen, in the excluded of the excluded middle. In addition, the LORD's name is in most creation myths a tautology, by definition true but not informative. Ungraspable is the ungraspable, which for that reason is free to do whatever it pleases; predictably unpredictable, inherently untrustable; always there to see, never to be seen, Bentham's panopticon in advance of itself. For what my eyes happen to catch depends both on where my body stands and on how my mind has been molded.

Then the scale, by definition the translation function that enables me to claim that this is this and that that is that. Yet I have repeatedly stressed that in the Realm of Power everyone and everything hops capriciously about, sometimes appearing as this sometimes as that. To put it bluntly, God (a term which to me functions as a pseudonym of power) does not operate according to the laws of logic. And therein lies my understanding the reason why the social sciences in comparison with the hard sciences have accumulated so little knowledge. If it is true, which I believe it is, that human action is structured like a tragedy – everything beautifully right in the beginning; everything horribly wrong in the end; no one to blame in between – then the social sciences are faced with a tremendous challenge, easier to state than to do anything about. To be precise: if human action actually *is* structured as a tragedy, how can we then rely on the principle of truth preservation for tying our premises and conclusions together? Surely the most common purpose of human action is to topple truth, not to preserve it, to falsify rather than retain what is now the case; less a matter of formal logic, more an instance of creative imagination. This to me is the problem of trust and verification, the real issue that the map-maker's scale is addressing.

Finally the *mappa*, the formation of the taken-for-granted, the painter preparing the canvas to ensure that the paint will not run off and the surface not crack, the glazier polishing the tain of the mirror. This is in effect what the unconsciously adopted socialization techniques are designed to do, making you and me obedient and predictable in the process, everything hidden yet everything faintly glimpsed.

★

As might be expected, a similar form of cartographic reason guided the thoughts-and-actions of the Greeks as well. Nowhere is this more evident than in Plato's *Republic* with its three figures of the Sun, which together with the concept of good-ness functions as the analyst's *fix-point par excellence*; the Divided Line which embodies the *scale* through which abstract ideas are turned into concrete things, degrees of truth corresponding to degrees of being; and the Cave Wall, the *mappa* of the surveyor's projection screen, the taken-for-granted background without which there would never be any shadows to observe, hence no maps to hide and seek.[13]

When the chips are down, this may be the real reason why the paragraphs of the Constitutional Law and the definitions of the Laws of Thought refuse to die away; the twin sisters of rhetoric and dialectics joined together in a struggle against indi-vidual madness, social chaos and political turmoil. Not everything forbidden should be allowed.

Once that warning has been issued, I also recall that it was Aristotle's teacher who noted that were it not for the Sun there would be no light and were it not for the light there would be nothing to see. Beware though! For after a minute of staring at the fireball your unprotected eyes are sure to be blinded. Illumination comes with a price.

## Notes

1 The fullest record of these explorations is in G. Olsson, *Abysmal: A Critique of Cartographic Reason*, Chicago: University of Chicago Press, 2007. In what follows I will draw exten-sively on passages from that minimalist tome, not because of laziness or as a case of self-plagiarism but in frustration over my inability to come up with phrases that are more precise. Since the appearance of *Abysmal* at least three dissertations have been explicitly devoted to the in-between: C. Abrahamsson, *Topoi/graphein*, Uppsala: Kulturgeografiska insitutionen, Uppsala universitet, 2008; M. Sand, *Konsten att gunga: Experiment som aktiverar mellanrum*, Stockholm: Axl Books, 2008 and P. Sangasumana, *Mapping Inbetweenness: The Cases of Conflict Induced Internally Displaced in Sri Lanka*, Colombo: Department of Geography, University of Sri Jayewardenepura, 2010.

2 S. Dalley, *Myths from Mesopotamia: Creation, the Flood, Gilgamesh, and Others*, Oxford: Oxford University Press, 1989, p. 235.

3 N.K. Sandars, *Poems of Heaven and Hell from Ancient Mesopotamia*, Harmondsworth: Penguin, 1971, p. 92.

4 Olsson, *Abysmal*, p. 6.

5 I am most grateful to the Reverend Pinnawala Sangasumana for opening the gates to the unknown.

6 Literally the "water-increasing-official."

7 E. Auerbach, *Mimesis: The Representation of Reality in Western Literature*, trans. W.R. Task, Princeton, NJ: Princeton University Press, 1953.

8 Exod. 20:3–17 and Deut. 5:6–21. All biblical quotations will be from *The New International Version*, Grand Rapids, MI: Zondervan Publishing House, 1973.

9 G. Olsson, "När Himmelbjerget kom till Mohammed," in Kjeld Buciek et al. (eds), *Rumslig praxis: Festskrift til Kirsten Simonsen*, Roskilde: Roskilde Universitetsforlag, 2006.

10 J.J. Stamm with M.E. Andrew, *The Ten Commandments in Recent Research*, London: SCM Press, 1967, p. 82.

11 Exod., cited in note 8, 33:19–23.

12 For more detailed analyses of these biblical stories, see G. Olsson, *Abysmal*, esp. the chapters entitled "Abr(ah)am" and "Peniel." Also J. Miles, *God: A Biography*, Simon & Schuster, 1995.

13 My own analyses appear throughout *Abysmal*, but the connections with Plato's *Republic* (Books 6 and 7) are primarily in the chapters entitled "In-Between," "Plato" and "Philadelphia."

# 5

# *CHOROS, CHORA* AND THE QUESTION OF LANDSCAPE

*Kenneth R. Olwig*

## Introduction

Landscape has long been a focus of geography's engagement with the humanities because it is both a core concept within some geographical traditions and an important topic in the visual and the literary arts. Some geographers have conceptualized landscape as primarily a form of region or place,[1] while others have seen it as primarily a pictorial, scenic form of representation[2] and yet others have emphasized the role of literature in the representation of landscape.[3] In many circumstances these differences can be explained in semiotic terms involving differing ideas of the relationship between a type of signifier (a pictorial or literary representation) and that which is signified (a place), and how this affects perception and behavior.[4] Vision, understood in this context, is a sensory modality that has a complementary relationship to other sensory modalities while providing its own valuable perspective on the seeing, imagining, and representation of the world.[5] There is, however, an influential Ptolemaic/Platonic strand of perspectival visual representation that inordinately complicates the under-standing of the relationship between a form of representation and that represented, and hence the understanding of what is meant by landscape. This form of representation is difficult to grasp because the subject of representation is first and foremost the very *space* that is not only the foundation of the representational form itself, but also of our idea of the subject represented – in this case landscape. It will be argued here that the relationship between these differing conceptualiza-tions of landscape can be clarified if the development of the concept of landscape is related to the parallel development of the concept of chorography, and its Greek root *choros/chora*.

## Chorography and landscape

*Choros* or *chora* is the root of *chorography*,[6] historically a branch of geography, but it is also a concept that, usually spelled as *chora*, plays an important role in philosophy. Chorography is a classic, often broadly humanistic branch of geography that was eclipsed by the mid-twentieth-century rise of geography as spatial science,[7] but which is enjoying renewed interest today.[8,9,10,11,12] The ancient Greek root of chorography, *choros* or chora,[13] literally means "a definite space, piece of ground, place" and makes particular reference to "the lower world," defining it among other things as "land or country."[14]

The Berkeley Geography Department of the University of California, which fostered such humanities-oriented geographers as Clarence Glacken, Yi-Fu Tuan, David Lowenthal, and Edmunds Bunkse, was founded by Carl Sauer, who located the roots of landscape geography in the geographical tradition of chorography, the study of region as place. "[T]he facts of landscape are place facts," as Sauer put it.[15] He identified *choros/chora*, which he spelled "*chore*," with landscape.[16] Landscape, as I have shown, is a concept which has had an areal meaning similar to *choros/chora* since at least the Middle Ages.[17,18]

In contrast to Sauer, Denis Cosgrove has argued that: "It is well known that in Europe the concept of landscape and the words for it in both Romance and Germanic languages emerged around the turn of the sixteenth century to denote a painting whose primary subject matter was natural scenery."[19] For him, landscape was primarily a form of pictorial representation and only secondarily a representation of a natural scene in pictorial space.[20]

Despite the difference between Sauer's and Cosgrove's conception of landscape, both nevertheless link their understanding of landscape to chorography. Sauer, however, traced his understanding of chorography to narrative chorography in the tradition of ancient Greeks such as Herodotus and Homer, or the Roman Strabo, as shall be seen later. Cosgrove, on the other hand, traces the advent of scenic landscape depiction to the perspectival representational techniques stemming from the Renaissance rediscovery of the cartography and chorography of the second century AD Alexandrian Greek astronomer, astrologer and geographer Claudius Ptolemy.[21]

## Plato, Ptolemy, chorography and *chora*

Ptolemy provided a set of cartographic techniques that were seized upon in the Renaissance to create the perspectival illusion of spatial depth. His cartographic instructions were contained in his *Geographike Uphegesis* meaning "guide to geography,"[22] which was rediscovered in the Renaissance. "Geography," for Ptolemy, was "the representation, by a map, of the portion of the earth known to us, together with its general features."[23] *Topography* was a subdivision of geography concerned with locations within the cartographic space of the map whereas chorography was concerned with regions: "geography differs from chorography in that chorography concerns itself exclusively with particular regions and describes each separately, representing practically everything of the lands in question ...."[24]

Whereas Ptolemy saw the construction of the map, and the plotting of locations upon it, to be matters for mathematics and science, he could not fit chorography into this exact scheme. He contrasted chorography to geography, stating that "no one can be a chorographer unless he is also skilled in drawing" whereas "by using mere lines and annotations it [geography] shows positions and general outlines." This leads to the conclusion that "while chorography does not require the mathematical method, in geography this method plays the chief part."[25]

In the Greek original Ptolemy described the -*graphy* in chorography as "*mimesis-dia-graphes*," meaning imitation/representation through a graphic form. In the process of translation into Latin this came to be rendered by some as "*imitatio picturae*" or, as it was put in English at the time, a "certaine imitation of paintinge."[26] This suggested to contemporaries that "Ptolemy himself wanted to establish some kind of relationship between the two disciplines."[27] The linkage between pictorial art and chorographic cartography was particularly notable in the case of the theater. It was thus common to call the cosmographic atlases of the time theaters, as in Ortelius' *Theatrum Orbis Terrarum* (theater of the lands of the globe).[28] A theater is different from a landscape painting or drawing because the perceiver is within the space of the theater and does not gaze into this space from without, as through a window, as is the case with a painting. Likewise, in a theater using the techniques of perspective scenography, which were developed in tandem with those of perspective art, the figures performing within the perspectival space of the stage scenery are not frozen painted images but moving embodied actors. From the creation of theater scenery it was thus but a short step to the architectural design of buildings and the planning of cities and rural scenery.[29]

The problem of how to translate Ptolemy turned largely on the interpretation of the meaning of the -*graphy* in three key interlinked terms in Ptolemy's discussion of geography: *geography*, *topography*, and *chorography*. Ptolemy defined geo*graphy* proper as the mapping (*diagraphos* – meaning to mark out by lines) of the areas of the globe as known to humankind (-*graphy* referring to *writing* or *representation*). There was a long tradition in Greek and Roman geography, traced back to Homer, Herodotus, and Strabo, in which geography and history were linked in chorography, and were primarily seen to be topics for writing, which is to say, language and discourse.[30,31] When, however, geography is redefined as the graphic practice of mapping, then geography, topography, and chorography become identified primarily with the use of graphic and pictorial techniques of representation rather than with writing. This emphasis upon the pictorial can be related to the influence of Platonism upon Renaissance interpreters of Ptolemy.

## Platonic cosmology and landscape

Ptolemy's geography fit perfectly the Platonic cosmology that fascinated the Neo-Platonists of the Renaissance. Ptolemy, who appears himself to have been influenced by Plato,[32] based his geography upon the science of mapping. Ptolemy's cosmography was earth centered and, like that of Plato, it separated the sublunar sphere of the earth from the surrounding sphere, or spheres, of the cosmos which revolved around the fixed point of the earth. For the Greeks the cosmos was not simply the universe, it was a principle of order, and the ideal geometric paradigm for that order was to be found in the heavens. According to Liba Chaia Taub, Ptolemy's astronomical writings represented "the culmination of a rather neglected form of Platonic ethical theory, with its special emphasis on astronomy."[33]

Cartography, for Ptolemy, was based upon the use of celestial coordinates to create a grid, or graticule (latitude and longitude), upon which could be plotted locations on the globe. This kind of map was thus composed of a geometric structure, analogous to the paradigmatic archetypal cosmic geometric forms or ideas that, in Plato's cosmology, constituted a higher *ideal* reality which was visible in cosmic space. The concept of the *idea*, that is, the root of *ideal* came from the Greek word *idein*, meaning "to see."[34] Ideas, in other words, were something one saw, as when one saw geometrical forms in the motions and shapes of the heavenly bodies, and when one saw the rationality of a geographical theorem via its pictorial representation as lines and points within a spatial framework. Sciences such as geometry made it possible for people to perceive, as mental images, archetypal ideal ideas expressed through their graphic representation, as in a geometric theorem or, as in Ptolemy's case, a map or globe. The Ptolemaic map thus can be seen as a representation of the idea of an ideal cosmic geometric nature (that of the earth as celestial globe) which undergirds and orders the chaotic phenomena of the sublunary realm of earthly nature. The perspectival representation of chorography as landscape likewise represents an ideal idea of landscape.

The role of representation in Platonic philosophy suggests that the *-graphy* in geo*graphy* provided a way of visually re-presenting archetypal ideas in Plato's cosmology. It therefore implied more than a simple graphic likeness. Ptolemy's *Geography*, in Latin translation, was often given the title "Cosmography." This mistranslation was not unreasonable, however, because the *Geography* can be perceived as being concerned with cosmography defined as "the constitution of the whole order of nature or the figure, disposition, and relation of all of its various parts."[35] Ptolemy's geo*graphy*, understood as cosmo-*graphy*, thus can be seen to form the graphic pictorial image that represents the underlying archetypal ideas behind Platonic cosmology. Cosmology was "a branch of systematic philosophy that deals with the character of the universe as a cosmos by combining speculative metaphysics and scientific knowledge; especially: a branch of philosophy that deals with the processes of nature and the relation of its parts."[36] It is thus in this sense that the perspectival representation of chorography was a representation of nature.

The Ptolemaic cosmography created a world picture that was first a representation of an ideal Platonic spatial framework within which was inscribed a world that, especially through linear perspective, created an illusion of the lived space of daily life. Ptolemy himself was aware of the ability of mapped space to create an illusory bounded whole, like that of a face:

The purpose of chorography is the description of the individual parts, as if one were to draw merely an ear or an eye; but the purpose of geography is to gain a view of the whole, as, for example, when one draws the whole head.[37]

This image of the globe as a head with a face was to reappear in pictorial form in a famous woodcut, illustrating Ptolemy's "cosmography," found in Peter Apian's influential *Cosmographicus Liber* from 1533.[38] By representing the land as a head and face the Renaissance cosmographers gave the landscape a mask-like personality – *persona* being the Latin for mask and, by extension, face – that was capable of being captured by painters, much as in a portrait. This face, however, is an ideal construction within a Platonic space. Gilles Deleuze and Félix Guattari call this sort of inversion, by which an objective environment takes on a subjective scenic face, "deterritorialization":

> We could say that it is an *absolute* deterritorialization: it is no longer relative because it removes the head from the stratum of the organism, human or animal, and connects it to other strata, such as signifiance and subjectification. Now the face has a correlate of great importance: the landscape, which is not just a milieu but a deterritorialized world. There are a number of face–landscape correlations on this "higher" level. Christian education exerts spiritual control over both faciality and 'landscapity' (*paysagéité*): Compose them both, color them in, complete them, arrange them according to a complementarity linking landscapes to faces.[39]

A landscape that is fundamentally a representation of an ideal space within which the world is given a scenic face and personality is describable as deterritorialized, but this, of course, is not the only meaning of landscape. Prior to the Renaissance deterritorialization of landscape via Ptolemaic/Platonic spatial and perspectival forms of representation the primary meaning of landscape, as well as the choros/chora in chorography, was a form of place territorialized as region.

## Space, place, region and choros/chora

The term "chorography" has come to be defined in purely Ptolemaic terms as: "The art or practice of describing, or of delineating on a map or chart, particular regions, or districts; as distinguished from *geography*, taken as dealing with the earth in general, and (less distinctly) from *topography*, which deals with particular places, as towns, etc."[40] The problem with this interpretation is that it does not capture the complexity of meaning of the ancient Greek usage. *Chora* is thus one of the most daunting concepts in Plato's cosmology as set forth in his *Timeaus*. Plato describes *chora* as being apprehended:

> …without the senses by a sort of bastard reasoning and hardly an object of belief. This, indeed, is that which we look upon as in a dream and say that anything that is must needs be in some place and occupy some room.[41]

Plato is describing *chora* as apprehended in relation to the cosmology put forth in the *Timaeus* and it is in this context that its reasoning is "bastard," without legitimate parents. *Chora* does not fit into Plato's cosmology and the ideas that are its legitimate progeny. As Jacques Derrida puts it, *chora* "is something which cannot be assimilated by Plato himself, by what we call Platonic ontology, nor by the inheritance of Plato."[42]

Derrida deals with the philosophical conundrum posed by *chora* by stepping outside the box of Platonic cosmology and examining the meaning of *chora/choros* as an ordinary term in ancient Greek in which it meant, among other things, "place," "region," and "country" in the sense of the land of a people.[43] He takes note that these are politically defined places and comments that "the ordered polysemy of the word [*chora*] always includes the sense of political place or more generally of *invested* place, by opposition to abstract space. *Chora* 'means': place occupied by someone, country, inhabited place, marked place, rank, post, assigned position, territory or region."[44] This observation is in agreement with the ancient Greek and Roman conception of *chorography* as practiced, for example, by Strabo according to the geographer Christiaan van Paassen. Strabo was concerned with the region and its contents qualitatively described. His chorography thus did not involve "a determination of length and breadth, no copying of the outline, but a narration of 'such are the conditions here.'"[45] Among the things that Strabo treated was therefore the "laws and political institutions of the various peoples."[46] The nineteenth century was a period when interest in the chorography of Strabo, and early predecessors such as Herodotus, was revived by humanities-oriented geographers after having been looked down upon because it did not have the quantitative, scientific, rational ambitions of Ptolemy.[47] It was this anti-Ptolemaic notion of chorography that attracted Sauer, who noted approvingly Alexander von Humboldt's observation that: "'In classical antiquity the earliest historians made little attempt to separate the description of lands from the narration of events the scene of which was in the areas described. For a long time physical geography and history appear attractively intermingled'" (von Humboldt quoted in Sauer).[48]

*Chora* was not just a philosophical concept; it also had a substantive meaning and this meaning, as Derrida tells us, had implications for "the discourse on places, notably political places."[49] The role of discourse in the constitution of *chora* is brought out in Derrida's description of the constitutive role of the *agora* as "the political place (*lieu*) where affairs are spoken of and dealt with."[50] The term *agora* is used to name the place where assemblies are held as well as to denote the people who make up the deliberative assembly, and even the speech pronounced at the assembly.[51] For an ancient Greek the essence of *choros* as a polity would have been expressed through the intangible discourse of the citizens gathered at the *agora*. The country would not have been defined, as in a modern state, by lines on a map, but by the discourse of the citizenry at the *agora* through which a polity was established. An *agora* is a place where Greeks aggregated for the purpose of a market and for the purpose of political exchange through discourse and it is derived from the Greek *ageirein*, meaning to assemble.[52]

The *agora*, which forms the core of the historical Greek *choros*, is literally a *place* in the etymologically original sense of the term: "Middle English, from Middle French, open space in a city, space, locality, from Latin *platea* broad street, from Greek *plateia* (*hodos*)."[53] It was in such a place that the Greek polity flocked together, gatherings took *place* and a polity was leveled, planted in a *place* that symbolized their political heritage.[54] Some theoreticians of place, such as Fred Lukermann[55] and Edward Casey,[56] have looked to *chora*, as defined by Plato, for a concept equivalent to place but the problem is that when place is defined in terms of *chora*, as understood by either Plato or Ptolemy, it inevitably is effected by the Platonic attempt to reduce *chora* to a spatial phenomenon. Echoing Plato, Lukermann thus writes that "choros technically means the boundary of the extension of some thing or things, it is the container or receptacle of a body." *Choros* therefore "may safely be translated in context as area, region (*regio*), country (*pays*) or space/place – if in the sense of the boundary of an area."[57] This, of course, resembles the Ptolemaic cartographic understanding of *chorography*, which has to do with the creation of a regional whole within the space of the map by connecting together locations along the boundary of the region. What one sees here thus is a kind of leakage from Plato's discussion of *chora*, where Plato struggles to assimilate this "bastard concept" into the spatial ontology of the Timaeus. But, as Derrida put it, *chora* "is something which cannot be assimilated by ... Platonic ontology."[58] *Chora* thus becomes, in Derrida's words, "a space that cannot be represented."[59] The *agora* centers on a holy core but has no defining spatial boundary. It is the gathering of people at the *agora*, not a bounded topographical space within which it meets, that defines the *choros* of its polity.

When one tries to assimilate *chora* to a Ptolemaic/Platonic spatial framework it leads to a concept of place that is inextricably bound up with location in space, and hence a notion of "space/place" that is, in Derrida's words, "a kind of hybrid being."[60,61] Such a hybrid is inevitably characterized, to borrow Nicholas Entrikin's term, by a "betweeness"[62] characteristic of the modernity originating in the Renaissance.[63] The elision of meaning from *choros/chora* as a region or land as the place of a polity to place as a location in space, that inevitably occurs when *choros/chora* is used in post-Ptolemaic discourse, suggests that it might be better to avoid simply translating *choros/chora* as place and rather to use a phrase such as place-region or landscape-region.[64]

## Conclusion

The interjection of the concept of *choros/chora*, and with it *chorography*, into the discourse on landscape helps clarify a number of confusions that have long plagued geographical discourse, and which are of particular relevance to the connection between geography and the humanities. One such confusion is that between landscape as a visual form of representation, identifiable with the so-called "new cultural geography" of Denis Cosgrove and Steven Daniels,[65] and landscape defined as an area or region identified with the "Berkeley" geography of Carl Sauer.[66] There is, in fact, no necessary opposition between landscape defined as a region or country and its pictorial representation as a prospect of that place. The one is simply a representation of the other, even if they may both inform and shape each other. There is no necessary connection, however, between the use of linear perspective and the visual representation of landscape. Just as humanist geographers criticized the idea of the mental map for assuming that human spatial perception is similar to the spatiality of the geographer's map,[67] the same argument can be applied to the idea that visual perception is similar to that of linear perspectival scenery, rooted in the space of the map. Not all landscape painters used linear perspective and not all paintings using linear perspective had landscape as their motif. What is important, as Cosgrove has shown, is that some used linear perspective to both represent and define landscape, and the visionary idea of landscape that this form of representation engendered was subsequently to have a major effect on ideals governing perception and shaping of the physical environment as idealized landscape scenery.[68] Whether this form of visionary spatial representation and its consequences for the planning and design of the physical environment are for better or worse is a vital subject of debate[69,70,71] which should not be confused with the question of whether landscape can or should be represented visually.

The *choros*, as Derrida and others have suggested, came about through the language and discourse of the polity at the *agora*. Similarly, *chorography* has a long history of narrative representation. The distinction between representation through language and by visual means is useful because it brings out the importance of language and narrative to a discipline that tends to be dominated by the visual and the spatial.[72,73] Language is as important to the humanities as visual modes of expression, and the study of language has long had strong ties to geography. It is thus difficult to imagine the study of a language such as Danish that does not include the study of Denmark's chorography.[74] Language is fundamental to any polity and Denmark would be incomprehensible without Danish. This broad humanistic study of language is termed philology, a concept that derives ultimately from the Greek meaning "love of argument, learning, and literature, from *philologos* love of words and learning."[75] This chapter is an example of a philological approach to chorography, landscape, and geography.

## Notes

1  C.O. Sauer, "The Morphology of Landscape," *University of California Publications in Geography* 2(2), 1925, 19–53.
2  D. Cosgrove and S. Daniels (eds), "Introduction: Iconography and Landscape," in *The Iconography of Landscape*, Cambridge: Cambridge University Press, 1988, pp. 1–10.
3  K.R. Olwig, *Nature's Ideological Landscape*, London: Allen & Unwin, 1984.
4  K.R. Olwig "'This is not a Landscape:' Circulating Reference and Land Shaping", in M. Antrop et al. (eds), *European Rural Landscapes: Persistence and Change in a Globalising Environment*, Dordrecht: Kluwer, 2004, pp. 41–66.
5  D. Cosgrove, *Geography and Vision: Seeing, Imagining and Representing the World*, London: I.B. Tauris, 2008.
6  *Oxford English Dictionary*, Oxford: Clarendon Press, 1989.

7 K.R. Olwig, "Nature – Mapping the 'Ghostly' Traces of a Concept," in C. Earl, K. Mathewson and M. S. Kenzer (eds), *Concepts in Human Geography*, Savage, MD: Rowman and Littlefield, 1996, pp. 63–96.

8 K.R. Olwig, "Landscape as a Contested Topos of Place, Community and Self," in S. Hoelscher, P.C. Adams, and K.E. Till (eds), *Textures of Place: Exploring Humanist Geographies*, Minneapolis, MN: University of Minnesota Press, 2001, pp. 95–117.

9 K.R. Olwig, "Has Geography Always Been Modern?: *Choros*, (non)Representation, Performance, and the Landscape," *Environment and Planning A* 40, 2008, 1843–61.

10 I.J. Birkeland, *Making Place, Making Self: Travel, Subjectivity and Sexual Difference*, Aldershot: Ashgate, 2005.

11 M.R. Curry, "Toward a Geography of a World Without Maps: Lessons from Ptolemy and Postal Codes," *Annals of the Association of American Geographers*, 95 (3), 2005, 680–91.

12 M. Pearson, "*In Comes I": Performance, Memory and Landscape*, Exeter: University of Exeter Press, 2006.

13 *OED*, cited in note 6.

14 H.G. Liddell and R. Scott, *A Greek–English Lexicon*, Oxford: Clarendon Press, 1940, p. 2016.

15 Sauer, "The Morphology of Landscape," p. 26.

16 Sauer, "The Morphology of Landscape," p. 46.

17 K.R. Olwig, *Landscape, Nature and the Body Politic*, Madison, WI: University of Wisconsin Press, 2002, p. 232.

18 Samuel Johnson, a chorographer and a lexicographer, gives as the first sense of landscape a definition similar to Sauer's understanding: "A region; the prospect of a country," S. Johnson, *A Dictionary of the English Language*, 1755, London: W. Strahan, 1968. The root *pays* in *paysage* also means a place, area, or region and this common European under-standing of landscape is reflected in the European Landscape Convention's definition of landscape as: "an area, as perceived by people, whose character is the result of the action and interaction of natural and/or human factors," Council of Europe, European Landscape Convention, Florence, 2000, CETS No. 176. The use of the modern German spelling "*Landschaft*" to indicate this sense of landscape as place-region can give the misleading impression that this is a specifically German sense of the term; D. Cosgrove, "Landscape and Landschaft" lecture delivered at the Spatial Turn in History Symposium, German Historical Institute, February 19, 2004, *GHI BULLETIN*, 35, 57–71.

19 D. Cosgrove, *The Palladian Landscape: Geographical Change and its Cultural Representations in Sixteenth-Century Italy*, University Park, PA: Pennsylvania State University Press, 1993, p. 9.

20 This conception fits Johnson's second sense of landscape as: "A picture, representing an extent of space, with the various objects in it." S. Johnson, *Dictionary of the English Language*, cited in note 18.

21 D. Cosgrove, "Prospect, Perspective and the Evolution of the Landscape Idea," *Transactions of the Institute of British Geographers N.S.* 1, 1985, 45–62.

22 C. Ptolemy, "Geographike Uphegesis (excerpts)," in I.E. Drabkin and M.R. Cohen (eds), *A Sourcebook in Greek Science*, Cambridge, MA: Harvard University Press, 1948, pp. 162–81.

23 Ptolemy, "Geographike Uphegesis," pp. 162–3.

24 Ibid.

25 Ptolemy, "Geographike Uphegesis," p. 164.

26 L. Nuti, "Mapping Places: Chorography and Vision in the Renaissance," in D. Cosgrove (ed.), *Mappings*, London: Reaktion Books, 1999, p. 91.

27 Ibid.

28 A. Ortelius, *Theatrum Orbis Terrarum*, no publisher, 1570.

29 Olwig, *Landscape*.

30 F. Lukermann, "The Concept of Location in Classical Geography," *Annals of the Association of American Geographers* 51:2, 1961, 194–210.

31 C. van Paassen, *The Classical Tradition of Geography*, Groningen, NL: J.B. Wolters, 1957.
32 L.C. Taub, *Ptolemy's Universe: The Natural Philosophical and Ethical Foundations of Ptolemy's Astronomy*, Chicago, IL: Open Court, 1993.
33 Taub, *Ptolemy's Universe*, p. 152.
34 J.H. Miller, *Topographies*, Stanford, CA: Stanford University Press, 1995, p. 286.
35 Merriam-Webster, *Webster's Third New International Dictionary of the English Language, Unabridged*, Springfield, MA: Merriam-Webster, 2000, cosmography.
36 Merriam-Webster, *Webster's Third New International Dictionary*, cosmology.
37 Ptolemy, "Geographike Uphegesis," p. 163.
38 G. Strauss, *Sixteenth-Century Germany: Its Topography and Topographers*, Madison, WI: University of Wisconsin Press, 1959, pp. 55–6.
39 G. Deleuze and F. Guattari, *A Thousand Plateaus: Capitalism and Schizophrenia*, London: Continuum, 1988, p. 172.
40 OED, chorography.
41 Plato, *Cosmology: The Timaeus of Plato*, trans. with a running commentary, Cornford, F.M., London: Routledge & Kegan Paul, 1937, p. 192.
42 J. Derrida and P. Eisenman, "Chora", in *Chora L Works*, ed. Thomas Lesser and Jeffrey Kipnis, New York, NY: Monacelli Press, 1997, p. 10.
43 Derrida and Eisenman, "Chora," p. 16.
44 Derrida and Eisenman, "Chora," p. 23.
45 van Paassen, *Classical Tradition of Geography*, p. 7.
46 van Paassen, *Classical Tradition of Geography*, p. 8.
47 Van Paassen, *Classical Tradition of Geography*, p. 3.
48 Sauer, "The Morphology of Landscape," p. 23.
49 J. Derrida, "Khora", *On the Name*, Stanford, CA: Stanford University Press, 1995, p. 104.
50 Derrida, "Khora," p. 23.
51 M. Hénaff and T.B. Strong, *Public Space and Democracy*, Minneapolis, MN: University of Minnesota Press, 2001, p.44.
52 Merriam-Webster, *Webster's Third New International Dictionary*, agora.
53 Merriam-Webster, *Webster's Third New International Dictionary*, place.
54 *Platea* is akin to "Latin *planta* sole of the foot" which is also the source of the word *plant*, deriving from the "Late Latin *plantare* to plant, fix in place" (Merriam-Webster, *Webster's Third New International Dictionary*, 2000, plant). *Choros* is akin to the Greek *chēros*, the etymological source of heritage, the *choros* being something that you have inherited through bereavement. Merriam-Webster, *Webster's Third New International Dictionary*, 2000, chor.
55 Lukermann, "The Concept of Location in Classical Geography."
56 E.S. Casey, *The Fate of Place: A Philosophical History*, Berkeley, CA: University of California Press, 1997.
57 Lukermann, "The Concept of Location in Classical Geography," p. 55.
58 Derrida and Eisenman, "Chora," p. 10.
59 Derrida and Eisenman, "Chora," p. 11.
60 Derrida and Eisenman, "Chora," p. 7.
61 This tendency is illustrated by the OED's statement on how chorography has become "a term, with its family of words, greatly in vogue in 17th c., but now little used, its ancient sphere being covered by *geography* and *topography* jointly." OED, chorography.
62 J.N. Entrikin, *The Betweenness of Place: Towards a Geography of Modernity*, Baltimore, MD: Johns Hopkins University Press, 1991.
63 Olwig, "Has Geography Always Been Modern?," 2008.
64 This approach was taken in a humanistic geography of Norden's landscape-regions edited by myself and M. Jones in M. Jones and K.R. Olwig (eds), *Nordic Landscapes: Region and Belonging on the Northern Edge of Europe*, University of Minnesota Press, 2008.
65 Cosgrove and Daniels, "Introduction: Iconography and Landscape."

66 Sauer, "The Morphology of Landscape."
67 Y.-F. Tuan, "Images and Mental Maps," *Annals, Association of American Geographers* 65:2, 1975.
68 Cosgrove, *The Palladian Landscape*, 1993 and *Geography and Vision*, 2008.
69 M. de Certeau, *The Practice of Everyday Life*, Berkeley, CA: University of California Press, 1984.
70 M. Foucault, *Discipline and Punish: The Birth of the Prison* (orig. title *Surveiller et punir*), Harmondsworth: Penguin, 1979 (orig. 1975).
71 H. Lefebvre, *The Production of Space*, Oxford: Blackwell, 1991 (orig. 1974).
72 Y.-F. Tuan, "Sight and Pictures," *Geographical Review* 69:4, 1979, 413–22.
73 Curry, "Toward a Geography of a World without Maps."
74 George Perkins Marsh, for example, was a geographer and Nordic philologist important to the development of Berkeley geography and a major inspiration to my own work as geographer and Nordic philologist. Olwig, *Nature's Ideological Landscape*, 1984.
75 Merriam-Webster, *Webster's Third New International Dictionary*, philology.

# 6

# THEMATIC CARTOGRAPHY AND THE STUDY OF AMERICAN HISTORY

*Susan Schulten*

What is the role of maps in history? In the relationship between geography and the humanities the map is the linchpin, for it is both a product of and an influence over historical change. Historical circumstances influenced the production of maps but at the same time these maps – and the geographical ideas behind them – became the lenses through which Americans understood the world. Admittedly it is much easier to investigate how maps were produced than it is to document their influence, for assessing the role of maps in history involves a fundamental problem: how do we know to what extent they actually *matter?* How can we know if a map reflects or shapes the past? To what extent is a map a product of the forces around it – such as knowledge of the landscape and printing techniques – and to what extent does it actually influence spatial knowledge?

One of the most fruitful recent approaches to discerning the meaning of maps in history has been to ask about the relationship between cartography and the rise of the nation. The nation flourished at a certain historical moment, yet depended on successful claims to *transhistorical* legitimacy. Once we see the nation as contingent, rather than inevitable, we can investigate how geographical and historical knowledge legitimated these claims to territory and tradition. In fact the development of a coherent nation – both in theory and in fact – made certain types of knowledge both possible and necessary. Since the publication of Benedict Anderson's *Imagined Communities* in 1983 there has been a heightened degree of sophistication paid to the historical study of nationalism. Especially exciting have been the ideas produced by those with a geographically informed perspective and who think about the map as influencing, rather than simply reflecting, the currents around it. For instance, Raymond Craib argues that a visual image of modern Mexico was a precondition for the development of coherent national loyalty among its citizens in the early twentieth century. Martin Bruckner's research into early American ideas about language and nationhood has led him to ask whether there can even *be* a nation without a map. In a related vein Ian Tyrell has identified strong evidence that the United States was a nation in sentiments and ideology before it developed a strong state apparatus. Each of these scholars frames the relationship between knowledge and nationalism in reciprocal terms.[1]

In the nineteenth-century United States the advance of the nation both facilitated and depended upon a new genre of knowledge: thematic cartography. Janusz Klawe defines thematic mapping by contrasting it to traditional cartography, where all elements of the landscape – including hydrography, relief, and borders – might appear without special emphasis on any single aspect. By contrast, thematic or "weighted" maps identify limited aspects of the physical or human environment. Simply put, thematic maps identify the distribution of a particular kind of information, such as the disease, weather, or wealth,[2] and are usually designed as tools of analysis and distribution rather than location or navigation.

To be sure, all cartography is in a sense thematic, for cartographers always choose to include particular information. And if the vigorous debates within the history of cartography have taught us anything it is that maps are always the product of choices about content and arrangement, inclusion and exclusion. Yet, generally speaking, prior to the nineteenth century most maps depicted physical features or political boundaries. Maps were generally tools of identification, description, and location. By contrast, the proliferation of thematic maps signals a fundamental shift in the meaning and purpose of cartography, for they had the power to concretize abstract relationships and to illuminate patterns that would otherwise remain hidden. Though very much taken for granted today, thematic cartography constituted a seismic shift in the organization of information in the nineteenth century.

Thematic cartography originated in the 1830s in Europe and exploded in popularity at mid century; by the 1880s it had become commonplace. Anne Godlewska has argued that the growth of thematic cartography in France reflected the more general movement of geographical knowledge from description to analysis. Prior to about 1830, French geographers largely focused on descriptive and classificatory work; Alexander von Humboldt enlarged geography to answer specific questions about relationships and patterns in the environment. This change produced new types of maps that would identify a specific phenomenon in pursuit of a particular question, and in turn these new maps sparked further questions. Earliest among these were maps of epidemic disease outbreaks in the antebellum U.S. In a desperate effort to explain – and thereby cure – yellow fever and cholera, medical men mapped the incidence of these diseases and their potential contributing factors. Through maps they could test theories about the etiology of these epidemics. While topographic maps were akin to mimesis, thematic maps were more like an argument and both reflected and facilitated the growth of analysis through cartography.[3]

It might seem odd that it took so long to think about cartography as a tool to analyze data in spatial terms. But thematic cartography depended upon large bodies of information, such as the Census. The systematic accumulation of data was itself partly a function of an expanding nation. In the United States the Constitution provided for the Census as a tool of apportionment. As Congress expanded the scope of the Census over time it yielded a growing body of information about the nature of the population as well as institutions. New scientific agencies – again usually within the Federal government – in the nineteenth century generated other bodies of information, such as environmental observations which were used to create the first weather maps. In these and many other ways there is a direct and interdependent relationship between the rise of the nation, the growth of information, and the development of thematic cartography.[4]

This type of cartography also grew in the nineteenth century because it suited the needs of the emergent disciplines, which pursued knowledge by asking questions that could be uniquely answered through graphic and cartographic means. Just as the Federal government sponsored the Census, it also pursued the natural and human sciences in order to develop laws and patterns of distribution or behavior. Individuals and emergent professions began to use thematic cartography for specific purposes – to illustrate particular census statistics or to profile the geological patterns of a particular region. Those physical sciences which advanced thematic cartography were geology, climatology and meteorology (also known as climatology); in the human sciences mapping data was undertaken most aggressively within epidemiology, demography, sociology, economics, and other fields exploiting data provided by the expanding Census and a growing medical bureaucracy.[5]

In the United States the formative development period of thematic cartography was from the 1850s through the 1870s. During these decades Federal scientific agencies experimented extensively with thematic maps in order to translate statistical patterns into more concrete and visual language. Because thematic cartography was used by disparate fields the maps themselves are not concentrated within a particular area but rather scattered in treatises and journals; in fact thematic maps were not bound together in any systematic fashion in the United States until the Superintendent of the Ninth Census undertook such a project in the early 1870s. Before that time – because developments were made in such disparate fields – thematic mapping was not seen as a genre in and of itself. Only after 1870 were individuals, such as Daniel Coit Gilman, arguing that this *type* of mapping would revolutionize information. Furthermore, at least in the United States, thematic maps were almost invariably produced first by government agencies, not commercial firms. In fact it is worth noting that in the United States thematic cartography was frequently pioneered by Federal agencies, first among them the Coast Survey, the Naval Observatory, the Smithsonian Institution, and the Census Bureau. In all of these agencies the technique of isolating and mapping particular classes of data was a boon to governance because it organized information in explicitly visual, geographical terms. The appearance of distribution patterns and trends in turn facilitated new understandings of causation and change.

Given the space limitations here, I want to concentrate on just a handful of the thematic maps generated by the US Census Bureau in the 1870s. After the Civil War, the Census Bureau was completely transformed by the arrival of Francis Amasa Walker as its Superintendent. Walker was the son of Amasa Walker, a leader of the Free Soil movement in Massachusetts in 1848, who immersed his son in political economy. He graduated from Amherst College and then took up law, but immediately enlisted after the outbreak of war in 1861. Upon leaving the military Walker tried his hand at teaching and journalism before settling in as the Chief of the Bureau of Statistics. By 1870 he had become the Bureau's Superintendent and in the following year was named Commissioner of Indian Affairs. In the 1870s he also held a professorship of Political Economy and History at Yale's Sheffield School, presided over the American Statistical Association, and in 1881 became the third president of the Massachusetts Institute of Technology.

The breadth of these positions gives us some sense of Walker's tremendous intellectual range, but behind all of these lay a common interest in the organization and use of information. This interest led to his experimentation with thematic mapping in the Ninth Census, the most extensive and varied Census enumeration up to that point in the nation's history. In the late 1860s Walker saw and was deeply impressed by a series of maps made by August Meitzen in Prussia. These maps, depicting population distribution, encouraged Walker to ask one of his staff to design a few preliminary maps of the aggregate population and its "principal constituent elements": immigrants, African-Americans, and native-born whites.[6]

Walker was the first to use these maps to compare two populations and his breakthrough was recognized by Daniel Coit Gilman, who eagerly used them to convert scientists and social scientists to the progressive potential of statistical cartography. In his annual address to the American Geographical Society in January 1872 Gilman held up two of Walker's maps, one depicting the density of immigrants and the other of African-Americans. Notice, Gilman remarked, the way the maps instantly conveyed certain "truths" that raw data would yield only reluctantly and after methodical study. The crucial advantage of thematic mapping, Gilman argued, was its ability to ask new questions and to integrate disparate areas of knowledge. Through statistical cartography the reader could see at a glance that "the foreign born population is thickest, where the Africans are not, and *vice versa*."[7]

Walker was similarly struck by the potential for this new form of knowledge. As he later recalled, these preliminary maps were crude and limited but he "had only begun to appreciate the capabilities of this method," which amounted to a fundamentally different way to think about data. Flush with the buzz of discovery after viewing these new maps Gilman – along with several other professors at the Sheffield School – petitioned the Secretary of the Interior to increase funding to map the Census Reports due later that year. By that summer, the Secretary of the Interior had requested and received support from Congress to depict the distribution of the population (particularly the "foreign population"), the location of industry and agriculture, the distribution of disease, and other areas of social and material inquiry undertaken by the Census.[8]

Before 1870 the Census Office had never mapped its work; in light of this the extensive use of cartography and graphic illustration in the 1870s is remarkable and testament to Walker's radically new vision of the Census and its role in American public life. The Ninth Census had been mired in conflict and accusations of inaccuracy, certainly nothing new in the ninety-year history of the process. But upon completion the Report was hailed as a breakthrough for its relative breadth and comprehensiveness.[9] In fact the maps of the Census Reports were so well received that Walker was authorized to undertake an even more ambitious project: a statistical atlas of the United States, which would include dozens of maps and charts based on the Ninth Census.[10] Using graphs and maps, the atlas dissected the nation in myriad ways and, in the process, introduced a new kind of visual language. As Gilles Palsky and Michael Friendly have put it, in the nineteenth century thematic mapping became an increasingly "autonomous language," one where "the priority of representation was, in effect, reversed: topographic detail moved into the background and special themes into the foreground."[11]

Walker's own introduction to the atlas included lengthy, detailed explanations for each of the complex statistical charts, yet the maps, he argued, spoke for themselves. As he wrote, "the Geographical illustrations, in general, require no verbal description and explanation, beyond what is given on their face. It is not the Compiler's intention to preach from them, as a text; nor does he assume that attention needs to be directed to their more obvious or their more recondite suggestions." This assumption of the transparency of the map – that it simply renders facts in graphic form – is significant and captures the tenacious assumption that maps should be regarded as scientific, rather than argumentative, documents.[12]

The *Statistical Atlas* is a visually compelling text. Page after page of large, beautiful, groundbreaking maps and graphics counted, measured, and profiled the nation in provocative ways. But the legacy of the *Atlas* was not just in this visual audacity. On a more subtle level the *Atlas* used cartography to pose new social questions. By visualizing the distribution of the Census data the maps emphasized spatial variation. In fact the *Atlas* took statistics to another level of analysis because, by mapping data, it allowed new questions to arise and, conversely, its spatial analysis *necessitated* cartographic display. This is so obvious to us today that we may fail to see how new it was in the nineteenth century.

That Walker knew he was on to something is apparent in his arrangement of data. His special concerns were anemic population growth and urbanization. In fact, in his mind, the two were linked: urbanization caused (or at least reflected) the declining birth rate. Thus he organized the entire atlas around not just maps of the population but also maps of population *density*; this choice is worth noting. For instance, earlier attempts to map statistics had usually used the state as the unit of enumeration, thereby deemphasizing urbanization by "leveling" the distribution across the state as a whole. Walker, however, adopted more sophisticated cartographic techniques that captured variations through continuous shading. When he began to map social statistics Walker again took these population density outlines and laid them in blue ink over individual maps charting the distribution of race, ethnicity, illiteracy, debt, wealth, taxation, sex, birth rate, and a variety of diseases. By doing this he was comparing population density with the incidence of disease, immigration, illiteracy, wealth, and the like. He allowed the viewer to connect and correlate these phenomena with the concept that he considered paramount: population density.

By integrating different types of information Walker was using the maps to pose new questions about social organization. As he himself asked when introducing the maps, Why don't foreign born suffer from fevers while colored do? Conversely, what of consumption? Is it related to geography? Stock? Breeding? Occupation? This interest in the cause of social problems also explains Walker's inclusion of physical maps, for these would incorporate variables such as elevation, climate, and rainfall. For instance, like Gilman before him, Walker could not help but notice that the "foreign" and "colored" elements of the population did not overlap; neither did their respective temperature and rainfall zones. Blacks, he assumed, were physiologically adapted to the moist, hot climates of the south. The problematic conclusions that Walker reached were less important than the questions he was asking. Such complex questions were not just *enhanced* by cartography but made possible by them. Without maps, patterns of distribution would be left invisible.[13]

Thematic maps also effectively tied the population to the territory in new, more direct ways. They made the population both a more concrete phenomenon – by identifying concentrations of wealth and poverty, education and illiteracy – but also more abstract, by treating it as an entity that could be counted and measured in various ways. At the same time, the nation's growth in turn fueled the *need* for this type of new cartography. That is, the nation both demanded and made possible a more spatial organization of knowledge. By showing Americans themselves organized along lines of race, ethnicity, climate, resource distribution, wealth, poverty, sickness, and education, thematic cartography both divided and unified the population.

Walker sponsored these maps in part to foster national sentiment and in a more basic way to substantiate the nation in terms of not just its territory and boundaries, but also its internal structure. More generally, we ought to see these thematic maps made by the Census as images that shape and give meaning to what might otherwise be incomprehensible statistics. Thematic cartography could also be considered the intellectual and conceptual antecedent for Geographic Information Systems. GIS has a complex history but most scholars date its origins to the 1960s, when the quantitative turn in geographical analysis joined with the emergent computer age to create a new way to organize and manipulate information along spatial lines. Yet the concept of GIS is that cartography, can be a tool of decision making and that data can be organized in spatial terms to help think through problems of urban planning, politics, marketing, or logistics, just to name a few. The current and potential applications of GIS are overwhelming, but behind all of them is the concept of spatial analysis, and it was in the nineteenth century that thematic cartography first demonstrated the potential to render data in spatial terms.[14]

Thematic cartography became an engine of inquiry and a tool to analyze geographically dependent phenomena. The genre advanced emergent social sciences, concepts of the nation, and governance generally. It suggests just one way that our understanding of history – and the humanities generally – might be greatly enriched by paying closer attention to this sea change in cartographic thought.

# Notes

1  M. Bruckner, *The Geographic Revolution in Early America: Maps, Literacy, and National Identity*, Chapel Hill, NC: University of North Carolina Press, 2006; and "Lessons in Geography: Maps, Spellers, and Other Grammars of Nationalism in the Early Republic," *American Quarterly* 51 n. 2, June 1999, 341, n.39; R. Craib, *Cartographic Mexico: A History of State Fixations and Fugitive Landscapes*, Durham, NC: Duke University Press, 2004; I. Tyrell, "Making Nations/Making States: American Historians in the Context of Empire," *Journal of American History* 86:3, 1999, 1015–44.

2  J.J. Klawe, "Population Mapping," *The Canadian Cartographer* 10:1, June 1973, 44.

3  A. Godlewska, *Geography Unbound: French Geographic Science from Cassini to Humboldt*, Chicago, IL: University of Chicago Press, 1999, p. 265 and *passim*.

4  The growth of thematic mapping in American life is the subject of my current research. See, for example, Susan Schulten, "The Cartography of Slavery and the Authority of Statistics," *Civil War History* 56, March 2010, 5–32.

5  On the origins and development of European thematic cartography see A.H. Robinson, *Early Thematic Mapping in the History of Cartography*, Chicago, IL: University of Chicago Press, 1982. Recently three books have taken an in-depth look at the effect of single thematic maps on history: T. Koch, *Cartographies of disease: Maps, Mapping and Medicine*, Redlands, CA: ESRI Press, 2005; S. Johnson, *the Ghost Map: the Story of London's Most Terrifying Epidemic – and How it Changed Science, Cities, and the Modern World*, New York, NY: Riverhead, 2006; and S. Winchester, *The Map that Changed the World: William Smith and the Birth of Modern Geology*, New York, NY: HarperCollins, 2001.

6  These were not the first maps based on the Census: in 1861 the Coast Survey had executed a map of the slave population of Virginia, followed by one of the southern states generally. See Schulten, "The Cartography of Slavery."

7  D.C. Gilman, "Annual Address: Geographical Work in the United States during 1871," delivered January 30, 1872, *Journal of the American Geographical Society of New York* 4, 1873, quote p. 141.

8  F.A. Walker, "Introduction," *Statistical Atlas of the United States*, New York: J. Bien, 1874.

9  "Ninth Census vol. 1. The Statistics of the Population of the United States Embracing the Tables of Race, Nationality, Sex, Selected Ages, and Occupations to Which are Added the Statistics of School Attendance and Illiteracy, of Schools, Libraries, Newspapers and Periodicals, Churches, Pauperism and Crime, and of Areas, Families, and Dwellings. Compiled from the Original Returns of the Ninth Census, June 1, 1870, Under the Direction of the Secretary of the Interior by F.A. Walker, Superintendent of the Census," Washington, DC: Government Printing Office, 1872.

10  *Statistical Atlas of the United States Based on the Results of the Ninth Census 1870 with Contributions from Many Eminent Men of Science and Several Departments of the Government; Compiled Under Authority of Congress by Francis A. Walker, M.A. Superintendent of the Ninth Census, Professor of Political Economy and History, Sheffield Scientific School of Yale College*, Julius Bien, Lith. 1874.

11  M. Friendly and G. Palsky, "Visualizing Nature and Society," in *Maps: Finding Our Place in the World*, Chicago, IL: University of Chicago Press, 2007, p. 220.

12  Walker, *Statistical Atlas*, p. 3.

13  Walker, *Statistical Atlas*, Introduction, "Political Geography and Statistics," in *Encyclopedia Britannica*, 9th edn, vol. 23, 1888, pp. 821–2. See also T.M. Porter, *The Rise of Statistical Thinking, 1820–1900*, Princeton, NJ: Princeton University Press, 1986.

14  For a twentieth-century history of GIS, see T.W. Foresman (ed.), *The History of Geographic Information Systems: Perspectives from the Pioneer*, Upper Saddle River, NJ: Prentice Hall, 1998.

# PART II
# Reflecting

# 7

# DO PLACES HAVE EDGES?

## A geo-philosophical inquiry

*Edward S. Casey*

I

The question in my title poses an odd choice between two incompatible alternatives. Either places self-evidently possess edges or else (if this is not true) edges belong to physical things alone. Either way, further discussion is closed off. On the first alternative the edgefulness of places is taken for granted, as if it were so obvious as to need no further description, much less justification. On the second option only things are presumed to possess edges, as if edges and things form an exclusive dyad. These antinomical positions not only cancel each other out, but they also declare subsequent treatment otiose – an exercise in futility.

I want to argue that places do indeed have edges, even if they have them in a special way that has not yet been fully recognized: short of this recognition, it is all too tempting to revert to the standard view that edges belong to only things. In what follows I shall lay out in brief compass the major way in which edges belong to places; my assumption is that places require edges if they are to be coherent and identifiable places at all, viable arenas for perception and action. In order to see how this is so, however, we must move beyond the restrictive but widely held assumption that edges are the exclusive property of material objects. My analysis will be philosophical, and more particularly phenomenological, in spirit but it should be of interest to geographers and landscape theorists as well.

By a "place" I refer to any spatial spread ranging from a bioregion or a national territory to a human settlement of any kind, whatever its exact size. Each such place calls for detailed examination of its intrinsic capacity for bearing and displaying edges. My view is not only that places have distinctive edges but that they are pre-eminent bearers of edges: they are distinguished by their edges and present them to the environing world. In a world of places, edges ensure that (as Walt Whitman said) "all goes outward and onward."[1]

In arguing for distinctive edges of places I shall be flying in the face of the usual view that *edges belong to things*. This view is as ancient as Socrates, who said that an edge is nothing but "the limit of a solid."[2] On this view edges properly attach to things regarded as solid substances – three-dimensional objects that are material in their constitution. I take this to be tantamount to the literal *reification* of edges, confining them to concrete things. This betrays an object-obsessed approach to edges: to believe that edges belong strictly to things is to conceive edges themselves as objects that we can quantify and measure, handle and use. Such an approach reflects a pervasive tendency in Western epistemology to turn experiences and its contents into "ob-jects": it is to participate in what Merleau-Ponty calls "the freezing of being."[3] In Heidegger's nomenclature, edges so conceived belong to the realm of the ready-to-hand (*zuhandenes*) and the present-at-hand (*vorhandenes*): they are either instrumental in character or they are "objective" in the normative sense of determinable in terms of constant, specific, numerable properties that are subject to prediction and control.[4]

As instrumental, edges are assessed in terms of properties like manipulability, ease of use, and suitability. Issues at stake include how the edges of different things fit together; how accessible they are to practical interventions of various kinds; and how they stand up to regular employment (in short, their "reliability"[5]). Think of teethed edges that are meant to lock into a commensurately notched set of edges: say, the top of my thermos as it screws tightly onto its base. The actual shapes need not be formally geometric; they can be informally "morphological" in Husserl's term for non-Euclidean forms that do not belong to classical geometries of shape.[6] Indeed, slight irregularities can contribute to greater utility, since they may contribute to the give-and-take of physical things as they enter into intimate inter-action with other things – as with a comb that sweeps through my hair: its edges are firmer than those of the hair, yet still pliable enough to allow for significant variations in shaping my hair into different styles. Or consider a belt as it surrounds the mid section of the human body; it must bend enough to reflect the organic curvature of the stomach and hips of the person it engirdles in order to fit the body of the person who wears it – whereas an iron belt would be not only inefficient and felt as "unfitting" but also as cruel in its rigidity.

In contrast, the edges of objective or present-at-hand things need to be altogether regular and unyielding. "The rigid 1 body," said Husserl, "is the normal body."[7] By "normal body" he meant the exactly measurable body of a material thing whose very essence, thus regarded, entails physical and mathematical fixity. The computer on which I now write has a screen size of exactly $15 \times 6$ : just that and not any more or less. Its edges are altogether precise; their metallic rigidity rein-forces their measurable precision. Such edges are valued not for their functional value – as are the keys of the same machine (notable for fitting the fingers well) and its working software (whose programs exhibit functional flexibility) – but for their ever-the-sameness, their being just what they are. For the screen regarded as nothing but a present-at-hand object, reliability is a secondary virtue: rigidity of material substance may certainly reinforce reliability but the rigidity as such is part of the strictly objective standing of the computer considered as a sheer physical object.

If the edges of ready-to-hand things are prized for their pliable practicability, those of present-at-hand particulars are esteemed for their measurable constancy. The difference is that between edges which suggest their actual or eventual use and edges which accrue to objects as their formally determinable parameters (taking "para-meter" in its original sense of "to measure alongside"). One kind of edge is extroverted, one introverted, as it were; one is adherent, the other inherent. But both, for all their differences from each other, remain resolutely edges of material things, to which they belong as an inseparable feature.

## II

But what if edges are not married exclusively to things taken in either of the two objectified modalities I have just outlined? What if they also accrue to something not of the nature of a physical object at all? What if we were to resist the endemic reification of edges that results in their effective *rigor mortis*?

I have already hinted at the importance of edges for places. Now this bare hint must be followed out, especially in terms of the relationship between the edge-world and the place-world, each construed in a sufficiently generous sense. Let me do this by first of all proposing that these two worlds are not separate realms but are *conterminous*. By this last term I have in mind the two primary ways in which place and edge interact – in which place is edged:

(i) *terminus a quo* ("limit from which"): from a place or set of places, edges begin; they spread out from there – not in the sense of instrumental outreach but in the very different respect that the edges of one place are sources for the edges of other places, generative contours for entire place-worlds. Instances come from natural as well as cultural situations. Creosote bushes in the southern Nevada desert open up whole regions of the local landscape; these bushes grow together up to a certain limit and then abruptly terminate. From the discernible line of termination a new region extends, one that lacks these bushes altogether and is comparatively barren. One place, that of dense creosote growth, gives rise to another – sometimes across a remarkably straight line – thanks to the edge of the first opening onto the prospect of the second. Only twenty miles away from this particular scene, a wall on the west end of Las Vegas marks the limit of housing construction; on its other side, there is only a dry gulch and, just beyond that, a small national park, an uncultivated reserve of hills and mountains of extraordinary coloration and configuration. An entirely artificial wall – placed where it is by the builder of a local subdivision in the sprawling suburbs of Las Vegas – abruptly gives way to a wild landscape, traversed only by a single access road. The edge of a bulldozed and built-over place becomes the opening edge for a natural landscape.

In these two examples from Las Vegas, then, the edges of two regions – one wholly natural, the other a matter of carefully calculated design – are in effect places of origin, *loci a quo*, for two wild regions that extend out from them as their base or source. Questions of utility or exactitude are not at stake; rather, the issue is that of *regioning* (to adapt another term of Heidegger's).[8]

(ii) *terminus ad quem* ("limit to or toward which"): edges extend *to* places as well; by this I mean that a given edge can just as well be seen as the end of one place as the beginning of another. A place comes to its own edge. If it did not come to *some* edge, it would not count as a place at all. Places require edges as much as things do. Since a place (unlike the early modern notion of space) is not infinite, it runs out in its own edge. This edge is at the same time the start of another place with which the first place is contiguous. However different in aspect or character it may be, such an edge acts to delimit the farther place or region[9] of which it is the opening edge.

Here is a different example from the same part of the place-world: now I am part of a silent vigil at the Nevada Test Site just north of Las Vegas – a notorious stretch of land devoted to nuclear and other testing of bombs, pockmarked by nearly 1,000 explosions, some above ground and some underground. The protesting group of which I am part is allowed to assemble only in one predesignated place, just outside the main entrance to the test site – an entrance ironically named "Mercury." (I say "ironically," since Mercury or Hermes was for ancient Greece and Rome the god of boundaries and was known for moving swiftly over crossroads marked by Herms, the signposts that consist in single phallic stone shafts.) As we gather together in a circle we are very aware of the place to which we are confined – confined primarily by the barbed-wire fence that asserts itself on the east side of where we are located; beyond this fence we glimpse the test site that is visible in the near distance. The merest glance takes me right up to the fence and then through it to the buildings and other structures of the test site.

It will be noticed that the artificial and the natural combine even more closely in this last instance than they did in the previous examples: for it is in the same situation that I witness both the intrusive fence along with the test site structures just beyond *as well as* the landscape on which they are both built. As Merleau-Ponty puts it, here "everything is cultural in us (our Lebenswelt is 'subjective') and everything is natural in us (our perception is cultural-historical)."[10] We can go still farther and say that everything inhabiting the edge of a place such as the Nevada Test Site is cultural and natural, both at once – though not necessarily in exactly equal measure.

The difference between the two edge circumstances can be put thus: in the second case (*terminus ad quem*), edges move us from one place to another; in the first (*terminus a quo*), each edge serves to contain, to define, to arrest movement of the eye if not the foot. The generativity of the first compares with the defensiveness of the second. Either I start out from an edge of one place or region to get to another place or region or else I stay confined in one place or region – stay within its edge, whether in vision or in bodily motion. Leading out here contrasts with being held in.

But the contrast is mitigated by the fact that it is often the *same edge* that serves in the two roles. The edge of the creosote growth in my first example is at once the *terminus a quo* of this growth and a *terminus ad quem* that opens onto an adjoining non-creosote region even as it marks the outer limit of the creosote bushes themselves. The wall at the Las Vegas city limit ushers in the natural places outside the city; but it is also a clear perimeter, at once arbitrary and artificial, for the city itself. So, too, the fence at the Nevada Test Site opens out onto the test site itself and the natural landscape in which both are embedded, while also terminating the place permitted for protesting the untrammeled testing of nuclear fission and fusion for military purposes. The Janusian character of these edges of places tells us something significant about the role of edges in the place-world. The fact that they are often themselves conterminous indicates at once their specialness as well as the close tie between the edge-world and the place-world – their convergence if not their complete coincidence.

But the basic question remains unanswered: what is the relationship between edge and place?

## III

Before we answer this question directly, let us take up a related question: how are edge and *event* related? From a discussion of this new question we may derive crucial clues as to the nature of placial edges. Notice, to begin with, the close relationship between event and place: events *take place* in a place. Moreover, both places and events are spatio-temporal, each thereby contesting the primacy of Space and Time, those twin colossi of early modern philosophy and science. Events subvert the hegemony of monolinear, successive time by opening out onto history – not just in the intense sense of "historical events" but insofar as every event is historical to some degree, even if not altogether world-historical in the manner of the Nevada Test Site. Similarly, places open onto regions and, finally, whole place-worlds, as we have seen from the example of the creosote plains of the Nevada desert. In other words, event and place join forces in deconstructing the constricted conceptuality of Space and Time by exhibiting a more expansive way of being spatial and temporal than being merely simultaneous or successive – and by the fact that event and place in their augmented avatars occur indissociably, each interbraided with the other.

Further, *an* event – not unlike Dewey's notion of "an experience"[11] – involves a factor of *surprise* in which edges have a major role to play. These edges are themselves spatio-temporal: they are situated temporally and spatially at once and indissociably. But in this dual capacity they play an occlusive role: the outer edge of an event prevents us from seeing what is to come, on the other side of this event – what is to come by way of things that will appear or happenings to follow. When these things or happenings arise they do so unexpectedly to some significant degree, since they are situated on the other side of the current event's edge: there where it ceases to be that event and is in the process of becoming other to itself: another event. And when the next event emerges, whether by the incursion of something external beyond a given edge or by an internal alteration within a defining edge, it is always to some degree unanticipated: it "takes us by surprise," where *sur-prise* means literally "taken over." The emergent event takes us over – and sometimes makes us over.

Places are like events in just this way: they, too, surprise us at their edges. Insofar as both places and events come to an end in their edges, what lies on the other side of these edges is not fully disclosed – not yet at least. A dramatic instance of this otherness of the edges of an entire region is found in the history of early navigation in the Mediterranean: what lay beyond the Pillars of Heracles (i.e. the Straits of Gibraltar) was unknown and unbounded, thus profoundly threatening to the inexperienced. Some aspect of this circumstance is found in every place, just as it characterizes every event as well. The fear (or, conversely, the allure) of the new is a phenomenon of the edges of places as of events – and of the two together when they are combined, as in the ancient exploration of the limits of the known world.[12]

## IV

By considering the edges of events, we are now better able to appreciate different features of the edges of places: their capacity to harbor surprise, their spatio-temporal character, above all their pivotal role in the change of one place into another (or its replacement by another). What holds for the edges of events holds true for the edges of places: which is what we should expect if places are not static entities, mere sectors of pre-established spaces. If place is eventmental to its core, then its edges will be dynamic in their interaction with what surrounds them. This dynamism is intensified by the spatio-temporality of placial edges, which operate in spatial outreach as well as in temporal distention. Surprise is also part of this same dynamism as one place confronts another, whether it gives way to the new place or holds its own but alters its internal constitution.

## V

We have still not figured out what kind of edges places possess. To begin with it is crucial to point out that several sorts of edge, not just one, conspire in the constitution of a single place. Indeed, to be a place at all is to possess just such a multivalency of edge. This is an indispensable feature of places as dynamically open-textured – in contrast with *sites*, which are spaces that are strictly delimited, determined in advance by overriding considerations ranging from issues of exact location and cartographic accuracy to the character of infinite space (e.g. homogeneity, isotropism, etc.): as a consequence, their edges tend to be standard or predictable.[13] Similarly, a material thing typically bears just one kind of edge (i.e. its outer rim). But a place normally supports a variety of edges. Here, then, is an already crucial if minimal distinction between edges of things and edges of places. If the edges of places and events closely resemble each other for the reasons I have just discussed, those of places and things (and sites) diverge markedly.

Moreover, the variety of placial edges obtains even for a single place: think only of a place such as Central Park in New York City. This celebrated park has low walls along much of its perimeter but it also sports various open gates through which pedestrian and automobile traffic flows. It also has internal edges that differentiate parts of the park from each other, not to mention the numerous footpaths that effect their own edgework in intricate patterns. Even though there are certainly more complicated places than Central Park, this particular place shows itself to be rife with edges of disparate sorts – so many, in fact, that a description of it could continue almost indefinitely.[14] A proliferation of edges attaches to any given place, more so than to a given material thing or a site: this much we may take as axiomatic.

Place is a peculiarly powerful catchment area of edges, absorbing and exhibiting a plurality of them – natural and artificial, conspicuous and subtle, fully manifested or only adumbrated. This, along with its dynamically changing character, reflects the fact that *place itself has no definitive edge*, no set limit. To have a definite edge is a basic feature of a site: e.g. a building site, a location on a map. But places are not so restricted. Consider how places that we designate by phrases such as "Gramercy Park" or "Battery Park" – or regions like "Mid-town Manhattan" or "Lower East Side" – refuse to be characterized by stopping or starting at altogether determinate points. By the same token, places can intersect and when they do so their edges intertangle in diffuse ways that defy definite, much less complete, description. South Harlem merges into the Upper West Side from the east across Morningside Park, which acts as a buffer zone between these two parts of New York, at once connecting and separating them. Yet no local inhabitant would be willing to say that South Harlem extends only to a precise given point – say, to Manhattan Avenue (i.e. the street on the east side of Morningside Park) but not one yard beyond. Harlem residents mix with Columbia University students in the same park. Even if no such pronounced buffer zone as a park were present, however, there would still be no exact delimitation between two places or regions in a dynamic place-world such as New York City exemplifies.

Despite the endemic diversity of edges that a particular place possesses, placial edges can be gathered together under one heading: that of *boundary*. Where the edges of sites can be considered *borders* – that is, strictly determined and demarcated edges that, tellingly, are often referred to as "border-lines" – the edges of place are boundaried. This means that they exhibit a porosity and vagueness that allows them to be at once ever-changing and yet stable enough to serve as identifiable edges of places. To be porous and vague is to allow, and sometimes to facilitate, movement across such edges: not any and all movement but that of certain animals and humans, indeed of basic elements, that are moving from one place to another. Morningside Park, considered as an extended boundary, permits people to walk back and forth between different locations in South Harlem and the Upper West Side. It provides many opportunities for entries and exits: from Harlem I can approach Columbia University by walking across the intermediate spaces provided by the park. It is as if this park-boundary were perforated with many points of access and egress – in direct contrast with sites whose very definition and existence depend on the establishment of tightly contained limits.

The ease with which I and other denizens of this part of the city walk back and forth across this one location is to be contrasted with the difficulties of passing over an officially recognized border such as that separating the United States from

Mexico. In the latter case I must show proper citizenship papers: if I do not possess these papers I will be refused admission to the other side. I will be *stopped at the border* – where the "at" indicates a precise point of disallowed passage. Even if I can produce the right papers I am still very conscious of just where the border-line is located: if not marked as such, it is felt palpably as I am waived through to the other side. If not visible to the eye, it is felt under foot. In less dramatic situations, for example entering a public building, a border is indicated by the sheer presence of something like the door of the same building: it marks unambiguously a definitive difference between being *in* or *out of* this building.

Borders and boundaries can combine forces in a single situation: although serving as a boundary for people living in (or passing through) New York City in the area above 110th St. and west of 8th Ave., Morningside Park has its own borders in the form of the streets that line its outer limits – and more particularly, the sidewalks and curbs associated with them. But this active collusion between borders and boundaries in no way undermines their basic distinction as two genres of edge. (I call these "genres" since each includes a proliferation of species or varieties – that is, ways by which they are concretely realized on the ground as fences or gates or sidewalks, walls or doorways or lines.)

## VI

Edges of places, then, are to be considered as boundaries or boundary-like, where those of sites are borders or border-like. The resolution of the dilemma with which I opened is indeed to affirm the places have edges – but a peculiar sort of edge: the boundary – and thus to deny that material things alone possess edges: places as well as events come fully edged, albeit in their own peculiar ways.

This is not to deny that boundaries, as we use this term conventionally, can be considered edges not only of places but also of other sorts of entity (e.g. artworks, colors, persons, groups, etc.). Similarly, borders obtain not only for sites but for other things as well. What distinguishes the various avatars of the two terms on which I have focused has to do with the manner in which particular edges inhere in their respective subject matters: the border of a painting is a two-dimensional margin that surrounds the primary image, the boundary of a concept concerns its limit of meaning or use. But these different applications of "border" and "boundary" share a common semantic core: borders of any kind are distinct edges that lend themselves to precise delineation and specification, while boundaries are edges that possess a certain indeterminate leeway and are absorptive in certain basic respects: they are osmotic, we might say.

Just as the distinction between boundaries and borders is not arbitrary but holds consistently across diverse object domains, so my case in this chapter for applying the first word to places and the second to sites is not merely taxonomic. I am drawing on two terms whose central sense in English (and its equivalent in other languages) regulates extensions of these terms to disparate phenomena in the everyday life-world: in the present instance, to places and to sites regarded as the two main modalities in which human beings undergo space.

In this brief essay, I have offered a modest contribution to the analysis of edges in particular and to spatial studies more generally, including those that bear on issues in geography, cartography, and landscape topography. My remarks are part of an ongoing project to offer a comprehensive phenomenology of the ways that human beings inhabit, embellish, and represent the place-worlds in which they live and move and have their being. In the pursuit of this phenomenology, edges figure much more significantly than is often assumed to be the case.[15]

## Notes

1  W. Whitman, *Song of Myself*, in *Leaves of Grass*, New York: Bantam, p. 28.
2  Plato, *Meno*, 76a, edge "is that in which a solid terminates, or, more briefly, it is the limit of a solid."
3  M. Merleau-Ponty, *Phenomenology of Perception*, trans. C. Smith, New York: Routledge, 2002, p. 63.
4  For the ready-to-hand and present-at-hand, see M. Heidegger, *Being and Time*, trans. J. Macquarrie and E. Robinson, New York: Harper, 1962, sections 15–18, p. 22.
5  On reliability (*Verlässlichkeit*), see "The Origin of the Work of Art," trans. A. Hofstadter, in *Poetry Language Thought*, New York: Harper, 1971, pp. 33–6.
6  See E. Husserl, "Descriptive and Exact Sciences," *Ideas Pertaning to a Pure Phenomenology and to A Phenomenological Philosophy*, trans. F. Kersten, Dordrecht: Kluwer, 1983, Book One, section 74.
7  E. Husserl, "The World of the Living Present and the Constitution of the Surrounding World External to the Organism," trans. F.A. Elliston and L. Langsdorf, in F.A. Elliston and P. McCormick (eds), *Husserl: Shorter Works*, Notre Dame: Notre Dame University Press, 1981, p. 239. Husserl here refers to the standard early modern, especially Cartesian, idea of bodies as bits of *res extensa*; his own view is that the human body is the crucial exception to this: as a "lived body," it is moving and supple.
8  On regioning, see M. Heidegger, "Conversation on a Country Path," in *Discourse on Thinking*, trans. J.M. Anderson and E.H. Freund, New York: Harper, 1966, pp. 65–90.
9  In this chapter, "place" and "region" are not rigorously distinguished. For their more precise distinction, see these terms as discussed in the Glossary to E.S. Casey, *Representing Place: Landscape Painting and Maps*, Minneapolis: University of Minnesota Press, 2002, pp. 347–55.
10  See M. Merleau-Ponty, *The Visible and the Invisible*, trans. A. Lingis, Evanston: Northwestern University Press, 1968, p. 253 (working note of May, 1960).
11  See J. Dewey, "Having an Experience," Ch. 3 of Dewey's *Art as Experience*, New York: Perigee Books, 1980; original edition, 1934.
12  Not just at sea, of course; the same obtains for expeditions into the *terra incognita* of whole regions and continents: witness Alexander invading Asia Minor and India.
13  For farther discussion of site vs. place, see E.S. Casey, *Getting Back into Place: Toward a Renewed Understanding of the Place-World*, Bloomington: Indiana University Press; 2nd edn, 2009, pp. 226, 267–80, 288; as well as the Glossary in *Representing Place*, cited in note 9.
14  I have analyzed Central Park in more detail in my essay "Borders and Boundaries: Edging into the Environment," in S.L. Cataldi and W.S. Hamrick (eds), *Merleau-Ponty and Environmental Philosophy: Dwelling on the Landscapes of Thought*, Albany: SUNY Press, 2007, pp. 67–92.
15  The present chapter is a preliminary sketch of one part of a larger project, to be entitled *The World on Edge*, in which the author is currently engaged.

# 8

# RACE, MOBILITY AND THE HUMANITIES

## A geosophical approach

*Tim Cresswell*

How is it possible to be a refugee? This kind of question is rarely the topic of accounts of refugees that ask how many there are or even how they are to be accommodated. Such a question requires the humanities and a focus on meaning. My argument here is for a methodological and epistemological approach to geography that is based in the humanities and enables an interpretive practice that sees the production of (geographical) meaning in a diverse array of sites of representation. It is an argument for a "critical geosophy." This approach, I argue, is marked by an attentiveness to meaning, and ways in which meanings travel from one sphere of knowledge to another, that is born in the humanities. Exposing meanings and their travels, often rooted in and routed through history, opens up perspectives on otherwise opaque geographical phenomena. It allows us to make critical, material, interventions in important issues. There is a long tradition of critical and materialist work in the humanities that I build on here. It stands against a perception of the humanities as apolitical, idealist and ineffectual that is held by some. I develop these ideas below in the context of the recent turn to work on mobilities which has characterized significant portions of both the humanities and the social sciences. My argument is that what some have called the "new mobilities paradigm" is rooted as much in the humanities as the social sciences and stands as a critique of and addition to overly technocratic and social science-based accounts of "transport" and "migration" (amongst other things).[1]

This approach brings together some unlikely sources of inspiration: primarily John Kirkland Wright, Edward Said and Ian Hacking. John Kirkland Wright is an unlikely source for critical and materialist geography. His notion of geosophy is just about as far from insistently critical and materialist scholarship as it is possible to go within a discipline dominated by the social sciences.[2] In 1947 Wright coined the term "geosophy" to describe what we might call the geographical imagination or geographical knowledge. It was one of the two or three key moments when a leading geographer argued (implicitly at least) for an engagement with the world of the humanities. At the time, he was making the argument that geographers could benefit from exploring the geographical knowledges of non-academic, everyday, folk – fisherman, lorry drivers, farmers, nurses – in order to understand how their ways of knowing the world influenced their everyday lives.

To provide an account of such ideas in individuals is interesting but not especially critical or materialist; perhaps even the opposite. But what if these ideas are in the hands of particularly influential or powerful people? Then a translation occurs from the world of ideas to the world of shared action. Geography becomes part of the bedrock of ideology – of meaning in the service of power. There is a difference between John or Jane on the street having geographical ideas and (say) a judge or novelist or planner or politician. If they have a geographical idea or are informed by a geographical imagination, it is more likely to be widely exported: to become influential. A "critical geosophy" would be an account of geographical ideas and the roles that they play in the production, reproduction and transformation of power. Such an account would have (at least) one foot planted firmly in the humanities as it would be an intervention in the interplay between the world of meaning and the material world.

This insistence on the materialization of the cultural world can be seen in the work of Edward Said,[3] who wrote that "the work of the humanist critic is to materialize rather than spiritualize the culture in which we live."[4] Edward Said made this claim, in a stirring account, to the role of cosmopolitan worldliness in the work of the humanist critic. He was reacting against the kind of formalist political and social disinterestedness that, to many people, in his view characterizes work in the humanities. Rather than the model of the pure, disinterested scholar, Said advocates a critical materialist perspective in humanities-based work. Work in the humanities, he argues, should involve an interest in the worldliness of the acts of reading and writing as well as the world of the text. Here I want to underline Said's insistence on the critical material capacity of humanities scholarship and to illustrate how such an approach has informed my engagement with the idea of mobility.

The third key figure behind this essay is the philosopher of science, Ian Hacking. Hacking has insisted on the necessity of thinking of the processes by which things are socially (and culturally) produced as more literal. In his arguments for an "historical ontology"[5] Hacking suggests that there is a need to delineate the ways in which social types come into being – to outline the kinds of requirements which are necessary for something to be possible. This is not, in his terms, simply a matter of saying something exists (a strict nominalism) but is a matter of providing a material infrastructure which makes such an existence both possible and meaningful. He illustrates his case with an account of the category "woman refugee." Thinking of the category "woman refugee" as a social construct, he suggests, is in some senses obvious.[6] As with many other social types (migrant, capitalist, plumber), it is hard to think of the category "woman refugee" as springing from some version of nature. But if it is obvious, why say it? Hacking's answer is that it becomes more interesting as the analysis becomes more literal. If something is being constructed then there must be agents implicated in the process of construction. A "woman refugee" in

Canada is a geographical and historically contingent category based on inter-national laws of refugee status and the codification of these laws into national law. The category "woman refugee" is clearly a social construct. So are the actual fleshy people who inhabit this category also social constructs? It is certainly the case that the application of the category to real bodies affects their lives in important ways. One body categorized in this way will be offered some forms of legal protection. Another body entering which is not categorized in this way will be deported and returned to a place where life will be difficult or impossible. And then there is the existence of what Hacking calls a "looping effect." People (unlike atoms or light bulbs) can, and frequently do, modify their behavior and even their sense of them-selves to become more like a category they wish to inhabit: a category that will benefit them. Categories are often active participants in the process of making up fleshy human beings. It becomes possible to be a different person.

But categories are not the only things involved in this process of making up. Being a "woman refugee" depends on the existence of a landscape of woman refugees – of borders and fences, of law courts, holding centers, documents, officials, newspaper stories, safehouses, social workers, lawyers, judges, security guards and nations. There is, in other words, a geography of meaning maintenance; a geography that makes it possible to be X. And, following Wright, there is also a deeper set of knowledges that inform the construction of categories in the first place. What would a refugee be without notions of national belonging, of threatening mobility, of the values of place? There are geographical imaginations at work in the work of social construction. This is the subject of critical geosophy.

To sum up: without a focus on the backcloth of institutionalized knowledge that produces effects of power, Wright's call for a geosophy could be simply a call for idiosyncratic and disconnected descriptions of a variety of mental maps. By insisting on the groundedness of knowledge (as Said or Williams might have done) we can avoid the pitfalls of idealism. But simply saying that individual imaginations and forms of knowledge exist in relation to a web of material relations is not to show how meaning is connected to materiality. Hacking's historical ontology provides a frame-work for linking meaning to materiality. We only know it is possible to be a female refugee if we can trace the actions of legal frameworks, forms of regulation, the exist-ence of detention centers and such like. And these in turn are informed by a set of geographical taken-for-granteds (as well as other, not so geographical, imaginations).

It is my contention that an approach such as this can form part of an humanities-based engagement with the world, with history and with humans as agents of change. I flesh this argument out in what follows in relation to work within the recent mobilities turn in the humanities and the social sciences. I focus, in particular, on the connections between mobility and race.

## Understanding mobilities

Until recently there have been very few substantial geography monographs about movement and mobility. One that stands out is *The Geography of Movement* by Lowe and Moryadas, published in 1975.[7] There was (and still is, even in the new, even bigger, edition) a (short) reference to mobility in the Dictionary of Human Geography that refers the reader to "migration" and "circulation." There were many books on transport planning and modeling and on migration. There were books on exploration and tourism.[8] There were, in other words, accounts of particular forms of mobility. There was no account, however, of mobility as a concept and fact in the world. This is despite the fact that mobility is, self-evidently, a concept at the core of human geography, alongside such notions as landscape, space, place or territory (all of which had several monographs, edited collections and textbooks dedicated to them).[9] Looking through the available texts it is striking how the idea of mobility has been taken for granted in ways that place (for instance) no longer is. Mobility has been people (or things, or ideas) moving and it was as simple as that. The single book on mobility (or movement anyway) was a book of classic spatial science, full of gravity models and spatial interaction theories. The time when movement was most central to the discipline coincided with the time at which we were at our most quantitative – at our furthest distance from the humanities.

It seemed that the history of our discipline had bypassed mobility. Many of the concepts from spatial science had been transformed by an encounter with humanistic and critical approaches to the discipline. Location, the dot on the map, had become place. Abstract and absolute space (unrecognized as such by the practitioners of spatial science) had become relative, relational and even third space. Mobility, until recently, has followed no such pathway and subjected to no such analysis.

The recent "new mobilities paradigm" has helped us to think of mobility across a range of ontological, epistemological and normative terrains.[10] It has successfully de-emphasized "transport" as a sole and technocratic focus of mobility studies. But its interests and foci are still often closely entangled with a social scientific approach. Issues such as the increased use of mobile phones, the space of flows inhabited by the kinetic elite or the problems and pleasures of automobility are central to work on mobilities. There is an increased sense of the way different mobilities, at different scales, are intertwined. These are all popular, important and timely advances. But study of them quickly becomes a world of representative samples, bullet points and generalizability. This is the place inhabited by government reports, grant applications which ask you to justify your sampling method and "policy relevance." This is frustrating for scholars rooted in the humanities. It says little, for instance about the mobility that inhabits a borderland of broken-down engines, trains tracks extending to the horizon, bumper-stickered pick-up trucks and marginal people – hobos, traveling salesmen, carnival performers and Robert Johnston figures strumming their guitars playing tunes that have the sound of travel in them – the sound of slavery and of Scottish border ballads sung by crofters displaced by land-grabs. How do we know this land? Most of us know it through the pages of novels, the tracks on CDs and the moving images of films and DVDs. I, for one, have no desire to rid myself of these imagined geographies. This mobile world is not easily translated into the tortured language of social science as it is normally understood. So we have to start somewhere else. To understand any of these crucial issues (mobile phones, the kinetic elite, international migration) fully we have to understand what mobility has been made to mean and how these meanings have been authorized. We have, in other words, to undertake a critical geosophy. The sense of freedom and individualism that accompanies owning a mobile phone or driving a car was not invented with these technologies. The rabid xenophobic responses to immigration are inexplicable without understanding how positive meanings have been attached to place and boundedness in ways which exceed the example of reactions to immigration. Understanding them involves a careful delineation of the process of meaning production. It involves an approach informed by the humanities.

For mobility to be understood within a geosophical framework it is necessary to trace the way that movement is turned into mobility within a range of knowledges and representations. We need to examine its geographies both at the level of material landscapes and at the level of the imaginations that inform those landscapes. We are not surprised to discover practitioners of the humanities busy working with novels or photographs. But the kinds of approaches used with novels or photographs (often, by the way, the hardest to define in a grant application under the "methods" heading) allow us "humanist" social scientists to look at other arenas of knowledge in similar ways. What happens when we turn the interpretive gaze on legal documents, the minutes of council meetings, medical texts or even the writings of spatial scientists? What if we look for some of the things we may find in a blues song, in a census table? Then the traditional objects of social science meet the interpretive gaze of the humanistic scholar and we find the production of meaning. And, as in novels or films, these meanings are often informed by a geographical imagination that, in turn, informs people and their actions. These meanings are part of the "looping effect" that Hacking writes of.

Take, for instance, medical textbooks: an unlikely place for explorations of the meaning of mobility you might think. Feminist scholars provide a useful corrective. Until recently medical textbooks described human reproduction in terms of a mobile and active sperm cell and a passive, stationary egg. Eggs, we were told, merely "drift" and are "transported" while sperms "deliver" after a journey of "incredible velocity" propelled by "whiplashlike motions" of their tails.

> The classic account, current for centuries, has emphasized the sperm's performance and relegated to the egg the supporting role of sleeping beauty. The egg is central to this drama, to be sure, but it is as passive a character as the Brothers Grimm's Princess. Now, it is becoming clear that the egg is not merely a large yolk-filled sphere into which the sperm burrows to endow new life. Rather, recent research suggests the almost heretical view that sperm and egg are mutually active partners.[11]

Even the driest of texts, the medical textbook, is a site for the production of meanings about mobility. Mobility here is insistent, assertive and, above all, masculine. And these textbooks are part of the landscape that produces meanings for mobility. They sit on the desks of medical students, in libraries and on bookshelves in specialist bookstores. They have no influence on what sperms and eggs actually do but they have a great influence on how doctors practice and how men and women are treated. They are implicated in the production of a lived world informed by geographical imaginations about mobility and stasis. Beliefs such as this are, of course, replicated in any number of spheres of social/spatial life. Work on arenas as diverse as journey to work, international migration and bodily comportment have made similar points.[12] But each is enriched when it is juxtaposed to the other. Maybe there is something going on that transcends each case study, something about the way ideas about mobility inform a multitude of knowledges and practices.

## Race, mobility and the humanities

Thinking through and with the humanities has an insistently critical edge. The focus on meaning that an interpretive, creative engagement with the humanities allows us to accomplish can uncover the previously unnoticed aspects of a world of measurements and mappings.[13] Consider three episodes in the mobile history of race in America.

The first episode is an obvious site for humanities scholarship – the poetry of the African-American poet Langston Hughes. It is full of deeply spatialized imagery of the black experience in early twentieth-century United States. Consider the following:

> Merry Go Round
> COLORED CHILD AT CARNIVAL
>
> Where is the Jim Crow section
> On this merry-go-round,
> Mister, cause I want to ride?
> Down South where I come from
> White and colored
> Can't sit side by side.
> Down South on the train
> There's a Jim Crow car.
> On the bus we're put in the back–
> But there ain't no back
> To a merry-go-round!
> Where's the horse
> For a kid that's black?[14]
>                     (Langston Hughes, 1942)

The poem provides an account of the experience of Jim Crow laws for a black child. Jim Crow laws were about both the place of black people in the South and the mobility of black people. It was when black people moved that they were asked to take their place at the back of the bus. Here Hughes reflects on this fact of black mobility. This is connected to the mobility of black migration from South to North with its subterranean history of the Underground Railroad, the Mississippi, and the transition from slave to free black person. The child has taken this particular ride and arrived up north, Harlem perhaps. And here we have a merry-go-round; a circular motion. There is no front or back in evidence and no obvious place to be marginalized in or to. And the merry-go-round becomes part to the carnival – part of the raucous mixing of the everyday, the market place and the fairground.[15] It becomes part of a mobile life where there is no "proper place." Mobility is mixed up here. It reaches across scales from the back of a bus to the bobbing horse of a fairground ride through the blood-soaked move from South to North. And the line between the title and the poem reminds us that this is a child. This child has clearly learned at a young age that he or she is supposed to have his or her place and is confused by the sudden disappearance of this place. Not surprisingly Hughes understood the painful links between mobility and race in the black American experience. Not everyone is so sensitive to the narratives and experiences that accompany the brute fact of movement.

Failing to recognize the meanings of mobility has consequences. Consider the Bus Riders Union (BRU) in Los Angeles. Unlike the poetry of Langston Hughes, their activism may not seem like a subject for humanities-based research. The Bus Riders Union is a group of people who are fighting to keep the Metropolitan Transit Authority (MTA) from spending large amounts of money on a commuter light rail system at the expense of the already existing bus system. The light rail would be fast and allow people to move relatively easily from the outside of the city into the center. Buses are not so fast and not so easy but allow relatively poor, predominantly minority, often female, citizens to move about the city (often to clean the houses of commuters). By proposing to spend millions on the rail system and diverting funds from the bus system the MTA was effectively proposing to enable the mobility of some at the expense of the mobility of others. The Bus Riders Union made its case in court (as well as on the street and in the buses) and stopped the MTA from diverting funds in this way. The nature of the arguments made by the transport planners of the MTA and the advocates of the BRU is instructive.

The MTA argued that this was, more or less, a planning issue. There was a technical problem and a technical solution. It was about moving people over large distances, quickly and efficiently. The MTA made their arguments based on an abstract notion of movement. The BRU, in turn, argued that mobility was social, that it was about access to resources and about power. Movement was not simply abstract and technical, they argued, it was political. Over and above everything else they insisted that issues of public transit were unavoidably issues of race. This insistence was met with denial by the MTA. Providing trains or buses was about planning and it was as simple as that. It was about technological problems with technological solutions. This denial of the link between transit and race was based on the denial, or ignorance, of history and of meaning. It was based on exactly those connections between race and mobility that Langston Hughes makes so clear. To the BRU activists the struggle over the buses of Los Angeles was part of a longer struggle that connected the Underground Railroad with Jim Crow laws, slavery with Rosa Parks. To say that public transit provision in the United States was not about race was absurd. How could transit not be about race? The BRU insisted on the significance of moving on buses and trains. It was more than about getting from A to B in an efficient way. It was about justice. And justice is both historical and geographical. Making these connections involves thinking about the myriad ways that mobility and marked moving bodies are given meaning and then exploring how that meaning is lived. It involves bringing into consideration all the factors that produce the possibilities of moving for people coded as raced, the Jim Crow laws and the poetry of Langston Hughes as much as the planning documents of the MTA. An approach based in the humanities encourages the thinking through of these connections.

My third short episode is the case of the racially marked events surrounding the mobilities of people escaping Hurricane Katrina and its human-aided onslaught on New Orleans and the surrounding area in late August 2005.[16] Over one thousand people were killed and hundreds of thousands left homeless. New Orleans residents who could not find shelter and food with relatives and friends were evacuated to locations across the American south. The flooding had been foreseen and an evacuation of the city ordered. Along with the detritus of human life that came to the surface in the days of early September were issues of race, poverty and mobility. Clearly there are valuable and legitimate social science stories to tell here. There are people to be counted, maps to be charted, pie-charts to be diagrammed. And these all have the potential to make radical interventions in these intolerable geographies. But something would be missing, something that the kinds of radical sensibilities engendered by the humanities might include.

Following the hurricane there was an extended discussion of the differential effects of the hurricane on black and white populations. As there had been no public evacuation procedure people had been asked to simply leave in private vehicles. Subsequently many affluent and white people had been able to leave while impoverished and mainly black people had found themselves immobilized in the city.

As with some of the reactions to the activities of the Bus Riders Union in Los Angeles there were attempts to make mobility a largely technical issue – to "de-socialize" and, particularly, "de-racialize" mobility. In contrast to these arguments, another story from the days following Katrina makes the link between race and mobility quite explicit.

In the days and weeks following Hurricane Katrina, stories began to emerge about the categorization of people who were displaced by the hurricane and the floods that followed it. The press has referred to them as "refugees" and the meanings associated with this mobile subject identity had offended some of the displaced people and some of the spokespeople for the African-American community. Both Jesse Jackson and Al Sharpton expressed opposition to this use of this term, claiming it was racist. The word *refugee*, it seems, was quickly associated with notions of blackness and foreignness. Even President Bush asserted that: "The people we are talking about are not refugees…. They are Americans and they need the help and love and compassion of our fellow citizens."[17]

The arguments over the use of the term "refugee" points to both the importance of meaning as a component of understanding mobility and to the ways in which mobility is entangled with race. The term "refugee" is freighted with meanings of subversive and threatening mobility enacted by those deemed other, from elsewhere. The word "*crisis*" often accompanies the word refugee, as do the words "foreign" and "immigrant." A critical understanding of the label "refugee" in this instance involves a longer historical understanding of the various forms of documents (including such apparently non-humanities kinds of documents as legal records and the Geneva Convention) and knowledges that have been implicated in the making up of refugees – documents and knowledges that build on specifically geographical imaginations about locatedness and mobility. It is through the history of such knowledges that the term "refugee" has become wrapped up in notions of being out of place, of being foreign and suspect. The term is heavily racialized because of a long history of negative representations of refugees as other, as being from somewhere else, as threateningly mobile. This should lead to the question of what the refusal of the label "refugee" in this case meant for all those who could be, apparently legitimately, called "refugee."

## Conclusion

In a posthumously published essay Said wrote that "Change is human history, and human history as made by human action and understood accordingly is the very ground of the humanities."[18] The humanities have the capacity to enchant, to animate and to spiritualize. But it would be a mistake to be happy with these fine achievements of humanities scholarship and leave it at that. The humanities also have the capacity to "materialize" the world that surrounds us and remind us of how and why a meaningful world comes into being, to ask how these meanings are authorized, by whom and for what purpose and what ends? As geographers who engage with the humanities we have a particular role to play here. The geographical imagination – the subject of geosophy – feeds into all kinds of ways of making the world up. Ideas about a notion as fundamental as mobility, for instance, can have very material origins and effects. They become embroiled in arguments about how the world is and how it should be and how we might get from here to there. Race is one social arena that such an investigation can illuminate and intervene in. Indeed, much of the best and most inspiring writing on issues around race is located in the humanities.[19]

## Notes

1   M. Sheller and J. Urry, "The New Mobilities Paradigm," *Environment and Planning* A 38:2, 2006, 207–26; K. Hannam, M. Sheller and J. Urry, "Mobilities, Immobilities and Moorings," *Mobilities* 1:1, 2006, 1–22; J. Urry, *Mobilities*, Cambridge: Polity, 2007; P. Adey, *Mobility*, London: Routledge, 2009.
2   J.K. Wright, "Terrae Incognitae: The Place of the Imagination in Geography," *Annals of the Association of American Geographers* 37, 1947, 1–15. See also T. Cresswell, "Making Up

the Tramp: Towards a Critical Geosophy," in P. Adams, S. Hoelscher and K. Till (eds.), *Textures of Place: Exploring Humanist Geographies*, Minneapolis, University of Minnesota Press, 2001.

3  Though it was also central to the foundational work of Raymond Williams in his insistence of the importance of "cultural materialism" and "structures of feeling." See R. Williams, *Marxism and Literature*, Oxford [Eng.]: Oxford University Press, 1977.

4  E. Said, *The World, the Text and the Critic*, Cambridge, MA: Harvard University Press, 1983.

5  He is, of course, informed by the work of Foucault here. See I. Hacking, *Historical Ontology*, Cambridge, MA, London: Harvard University Press, 2002.

6  I. Hacking, "Making Up People," in T. Heller, M. Sosna and D. Wellbery (eds.), *Reconstructing Individualism: Autonomy, Individuality, and the Self in Western Thought*, Stanford: Stanford University Press; I. Hacking, *The Social Construction of What?*, Cambridge, MA London: Harvard University Press, 1999.

7  J. Lowe and S. Moryadas, *The Geography of Movement*, Boston, MA: Houghton Mifflin, 1975.

8  See, for instance, A. Blunt, *Travel, Gender and Imperialism: Mary Kingsley and West Africa*, New York: Guilford, 1994; F. Driver, *Geography Militant: Cultures of Exploration in the Age of Empire*, Oxford: Blackwell, 1999; A.R. Pred, "The Choreography of Existence: Comments on Hagerstrand's Time-Geography and its Usefulness," *Economic Geography* 53, 1977, 207–21; H.P. White and M.L. Senior, *Transport Geography*, London: New York: Longman, 1983; P.J. Boyle and K. Halfacree, *Migration into Rural Areas: Theories and Issues*, Chichester; New York: Wiley, 1998.

9  Y.-F. Tuan, *Space and Place: The Perspective of Experience*, Minneapolis: University of Minnesota Press, 1977; D. Cosgrove, *Social Formation and Symbolic Landscape*, London: Croom Helm, 1984; R. Sack, *Human Territoriality: Its Theory and History*, Cambridge: Cambridge University Press, 1986; N.J. Thrift, *Spatial Formations*, London; Thousand Oaks: Sage, 1996.

10  M. Sheller and J. Urry, "The New Mobilities Paradigm," *Environment and Planning* A 38:2, 2006, 207–26; K. Hannam, M. Sheller and J. Urry, "Mobilities, Immobilities and Moorings," *Mobilities* 1:1, 2006, 1–22.

11  See E.S. Heller, "Gender Language and Science," The 1996 Templeton Lecture, University of Sydney, HTTP: <http://www.scifac.usyd.edu.au/chast/templeton/1996templeton/1996lecture.html>.

12  T.P. Uteng, and T. Cresswell (eds), *Gendered Mobilities*, London: Ashgate, 2008.

13  D.E. Cosgrove, *Mappings*, London: Reaktion Books, 1999.

14  Langston Hughes [1942] in L. Hughes, A. Rampersad and D.E. Roessel, *The Collected Poems of Langston Hughes*, New York: Vintage, 1995.

15  M. Bakhtin, *Rabelais and his World*, Bloomington: Indiana University Press, 1984.

16  An extended account can be found in T. Cresswell, *On the Move: Mobility in the Modern Western World*, New York: Routledge, 2006.

17  This quotation is from Associated Press online at HTTP: <www.wwltv.com/stories/ww;090605refugees_.306635fl.html> (accessed 5 October 2005).

18  E.W. Said, *Humanism and Democratic Criticism*, New York: Columbia University Press, 2004.

19  B. Honig, *Democracy and the Foreigner*, Princeton: Princeton University Press, 2001; K. McKittrick, *Demonic Grounds: Black Women and the Cartographies of Struggle*, Minneapolis; London: University of Minnesota Press, 2006.

# 9

# THE WORLD IN PLAIN VIEW

*J. Nicholas Entrikin*

As befits two disciplines, neither of which is clearly defined and both of which address themselves to the whole of human life and thought, anthropology and philosophy are more than a little suspicious of one another. The anxiety that comes with a combination of a diffuse and miscellaneous academic identity and an ambition to connect just about everything with everything else and get, thereby, to the bottom of things leaves both of them unsure as to which of them should be doing what. It is not that their borders overlap, it is that they have no borders anyone can, with any assurance, draw. It is not that their interests diverge, it is that nothing, apparently, is alien to either one of them.

*Clifford Geertz*[1]

## Introduction

Clifford Geertz's comment on the ambiguous relationship between philosophy and anthropology could, with very little editing, be used to describe the relationship between philosophy and geography. Perhaps even more audaciously than anthropologists, geographers claim the entire world as their field of expertise. Yet to characterize philosophy and geography as "suspicious" of one another would be inaccurate. More often than not the former is simply unaware of the latter, and the latter, while keenly aware of the former, treats it as little more than a source of intellectual legitimacy or political ammunition. Even so, from time to time philosophy and geography have engaged in productive dialogue. This complicated, intermittent relationship represents one occasionally controversial aspect of the broader connection between geography and the humanities.

In what follows I will identify particular areas in a broadly defined geographical project where this seemingly unnatural disciplinary coupling becomes significant. Of course, all geographical concepts are potential objects of philosophical reflection. However, one area in particular, the human construction of worlds out of nature through place making and landscape creation, leads to potentially fruitful engagement of these two seemingly divergent perspectives on the world.

Geographical agency through place making involves often complex moral decisions about our relationship to others and to the natural world. Reflection on the moral and political dimensions of place making is a fairly recent turn in geographical thought and, important to the overarching theme of this volume, it is a turn that connects geography to contemporary research in the humanities.

## Geography and philosophy

For many, geography and philosophy are incongruous endeavors. To refer to oneself as a "geographical philosopher" or "philosophical geographer" is to invite puzzlement or mild amusement rather than understanding or curiosity. Such titles seem obfuscating and intellectually pretentious. No philosopher has embraced the first, although among contemporary philosophers Edward Casey and Jeff Malpas would easily fit such a description.[2] On occasion, the second title has been applied to those who work in a philosophically informed history of geographical thought. However, "philosophical geographer" is not a term that those working in this area would choose for themselves. Within geography, those who work with philosophy are more likely to describe themselves as "historians of geographic thought" or "geographic theorists," even though these descriptions may not accurately portray their work.

In the modern era of disciplinary specialization the initially limited engagement between philosophers and geographers came almost exclusively from the side of geography and concerned questions of scientific logic. Is geography a science? And if so, what kind of science is it? One sees the origins of these questions in nineteenth-century discussions of the classification of sciences: August Comte's hierarchy of sciences moving from the concrete to the abstract, and the distinction of the late nineteenth-century neo-Kantians, Wilhelm Windelband, and Heinrich Rickert, between idiographic and nomothetic sciences.[3] Geography, when discussed at all, was usually put into the concrete and idiographic categories.

Opposition to a loosely defined and often poorly understood "idiographic geography" is what united various groups of geographers in support of spatial analysis, the mid-twentieth century ancestor of geographical information sciences. To buttress their claims to being geography's "true" scientists, spatial analysts turned to the seemingly powerful analytical tools of mathematics and the conceptual framework of logical empiricist philosophy of science. The subsequent challenge to this model of geographical science in the latter part of the twentieth century came as a criticism of logical empiricism, a revolt that found sources in a diverse set of philosophical positions from phenomenology, existentialism, and the dialectics of Hegel and Marx, to the post-structuralism of Foucault.

What is notable in these shifting currents of geographical thought is that they all sought to establish their claims to intellectual legitimacy on their interpretation of philosophical texts. This represents a profound shift from earlier eras in which legitimacy and authority resided in the accurate interpretation of works written by the disciplinary pantheon. The newer strategy of philosophical "grounding" continues today in feminist, post-colonial, and post-structural geographies. Good examples of this work maintain the critical spirit of philosophical practice in exploring the conceptual foundations of geographical understanding. Too often, however, the writings of philosophers have been used by geographers simply to add hoped-for legitimacy to strongly held opinions and beliefs, or to de-legitimize the opinions and beliefs of others. Both good and bad examples combine to provide a picture of late-twentieth- and early twenty-first-century geography as more explicitly engaged with philosophical texts than during any other period since its professionalization as a university discipline.

The boundaries are more difficult to draw before this period. One could consider the shared origins of these fields in classical Western thought. Indeed, the most important and influential work by a modern geographer in philosophy and the history of ideas, Clarence Glacken's *Traces on the Rhodian Shore*, explores precisely these origins.[4] For example, the ancient Greek geographer Strabo wrote in his introduction to his multi-volumed *Geography* that the science of geography "described the parts of the earth."[5] But this cataloguing carried great import for Strabo, who stated that geography "as much as any other science," was the "concern of the philosopher," a statement that reads two millennia later as more hopeful than prophetic, and which no doubt reflects issues of translation and how much the meaning of the term "philosopher" has changed over the centuries.

The philosopher who discusses geography is rare in the history of modern Western thought. The most notable exception is Immanuel Kant, who lectured on physical geography and anthropology. These lectures have for a long time been largely ignored but more interest now exists in trying to relate them to his philosophy. The physical geography of Kant has been widely dismissed as intellectually inconsequential or worse. For example, David Harvey has referred to Kant's empirical work as "nothing short of a political and intellectual embarrassment," in part for its alleged racism and Eurocentrism, views that represent the antithesis of Kant's cosmopolitan ideal.[6] (Harvey is equally critical of Kant's anthropology.) He uses Kant's writings on geography and anthropology to make a sweeping claim about the weakness of current forms of Kantian cosmopolitanism, as represented, for example, in the work of Martha Nussbaum.[7] Although Harvey attributes the positive outcome of a "general revival of interest in geographical knowledges in recent times" to the renewed scholarly interests in a Kantian cosmopolitanism, it is not a lineage of which he is proud. For him, the "nobility of Kant's (and our) ethical vision needs to be tempered by reference to the banality of his (our) geographical knowledges and prejudices."[8]

The political philosopher-turned-geographer Stuart Elden has recently offered a more nuanced, less polemical view of this Kantian heritage.[9] Elden notes philosophy's understandable silence about Kant's physical geography. There are questions about the reliability of the sources (they were largely class notes) and, more importantly, the fundamental disproportion between a modest set of ideas expressed in a class and Kant's core philosophical work through his *Critiques*. Elden offers other reasons as well that are more internal to philosophy. He notes the important difference between the ways in which the two major traditions of philosophy treat history and biography. The analytic tradition, long dominant in the Anglo-American philosophical sphere, generally separates the biography of the philosopher from the significance of the work. On the other side, the Continental tradition pays more heed to the contextual elements of philosophy and philosophers. He, himself, has favored the latter tradition and claims that geographers have as well. One might contest this latter point but he makes this argument to express what he sees as the important connection of geography and anthropology to Kant's larger philosophical project. He follows the path outlined by other Kant scholars in arguing that Kant's aim was to help his students develop a cosmopolitan, global vision.[10] Geography and anthropology were fundamental to this task. The geographer offers awareness of the world's variety, its differences from place to place, the various forms of life, and the constant adaptation of cognitive and moral agents to changing environmental conditions. The philosopher possesses powerful tools of critical and analytic reflection and seeks knowledge beyond the contingent and the particular. The conflict internal to Kant's corpus is therefore reflected in the differing goals of the geographer and the philosopher. Elden is not an apologist for Kant but he does see the value of taking Kant's geography and anthropology more seriously and not simply as an excuse for dismissing modern-day work that has drawn inspiration from Kant.

In the modern era, beginning with the professionalization of disciplines in the late nineteenth century, philosophy and geography seemed first closer and then farther apart, as both of these two fields of inquiry were transformed into more narrowly defined and specialized disciplines. Not surprisingly they were closer at the dawn of the era of specialization when in American universities knowledge of the humanities meant knowledge of the classical world, which included both studies in philosophy and geography. For example, the first classes in geography at the University of California were courses on the ancient Mediterranean world and were part of the classical training of late-nineteenth-century students in the philosophy, literature, and languages of ancient Greece and Rome. This conception of the humanities gave way in mid-twentieth-century American academic life to less historically specific studies of languages, literature, philosophy, and the arts.[11] Geography was only rarely grouped in the humanities and dialogue with philosophers was basically non-existent.

Where these disciplines were occasionally joined in the early era of professionalization, however, was in support of liberal education as a means of building an educated citizenry for a democratic civil society. In this tradition one relatively neglected discussion of geography by a philosopher can be found in the now classic 1916 text *Democracy and Education* by the American pragmatist philosopher John Dewey.[12] Dewey placed geography alongside history as the core fields of a cosmopolitan civic education. Each discipline functioned to expand the vision of students beyond the here-and-now and to broaden their imaginations in ways that Dewey thought critical for an enlightened democratic citizenry. Dewey's discussion of geography, however, was the exception rather than the rule. Much now is made of the geographical implications of Martin Heidegger's ontology of "dwelling," seemingly central to the geographer's concern with how humans occupy the earth.[13] But the professional geography of Heidegger's era was not open to this then-radical possibility.

With very few exceptions, mid-century writings on what might be called the philosophy of geography were part of a general disciplinary concern with exploring and in some cases defending the scientific character of the field. For these authors the descriptive term most favored was "methodology." For example, David Harvey in his influential 1969 text *Explanation in Geography* introduces his argument by drawing a distinction between the philosophy and methodology of geography.[14] For Harvey, the philosophy of geography is a matter of opinion and belief – more subjective than the exploration of the logic of geographical science or methodology. He made emphatically clear that his approach was that of the methodologist.

Most explicitly philosophical work in geography in the mid to late twentieth century was strongly linked to the spatial scientific project. For example, *Philosophy in Geography*, edited by Stephen Gale and Gunnar Olsson, was largely a genuflection to the methodological clarity and theoretical potential of spatial science.[15] The contributors were primarily spatial analysts and regional scientists and the volume appeared in a series devoted to examining issues in social and behavioral scientific methodology and epistemology, especially questions about scientific inference. Some of its intellectual lineage was to be found in earlier conversations between philosophers and geographers on matters of the logic of science, dating as far back as Alfred Hettner's use of Rickert and Windelband, but evident as well in less visible connections between geographer Fred Lukermann and philosopher May Brodbeck, geographer Fred Schaefer and philosopher Gustav Bergmann, the Hartshorne brothers, the geographer Richard and the philosopher Charles, and others that were often the consequence of personal or collegial ties. Even the most trenchant criticisms of the excesses of spatial science, such as those found in the work of Robert Sack, come out of these same epistemological sources.[16]

The most recent explorations of philosophy by geographers have been marked by a shift away from these matters of method and epistemology and toward issues of morality and ontology. However, works in these latter areas often take very different directions, based on broad issues of humanist scholarship, such as the distinction between the centered versus decentered self, and more specifically geographical matters of the degree of engagement with core concepts and the related distinction between what might be called indigenous versus imported theory. An example of a geography-centric "geophilosophical theory" is found in the work of Robert Sack, especially his book, *The Geographical Guide to the Real and the Good*.[17] Sack's theory offers a geographical equivalent of a philosophical anthropology, an argument about the nature of human existence and what it means to be human. For Sack, humans are geographical beings who "are incapable of accepting reality as it is, and so create places to transform reality according to the ideas and images of what we think reality ought to be."[18] This quality he describes as the geographical problematic. In this view places become tools for human projects. The relative goodness of places can be measured either as instrumental value, allowing humans to reach particular goals, or intrinsic value by approaching a universal good without connection to achieving particularistic ends. In Sack's theory, humans are autonomous moral agents who through their projects continually reshape natural and cultural worlds to achieve desired ends.

This geographical theory of morality and its autonomous agents can be contrasted to the decentered geographical subjects in the "geophilosophy" based in the works of Gilles Deleuze and Félix Guattari, a lineage that traces back to Friedrich Nietzsche.[19] Deleuzian arguments have inspired geographical work on the contingent, interconnected, or "rhizomatic" flows of human and natural processes and have formed the basis of what has been referred to in geography as "non-representational theory."[20] The geographer Mark Bonta and the philosopher John Protevi express optimism for the potential of geophilosophy and Deleuzian-inspired geography, stating:

> Perhaps it's not too much, then, in our view, to say that Deleuze – once his work is fully understood – can be the Kant of our time. By this we mean not the author of mediocre and racist geographical works replete with common prejudices of his age, but the great philosopher who provided the philosophical "grounding" of classical modern science....We see *ATP* [*A Thousand Plateaus*] as the key to a new materialist geophilosophical paradigm because it combines Kant's insistence on the importance of the earth sciences and Descartes' attempt to provide a new cartography and coordinate system.[21]

This new paradigm is rooted in complexity theory, which engages the emergent qualities of material systems and that addresses "fragmented space," "twisted time," and "disequilibrium." Despite great interest in the work of Deleuze among a subset of geographers, however, no one has yet assumed the working title of "geophilosopher." In part this reflects the absence of an explicit engagement by Deleuze and Guattari with the geographical – that is to say, with issues of landscape, place, and space.

To at least one commentator on the geophilosophical project, trying to blend philosophy and geography through Deleuzian arguments may be of limited potential. In a review of Bonta and Protevi's book, the philosopher Dylan Trigg writes, "Given that the hybrid between geography and philosophy has been principally (and successfully) adopted through phenomenological work, the authors' attempt to resituate this convergence in terms of complexity theory has limited scope."[22] In referencing phenomenology, Trigg points in the direction of the most explicitly philosophical orientation in late-twentieth-century geography, so-called "humanistic geography."

In one of the first systematic statements on humanistic geography, Yi-Fu Tuan defines this orientation as a direct engagement with humanistic disciplines, including philosophy.[23] Edward Relph made clear reference to phenomenology and existentialism in his book on the experience of modern places and drew inspiration from the then largely unknown French geographer Eric Dardel, who was one of the first geographers to discuss the philosophy of geography as addressing the human experience of nature.[24] Histories of geography make reference as well to the influence of sources such as J.K. Wright's widely read *Annals* paper of 1947, "Terrae Incognitae: The Place of the Imagination in Geography," and his neologism, "geosophy," as marking a significant turn in the self-conscious engagement of geography as a form of knowledge. Dardel's book and Wright's paper may be seen as harbingers of a greater curiosity among geographers about the humanistic roots of their discipline – about matters of experience and interpretation, themes that would draw geographers' attention to philosophical traditions such as phenomenology and, later, pragmatism.

Evidence that this curiosity persists may be seen in several current projects. For example, the journal *Philosophy and Geography* demonstrates a commitment to cooperative and collaborative practice between the two disciplines. The journal began modestly under the leadership of the geographer Jonathan Smith and the philosopher Andrew Light as an annual publication. It has since grown to a tri-annual publication and joined with another journal in 2005 to re-emerge (with the same editors) as *Ethics, Place & Environment: A Journal of Philosophy and Geography*. Its goal is to stimulate and develop the growing overlap between the two disciplines through the examination of environmental ethics, understood broadly to address the natural and built environment. The same team has also collaborated to form *The Society for Philosophy and Geography* in 1997 and has been successful in combining the two fields in ways that are both substantive and institutional.[25]

Even with the relative success of this and other collaborative efforts, they reach only a relatively small portion of the two disciplines. Many geographers still hold firm beliefs about their field's status as a science and about the fundamental difference between science and philosophy. Why, they ask, would anyone want to connect a straightforward empirical discipline such as geography with a reflective enterprise that seeks to uncover the basic structures of human thought? Or, to turn the tables, why would one take a discipline as practical and useful as geography and connect it to something as impractical and unworldly as philosophy? After all, the critical and speculative examination of the human condition that defines scholarship in the humanities has, at its source, philosophical reflection. Can one say anything like that about geography? How does one combine this core concern of the humanities with a field often described as an earth science?

## The relational turn: self and place

Despite entrenched opposition, the emergence of widespread concerns with the "self" or "subject" in geographical inquiry has allowed for a greater opening to the humanities. Such "subjects," whether centered or decentered, are reflexive, fully dimensional geographical agents, unlike the one-dimensional geological agent of early-twentieth-century geography or the economically rational actor of mid-twentieth-century spatial analysis. As a social, moral, and aesthetic being, the geographical agent acts in the world not only in terms of how the world is, but also with ideas of what it could become. Human geographers now examine the role and position of the active subject who interprets the world through narrative, performance, and graphical representation. With this change has come a greater self-awareness about the role of the geographer as researcher. The image of geographers remotely gazing from above or as disembodied observers in the field has changed to geographers as subjects in the world narrating their experiences.

This more complex concept of the geographical actor has coincided with changed conceptions of place and landscape. The dominant meaning of place has changed from a location or a position in space to a complex relation of self and environment. Within this perspective, cartography and regional geography have become modes of representation as opposed to mirrors of nature; as modes of representation they have blurred the line separating actual and imagined geographies. Landscape, while maintaining the key element of the visual, has been explored as a "way of seeing" that presumes the social and political context of actors who engage in and transform the material world.[26]

This fully dimensional geographical agent acts in the world, either individually or collectively, as a place maker and transformer of landscapes. Each role draws together subject and object, culture and nature; each transforms space into place and nature into human landscapes. These worlds are created out of natural environments by imposing social rules and meanings. Homes, cultivated fields and factories, universities and prisons, city parks and wilderness preserves, civic monuments and nuclear waste sites, national territories and international trade zones are all places created as tools for human projects.[27] Their fluid spatial scales are adapted to collective and individual projects and in an ideal world these places are remade and transformed as these projects change. Such projects do not occur in empty space; they must adjust to already existing symbolic and material conditions and are frequently altered or undermined by conflicting social and natural processes as well as influenced by existing webs of power relations. They are created by authoritarian decree as well as by democratic consensus, by brute force as well as by the ethics of care.

The geographer who is drawn to philosophy is drawn not as a philosopher, but as someone who wants to capture this place-making process in its full dimensionality. In the end, geography is more about practice than reflection, although both are necessary. The geographer's practice is one that always engages a material world, even if that world is only known through symbols. These concerns are what have drawn geographers toward debates about epistemology and ontology. One could argue, however, that geography is more fundamentally connected to questions of moral and political philosophy. After all, place making requires not only empirical, instrumental knowledge but also moral and political judgment. It should not be surprising, then, that some of the most vital areas of connection between philosophy and geography concern issues of environmental ethics, social justice, and democratic community. By investigating humans' relations with nature, issues of inclusion and exclusion, and who decides how places are transformed, scholars working in these areas address the fundamental task of place making. New technologies tend to complicate rather than change these basic human concerns.

## Conclusion

In his *Philosophical Investigations*, Ludwig Wittgenstein describes science as "a shifting set of methods" for learning "something new about the world."[28] Philosophy, for Wittgenstein, has a different task, and that is "to understand something that is already in plain view." Geographers, like Geertz's anthropologists, are sufficiently diverse in their interests to engage both sides of this distinction. Indeed, as I have argued, twentieth-century concerns about that "shifting set of methods" are what led many geographers to read philosophy and then to write as "methodologists." The best humanities-inspired studies of space, place, and landscape, however, move closer to the latter perspective, of helping to understand what is or was in plain view. In such projects, the works of philosophers, literary scholars, and artists all become potential sources of inspiration for knowing better the worlds that humans create and transform, destroy and rebuild, through the places and landscapes that form the necessary infrastructure of everyday life.

## Notes

1 C. Geertz, *Available Light: Anthropological Reflections on Philosophical Topics*, Princeton, NJ: Princeton University Press, 2000, p. ix.
2 E.S. Casey, *Getting Back into Place: Toward a Renewed Understanding of the Place-World*, Bloomington: Indiana University Press, 1993; E.S. Casey, *The Fate of Place: A Philosophical History*, Berkeley: University of California Press, 1997; E.S. Casey, *Representing Place: Landscape Painting and Maps*, Minneapolis: University of Minnesota Press, 2002; J. Malpas, *Place and Experience: A Philosophical Topography*, New York: Cambridge University Press, 1999; J. Malpas, *Heidegger's Topology: Being, Place, World*, Cambridge, MA: MIT Press, 2006.
3 A. Comte, *Positive Philosophy, Volume I*, trans. H. Martineau, London: G. Bell and Sons, 1896; H. Rickert, *The Limits of Concept Formation in Natural Science: A Logical Introduction to the Historical Sciences*, trans. G. Oakes, Cambridge: Cambridge University Press, 1986; W. Windelband, *History and Natural Science*, trans. G. Oakes, *History and Theory* 19, 1980, 169–85.
4 C. Glacken, *Traces on the Rhodian Shore: Nature and Culture in Western Thought from Ancient Times to the End of the Eighteenth Century*, Berkeley: University of California Press, 1967.
5 Strabo, *The Geography of Strabo*, trans. H.L. Jones, London: Heinemann, 1917.
6 D. Harvey, "Cosmopolitanism and the Banality of Geographical Evils," *Public Culture* 12, 2000, 532.
7 M. Nussbaum, "Kant and Stoic Cosmopolitanism," *Journal of Political Philosophy* 5, 1997, 1–25; *For Love of Country?*, Boston, MA: Beacon Press, 2002.
8 Harvey, "Cosmopolitanism," p. 536.
9 S. Elden, "Reassessing Kant's Geography," *Journal of Historical Geography* 35, 2009, 3–25.
10 R. Louden, *Kant's Impure Ethics: From Rational Beings to Human Beings*, Oxford: Oxford University Press, 2000; R. Louden and M. Kuehn (eds), *Kant: Anthropology from a Pragmatic*

*Point of View*, Cambridge, England: Cambridge University Press, 2006; H.L. Wilson, *Kant's Pragmatic Anthropology: Its Origin, Meaning, and Critical Significance*, Albany: State University of New York Press, 2007.

11 S. Marcus, "Humanities from Classics to Cultural Studies: Notes Toward the History of an Idea," *Daedalus* 135, Spring 2006, 19.

12 J. Dewey, *Democracy and Education*, New York: Macmillan, 1944 [1916].

13 M. Heidegger, "Building, Dwelling, Thinking," in *Poetry, Language, Thought*, New York: Perennial Classics, 1971.

14 D. Harvey, *Explanation in Geography*, London: Edward Arnold, 1969.

15 S. Gale and G. Olsson (eds), *Philosophy in Geography*, Dordrecht: D. Reidel Publishing Company, 1979.

16 R.D. Sack, *Conceptions of Space in Social Thought*, London: Macmillan, 1980.

17 R.D. Sack, *A Geographical Guide to the Real and the Good,* New York: Routledge, 2003, p.ix.

18 R.D. Sack, *Geographical Guide*, p. 4.

19 G. Deleuze and F. Guattari, *What Is Philosophy?*, New York: Columbia University Press, 1994, p. 104.

20 For a geographical introduction to geophilosophy see M. Bonta and J. Protevi, *Deleuze and Geophilosophy: Guide and Glossary,* Edinburgh: Edinburgh University Press, 2004.

21 Bonta and Protevi, *Deleuze and Geophilosophy*, pp. vii–viii.

22 D. Trigg, review of *Deleuze and Geophilosophy: Guide and Glossary* by M. Bonta and J. Protevi, *Ethics, Place & Environment: A Journal of Philosophy and Geography* 10, 2007, 251.

23 Y.-F. Tuan, *Space and Place: The Perspective of Experience*, Minneapolis: University of Minnesota Press, 1977.

24 E. Relph, *Place and Placelessness*, London: Pion, 1976; E. Dardel, *L'homme et la terre*, Paris: Presses Universitaires de France, 1952.

25 A. Light and J.M. Smith (eds), *Philosophy and Geography II: The Production of Public Space*, Lanham: Rowman & Littlefield Publishers, 1998; A. Light and J.M. Smith, *Philosophy and Geography III: Philosophies of Place*, Lanham: Rowman & Littlefield Publishers, 1998.

26 D.E. Cosgrove, *Social Formation and Symbolic Landscape*, Madison: University of Wisconsin Press, 1998; D.E. Cosgrove and S. Daniels, *The Iconography of Landscape: Essays on the Symbolic Representation, Design and Use of Past Environments*, New York: Cambridge University Press, 1989.

27 R. Sack, *Homo Geographicus: A Framework for Action, Awareness and Moral Concern*, Baltimore: Johns Hopkins University Press, 1997.

28 L. Wittgenstein, *Philosophical Investigations*, trans. G.E.M. Anscombe, Malden: Blackwell Publishers, 2001 [1953], quoted in S. Cavell, *Philosophy the Day after Tomorrow*, Cambridge, MA: Belknap Press of Harvard University Press, 2005, p. 112.

# 10

# COURTLY GEOGRAPHY

## Nature, authority and civility in early eighteenth-century France

*Michael Heffernan*

Louis XV, the physical embodiment of French cultural and political authority for most of the eighteenth century, was called many names, not all of them complimentary, during his long and not entirely illustrious reign. The list of epithets has expanded ever since but Louis has never, as far as I know, been described as a fluvial geomorphologist, even though a very slender case can be made for considering France's penultimate pre-Revolutionary monarch in precisely these terms. Nestling among the 12,300 printed laws, edicts, and ordinances misleadingly attributed to the author name "Louis XV" in the electronic catalogue of the Bibliothèque Nationale de France is a printed volume whose title sets it apart from this mountain of official documentation. The book in question turns out to be, on closer inspection, a short geographical treatise entitled *Cours des Principaux Fleuves et Rivières de l'Europe*. This seems, on the face of it, to have been Louis's one and only proper publication.

The *Cours*, as it will henceforth be referred to here, received the occasional comment in eighteenth-century accounts of Louis's life and times, and is mentioned in passing in a few of the more comprehensive recent histories, but it has never been discussed in detail. This silence is partly explained by the book's publication date – 1718. In February of that year, Louis XV celebrated his eighth birthday, an orphan King who had unexpectedly inherited the throne three years earlier following a calamitous sequence of deaths that carried off his great-grandfather, Louis XIV, as well as his grandfather, father, mother, uncle, and both his older and younger brothers. The *Cours* is therefore neither a learned scientific treatise, nor a mature work of political economy. It is, rather, a short commentary, unfettered by nuance or detail, and written in the kind of simple, unadorned prose that an inquisitive, well-educated young boy might have used under the watchful instruction of his tutors.

We do not know how many copies of the *Cours* were printed but it was certainly a small number. Voltaire, who was released from a short spell in the Bastille a few weeks before the *Cours* was printed, later suggested that only 50 were produced, almost certainly an underestimate but not by much.[1] Reports that 150 copies, previously seized by the revolutionary authorities, went on sale in Paris in 1805 are plausible, this cache apparently comprising unbound sheets acquired by a Parisian book-seller who destroyed all but a handful to increase the value of those that remained.[2] Only 17 library copies of the *Cours* survive, 11 in France, four in the USA, and two in the UK. All are quartos with leather bindings varying in color and detail, most showing the Bourbon arms on front and back.

The book's 71 large-font pages describe the routes of 47 rivers, 21 in France and 26 in the rest of Europe, the text moving in accordance with current educational theory from the familiar and the known to the unfamiliar and the unknown.[3] The 21-page opening section on France is divided into four sub-sections devoted to the four principal rivers invoked since Medieval times to define the limits of the Frankish realm: the Seine, the Loire, the Garonne, and the Rhône. The size and direction of these rivers are discussed from source to sea, with comments on their towns and notable bridges, navigational capacities, commercial significance, and surrounding landscapes. Similar commentaries are provided on two tributaries of the Seine and three tributaries each of the Loire, the Garonne, and the Rhône. The opening section concludes with ten pages on the Escaut, the Meuse, the Moselle and their tributaries, the three river systems that flow across France's north-eastern frontier, the configuration of which had been defined only five years earlier by the Treaty of Utrecht. The 45-page second section describes the European rivers in the same simple terms, beginning with Germany (Rhine, Weser, Elbe, Trave, Oder, and Danube) and moving through Spain (Douro, Tagus, Guadiana, Guadalquivir, Ebro, and Segre), England (Thames and Severn), Poland (Vistula, Dziwna, Dnieper, Dniester, and Neman), Russia (Volga, Dvina, and Ob), and Italy (Po, Adige, Tiber, and Ofanto).

Posterity may have judged little Louis's little book to be no more than a charming piece of juvenilia but this chapter suggests that this long-forgotten volume is worthy of more serious consideration as text and artefact. The *Cours* was a far more complex work than its disarmingly simple prose suggests because it illustrates how and why geographical knowledge came be to regarded as integral to the emerging educational and cultural arena we now recognize as the humanities. This is a large claim, to be sure, particularly for a modest volume prepared as a vanity publication within the closed world of the early eighteenth-century French court. But this suggestion is endorsed by recent work on the history of geography that has demonstrated why is it is both possible and necessary to consider the printed and textual formats of geographical texts of all kinds as constitutive of much larger expressive meanings.[4] The argument is also justified by reference to the pioneering work of the sociologist Norbert Elias on the nature and significance of courtly practices and etiquette. In *The Civilizing Process*, first published in German as two separate volumes in the late 1930s, Elias famously examined the "socio-genesis" of the concepts of "culture" and "civilization," the foundations of the modern humanities.[5] According to Elias, the apparently universal sentiments, sensibilities and tastes that now define these concepts were actively created in early-modern Europe by a "civilizing process" involving shifting regimes of shame and repugnance, honor and duty that ultimately internalized learned practices as "habitus" or "second nature." The "civilizing process" involved a disciplinary matrix of social customs, conventions, and practices that regulated, in the face of many challenges and transgressions, the entire range of potentially disruptive corporeal activities, including eating, drinking, excreting, sex, dress, speech, comportment, gesture, behavior, and the use of violence. The ultimate outcome was the self-aware, self-restraining, self-governing citizen, the consensual building block of modern society and the centralized nation-state.[6]

Elias was quite precise about points of departure. In his view, the arcane rituals and codes of etiquette developed in the royal courts of early-modern Europe provided the exemplary regulatory regimes from which these wider processes continued in other social settings long after the closed, rigidly hierarchical courtly societies had been swept away by political change.[7] Elias's discussion of the mimetic processes through which courtly practices were diffused and modified across space and time was the least convincing aspect of his argument, partly because he said remarkably little about the formal educational strategies developed within royal courts, even though these educational tactics exemplified his wider argument and provided the tangible foundations on which educational programs were subsequently developed in other social contexts based on similar internalized social norms and conventions.[8] This chapter re-examines Louis XV's neglected text in the light of these ideas as a case study that illuminates how the discipline of geography emerged in the early-modern period as an educational technique for describing and knowing the world and how geographical knowledge became an integral component in a wider "civilizing process" that ultimately determined how the modern humanities developed as a field of intellectual inquiry.[9]

So what do we know about this little book? The *Cours* was first and foremost an attempt to educate the young King in the rudiments of printing and book-binding, the text specially prepared by his tutors so that he might oversee its printing and thereby reinforce other lessons he was receiving in spelling, grammar, and composition.[10] This decision was certainly endorsed by the boy's uncle Philippe, the Duc d'Orléans, who ruled as Regent from 1715 to 1723, and his principal tutor, the Bishop of Fréjus, better known by his subsequent title, Cardinal Fleury. During the Regency, the French Court and its attendant ministries operated not from the opulent splendour of Versailles but from the Duc's official residence, the Palais Royal in central Paris, adjacent to the other state offices in the Louvre where young Louis resided, surrounded by his nurses, tutors, and servants, in private apartments in the Palais des Tuileries.

To facilitate the King's education in the gentle art of printing, a fully equipped atelier, complete with a small printing press and the necessary book-binding equipment, was installed in April 1718, directly below the royal apartments. The Imprimerie du Cabinet de Sa Majesté, as the workshop was grandly titled, produced 25 separate published documents, many less than one page in length, during its first year of operation and perhaps 50 before it was closed in 1727. As the *Cours* was the only bound volume printed in the workshop, we can safely assume it was regarded as the primary objective of the exercise. The statement on the book's title page indicating that the book was "composé & imprimé" by Louis is therefore admirably precise (Figure 10.1). Working under the avuncular guidance of an elderly Parisian printer, Jacques Collombat, Louis composed the plates, printed the gilt-edged quarto pages, and prepared the variously colored leather-backed bindings, each bearing the royal arms. The final pages of the two sections indicate they were printed separately in June and September 1718.[11]

The decision to allow young Louis to wrap himself in an artisan's leather apron and smudge his pale cheeks and delicate hands with printers' ink was informed by some fairly obvious cultural and political calculations. The printed word was the

# COURS
## DES PRINCIPAUX
# FLEUVES
# ET RIVIERES
## DE L'EUROPE,

*Composé & imprimé*

### Par LOUIS XV. Roy de France
### & de Navarre.

*En* 1718.

### A PARIS,
### Dans l'Imprimerie du Cabinet de S. M.
*DIRIGE'E*
Par J. Collombat Imprimeur ordinaire du Roy, Suite, Maison, Bâtimens, Arts & Manufactures de Sa Majesté.

M. DCC. XVIII.

**FIGURE 10.1**   Title page of the *Cours*. British Library G.974.

bedrock of an enlightened, rational civilization but no one, least of all the Duc d'Orléans, had any doubt that the security and prestige of the monarchy would be influenced, even determined, by its ability to control the production and distribution of printed material. The young King therefore needed to learn the skills and crafts of publishing, the pre-eminent cultural industry of the age that enabled the state to promulgate its laws, decrees, and regulations across the realm but which also had the potential to undermine the carefully constructed mystique of royal authority.

The French Court was a closely regulated world and it can reasonably be assumed that the contents of the *Cours* would have been carefully vetted by Fleury and the Duc d'Orléans. The fluvial theme certainly resonated with the general belief that a sound, practical knowledge of topography was necessary for the elaborate rituals associated with hunting, a practice central to the symbolic and highly politicized territoriality of the French Court at Versailles.[12] The importance of topographic knowledge had been repeatedly asserted in standard manuals defining the pros and cons of courtly behavior, beginning with the classic sixteenth-century statements by Thomas Elyot, Baldassare Castiglione, and Niccolò Machiavelli. This largely explains why most European courts established official positions for geographers and cartographers, the first French incumbent being Nicolas de Nicolaï (1517–1583).[13] Royal geographers were required to oversee the official court map collections that described the King's domain, both actual and potential, as it radiated outwards from the court itself to the regional, national, and international scales. In some instances, royal geographers acquired considerable political power, notably so in the case of Nicolas Sanson (1600–1667), France's most prolific seventeenth-century map-maker who was the first to provide formal lessons in geography and map interpretation in the French Court, his pupils including both Louis XIII and Louis XIV, his value as military and geostrategic adviser eventually recognized by promotion to Conseilleur d'État under the latter monarch.[14]

The contents of the *Cours* were also appropriate in view of the significance of water in the carefully fabricated elemental mythology of the Sun King. Louis XIV's transcendent, Apollonian image had been defined by his apparent mastery of rivers and water courses, beginning with his crossing of the Rhine at the head of the French armies invading the Low Countries in 1672, an event celebrated in the paintings, murals, dances, and spectacular festivities of Versailles.[15] The construction of Pierre-Paul Riquet's Canal du Midi, a work of art as well as a feat of engineering that realized Colbert's mercantilist dream of direct water-borne traffic from the Atlantic to the Mediterranean, affirmed the Sun King's ability to regulate the national river system, to bend nature to his will.[16] The same could also be said of the massive, ultimately less successful attempts to divert waters from the river basins surrounding the ever-expanding complex at Versailles itself to ensure that the countless fountains and water-falls that animated the epicenter of absolutist power could function as the King wished.[17]

Viewed in these terms, the *Cours* can be read as a continuation of a well-established, intensely political tradition of courtly learning in topography and geography; as a straightforward piece of absolutist propaganda designed to reassert the territorial ambitions of the boy's immediate predecessor and to reinforce a central theme of French absolutism – the conviction that the new French monarch ruled not only by divine right but also by actively creating his own, expanding Kingdom in the image of his own, growing body, an outward topographic manifestation of the royal *corpus mysticum*.[18] In this reading, the interconnected river systems described in the *Cours* might have represented the still developing sinews and blood vessels of the young King himself, the physical embodiment of a re-born absolutist monarch surging with new life in a new century.

But the simple, studiously factual text of the *Cours* resists such an obvious interpretation. The book says nothing about the preceding, epoch-defining attempts to regulate the fluvial environment under Louis XIV and focuses instead on rivers as natural route ways, the conduits of peaceful commerce. Even the Rhine, the symbol of Louis XIV's grandiose territorial ambitions, is defined unproblematically as a German river, while Russia, whose European credentials were by no means universally acknowledged, is accepted as an integral part of the European arena, even as far east as the Volga and the Ob, central Asian river basins almost never defined as European at the time, and rarely afterwards, even within the Russian Court.[19]

It might be argued that the *Cours* is in fact a subtle critique of the excesses and hubris of French absolutism under Louis XIV, a manifestation of a new, eighteenth-century anxiety about the relationship between royal authority and the natural world. Under the absolutism of Louis XIV, divinely ordained royal authority had sought to reveal itself by dominating nature with the assistance of militarized science and engineering. By the time Louis XV succeeded to the throne, after more than half a century of systematic scientific investigation, the natural world was more commonly invoked as an independent source of authority, superior to temporal and spiritual powers, and hence the basis for most critiques of the monarchy and the clergy.[20]

The idea that royal authority derived from scientific understanding of the natural world, rather than a domination of it, was strongly endorsed by Guillaume Delisle (1675–1726), the King's geography tutor and the man whose lessons formed the basis of the text printed in the *Cours*. Guillaume Delisle, who had been recruited as a royal tutor shortly before Louis XIV's death, was the eldest son of Claude (1644–1720), a lawyer and history tutor in the royal household under Louis XIV whose interests in geography and cartography had developed through his friendship with Nicolas Sanson.[21] In contrast to Sanson and earlier court geographers, whose role had been strategic and military rather than educational and scientific, Guillaume Delisle was a product of the new institutions of state-sponsored science, most notably the Académie Royale des Sciences, established in 1666 but significantly reorganized in 1699 into what would become the epitome of Enlightenment science, a state-funded agency directed by a salaried, professional cadre of *savants* representing the full range of scientific disciplines.[22] Like his equally distinguished younger half brothers, Joseph-Nicolas (1688–1768) and Louis (1692–1741?), Guillaume had been elected to the Académie as an astronomer and worked in that capacity with both Jean-Dominique (1625–1712) and Jacques Cassini (1677–1756), successive directors of the Royal Observatory and the architects of the great carto-graphic project that eventually produced the *Carte de France*, the first accurate, triangulated map of the French national space.[23]

Delisle was also proprietor of a renowned, although barely solvent map-making and publishing business on the Quai de l'Horloge, a short walk from the royal apart-ments in the Palais des Tuileries.[24] By moving from the bourgeois world of the *savant* and *commerçant* into the heart of Court society, Guillaume was following in his father's footsteps, to be sure, but whereas Claude saw geography as an essentially descriptive introduction to history, Guillaume insisted the subject was a properly scientific disci-pline, rooted in mathematical theory and based on the technical measurement and observation of the earth's physical features, a task that naturally culminated in accurate maps and surveys.[25]

Delisle's rigorously scientific approach was endorsed by the Duc d'Orleans, who had himself been tutored by Delisle's father. During the early, liberal phase of his Regency, the Duc, a self-proclaimed atheist who fancied himself as a man of science, provided substantial funds for a major survey of France's natural resources and gener-ously supported the first triangulation of the Paris meridian in 1718, the foundation on which the Cassini map would be constructed.[26] So impressed was the Duc by Delisle's teaching of the young King that he promoted the tutor to a newly created position as First Geographer to the King on a generous state pension of 1,200 *livres* per year. The *brevet* or licence for this position was issued to Delisle in August 1718, between the printing of the first and second sections of the *Cours*.[27] Delisle would henceforth stand apart from the dozen or so "géographes ordinaires du Roi," the non-stipendiary, "by appointment" designations awarded to commercial cartographers who supplied maps to the royal collections. His *brevet* as First Geographer suggests Delisle had been given a different kind of licence by the Duc d'Orléans to instruct the young King that the natural world should be understood scientifically and objectively rather than dominated politically and militarily. Although the *Cours* was never intended to be read widely, of course, one might speculate that its neutral tone and avoidance of recent, still contentious events involving Europe's rivers was designed to communicate, at least to those who encountered the book within the Court, that France's intentions under the new King were peaceful and fully in accordance with the terms of the Treaty of Utrecht, signed in the spring of 1713. If the *Cours* also demonstrated that young Louis, whose succession was itself the subject of on-going controversy, was being responsibly educated as a modern, enlightened monarch, then so much the better.

The story does not end there, however, for the *Cours* also demonstrates how geographical knowledge was implicated in other ways in the "civilizing process," particularly through its visual content and the subsequent status it acquired within the Court as a cherished object, a childhood memento, and a royal gift, owned and possibly read by a highly selective readership who would have known in some detail precisely how the book came into existence. The main visual component of the *Cours* is an image of Louis XV created by Jean Audran, a Court engraver responsible for some of the most dramatic and theatrical images of the Sun King (Figure 10.2).

The young Louis is depicted as Court officials presumably wished him to appear at the time, his chubby face (with the right eye disconcertingly larger than the left) and boyish upper body poised in a mirror frame above the Bourbon arms. Audran's engraving is one of the earliest portraits of Louis standing alone unaccompanied by other family members or courtiers and strongly suggests that the image was designed to reinforce the book's significance as a *rite de passage*, a demonstration of the King's precocious intelligence and growing maturity. The artfully withdrawn curtain suggests a regal entrance, reinforcing the impression that the young King would soon be ready to assume the full responsibilities of his position.[28]

Tom Conley's psycho-analytical reading of "cartographic writing" in early-modern France provides a useful framework for exploring the significance of this image and the seemingly innocuous text within which it is located.[29] According to Conley, "cartographic writing," which he defines as maps and associated geographical texts, was critically important in fostering the modern sense of the "self."

**FIGURE 10.2**   Jean Audran's image of Louis XV in the *Cours*. British Library G.974.

An individual encountering a map or a geographical text, argues Conley, establishes a new *relationship with unknown* space based on *perspectives* at once personal and relative and new forms of *pictographic memory* that conflate language and image and allow the continual re-imagining of the self in locations yet to be encountered, while also facilitating acts of symbolic possession through *signatures* or other personalized markers of ownership on maps and texts.

Viewed in these terms, Audran's image acquires new layers of meaning. In his guise as a humble apprentice printer, young Louis had prepared multiple images of his own, more regal self and had carefully positioned these representations within the gilt-edged pages of a simple geographical text about the rivers of France and Europe, the natural waterways that shaped the topography of a Kingdom about which he knew little and a continent about which he knew even less. The book, and the image of the young King it contained, was therefore designed to communicate a subtle message – the pages containing the young King's image were printed by the King himself, implying that a critical moment of personal self-identification and self-promotion had taken place, and not only for the young Louis because every person who subsequently encountered a copy of his book would have known that both the image and the accompanying text had been brought into existence by the King himself. The signature image of the young King not only underlined his complete ownership of the book, as text and artifact; it also demonstrated his ability to fashion his own image, reinforcing the associated conceit that he was the active geographical agent connecting the courtly reader of the text with the unknown spaces described.

These more complex cultural meanings can also be explored by reference to the "afterlife" of some of the surviving copies of the *Cours*. Most copies were presented by Louis as gifts to favored courtiers while he was still a young man awaiting his majority, or later as mementos of a cherished childhood memory. This is confirmed by the only copy of the *Cours* in the British Library, inherited by that institution as part of the library of Thomas Grenville (1755–1846) who almost certainly acquired this copy from someone in the French Court during his short tenure as British minister in Paris in 1782, long after Louis XV's death in 1774. There is a note attached to the inside front cover of this copy, possibly in Grenville's own hand, which states that "the Royal author of this trifling work disposed of the few copies which he had printed as presents."[30]

The continuing significance of the book to the adult King is revealed by a closer examination of one of the four copies of the *Cours* in the Bibliothèque Nationale. This example was inherited from Madame de Pompadour (1721–64) and was listed in the catalogue of her impressive library published shortly after her death. It was presumably a gift from the King shortly after Madame de Pompadour became his mistress in 1745. The bindings depict the royal arms on the front and Madame de Pompadour's on the back, implying that copies of the text were re-bound later in the eighteenth century. The text is itself embellished with a signed dedication in the King's hand.[31] Madame de Pompadour's copy of the *Cours* may have reflected, and perhaps even initiated, her well-known interest in geography and cartography, an enthusiasm that the King would certainly have known about and which later prompted her decision to provide financial support from her own funds for the *Carte de France* and to subscribe to Robert de Vaugondy's *Atlas Universel*.[32] So delighted was Madame de Pompadour by the globe presented by Didier Robert de Vaugondy to the King in 1750 that she commissioned her own pair of globes from the same man the following year, much to the ribald amusement of the pamphleteers. During the Seven Years' War, she insisted on receiving regular reports about military maneuvers and took it upon herself to embellish maps of the Americas in the King's library, some of them produced by the aforementioned Delisle, using beauty spots to indicate troop deployments.[33]

The Pompadour copy of the *Cours*, the only one that has ever been displayed in a public exhibition, also highlights some of the wider themes discussed at the beginning of this chapter.[34] As a small, intensely personal gift of symbolic rather than monetary value, one that allowed the recipient privileged access to the private, childhood memories of the giver, the Pompadour copy of the *Cours* recycles some of the oldest traditions in the arts of courtly love and demonstrates how favor, affection, intimacy, and charm, the chief characteristics of modern "civilité" and idealized forms of sociability, were established as the ground rules governing personal and gender relationships within the eighteenth-century French court, and how childhood was cultivated as an era of tender and exquisite memory defined by, and dependent upon, the existence of carefully preserved artifacts, mementos, and souvenirs.[35] As Elias and others have argued, similar rules would ultimately determine courtship practices, the giving and receiving of gifts, and the remembrance and celebration of childhood beyond this carefully regulated world as part of a wider "civilizing process."

The *Cours des Principaux Fleuves et Rivières de l'Europe*, prepared in the early years of his reign by the eight-year-old Louis XV on the basis of the geographical education he received from his tutor Guillaume Delisle, provides a self-consciously child-like and deceptively naïve account of a key component of Europe's topography and has hitherto been regarded as no more than an historical curiosity of interest only to bookish antiquarians or those with a special interest in the relics of Europe's doomed royal households. This short chapter has developed the scarcely unheralded claim that long-forgotten geographical texts, including those constructed by and for children, have a richer and more complex story to tell than their often simple structure and prose might suggest, not only about the history of geography as an educational tactic within the humanities, but also about the wider cultural role that geographical knowledge has played in the creation of modern manners, customs and social conventions.

## Notes

1  Voltaire mentions the *Cours* at least twice, though with incorrect titles and dates. His print run estimate appears in *Œuvres complètes de Voltaire. Vol. XIX: Précis du siècle de Louis XV*, Paris: A.A. Renouard, 1819, p. 443.

2  A.A. Renouard, *Catalogue de la bibliothèque d'un amateur, avec notes bibliographiques, critiques et littéraires. Vol. IV: Histoire*, Paris: A.A. Renouard, 1819, pp. 12–13.

3  See, for example, N. Lenglet du Fresnoy, *Méthode pour étudier la géographie*, 4 vols, Paris: C.-E. Hochereau, 1716.

4 See, for example, R.J. Mayhew, "Materialist Hermeneutics, Textuality and the History of Geography: Print Spaces in British Geography, c. 1500–1900," *Journal of Historical Geography* 33, 2007, 466–88; C.W.J. Withers, "Eighteenth-Century Geography: Texts, Practices, Sites," *Progress in Human Geography* 30, 2006, 711–29; and more generally M. Ogborn, *Indian Ink: Script and Print in the Making of the East India Company*, Chicago: University of Chicago Press, 2007 and M. Ogborn and C.W.J. Withers (eds), *Geographies of the Book*, London: Ashgate, 2010.

5 N. Elias, *The Civilizing Process: The History of Manners and State Formation and Civilization*, Oxford: Blackwell, 1994. For a sympathetic criticism, see R. Chartier, "Social Figuration and Habitus: Reading Elias," in *Cultural History: Between Practices and Representations*, Ithaca, NY: Cornell University Press, 1988, pp. 71–94.

6 S. Mennell, *Norbert Elias: Civilization and the Human Self-Image*, Oxford: Blackwell, 1989.

7 A classic analysis of the French court is provided by O. Ranum, "Courtesy, Absolutism, and the Rise of the French State, 1630–1660," *Journal of Modern History* 52, 1980, 426–51.

8 J. Duindam, *Myths of Power: Norbert Elias and the Early Modern European Court*, Amsterdam: Amsterdam University Press, 1994.

9 For related studies of geography and court society, see L.B. Cormack, *Charting an Empire: Geography at the English Universities, 1580–1620*, Chicago: University of Chicago Press, 1997, 207–23; F. Fiorani, *The Marvel of Maps: Art, Cartography and Politics in Renaissance Italy*, New Haven: Yale University Press, 2005; and C.W.J. Withers, *Geography, Science and National Identity: Scotland Since 1520*, Cambridge: Cambridge University Press, 2001, pp. 30–111.

10 H. Omont, "L'Imprimerie du Cabinet du Roi au Château des Tuileries sous Louis XV (1718–1730)," *Bulletin de la Société de l'Histoire de Paris et de l'Ile-de-France* 18, 1891, 35–45.

11 Collombat had previously printed a brief guide for his young pupil in 1718, using the workshop press, entitled *Principaux termes de l'art topographique ou de l'imprimerie, avec le nom et usage des instrumens, pièces et utiles qui sont renfermées dans l'Imprimerie du Cabinet du Roy*. See Bibliothèque Nationale de France, MSS. Fr. 22106, fo. 80.

12 C. Mukerji, *Territorial Ambitions and the Gardens of Versailles*, Cambridge: Cambridge University Press, 1997; R.W. Berger and T.F. Hedin, *Diplomatic Tours in the Gardens of Versailles under Louis XIV*, Philadelphia: University of Pennsylvania Press, 2008.

13 R.W. Karrow, *Mapmakers of the Sixteenth Century and Their Maps*, Chicago: Newberry Library, 1993, pp. 435–43.

14 See the introduction by M. Pastoureau, in Nicolas Sanson d'Abbeville, *Atlas du Monde 1665*, Paris: Sand & Conti, 1988, pp. 11–46.

15 P. Burke, *The Fabrication of Louis XIV*, New Haven: Yale University Press, 1994; S. du Crest, *Des fêtes à Versailles: les divertissements de Louis XIV*, Paris: Aux Amateurs de Livres, 1990.

16 C. Mukerji, *Impossible Engineering: Technology and Territoriality on the Canal du Midi*, Princeton, NJ: Princeton University Press, 2009.

17 The definitive work is still L.A. Barbet, *Les grandes eaux de Versailles: installations mécaniques et étangs artificiels, descriptions des fontaines et de leurs origines*, Paris: H. Dunod et E. Pinet, 1907; but see also I. Thompson, *The Sun King's Garden: Louis XIV, André Le Notre and the Creation of the Gardens of Versailles*, London: Bloomsbury, 2006.

18 The classic work being E.H. Kantorowicz, *The King's Two Bodies: A Study in Mediaeval Political Theology*, Princeton, NJ: Princeton University Press, 1957.

19 L. Wolff, *Inventing Eastern Europe: The Map of Civilization on the Mind of the Enlightenment*, Stanford: Stanford University Press, 1994.

20 L. Daston, "Attention and the Values of Nature in the Enlightenment," in L. Daston and F. Vidal (eds), *The Moral Authority of Nature*, Chicago: University of Chicago Press, 2004, pp. 100–26.

21 N. Martin Dawson, *L'Atelier Delisle: l'Amérique du Nord sur la table à dessin*, Sillery, Québec: Éditions du Septentrion, 2000.

22  R. Hahn, *The Anatomy of a Scientific Institution: The Paris Academy of Sciences, 1666–1803*, Berkeley and Los Angeles: University of California Press, 1971; D.J. Sturdy, *Science and Social Status: The Members of the Académie des Sciences, 1666–1750*, Woodbridge, UK: Boydell Press, 1995.

23  M. Pelletier, *Les cartes des Cassini: la science au service de l'État et des régions*, Paris: Éditions du Comité des Travaux Historiques et Scientifiques, 2002.

24  M. Sponberg Pedley, *The Commerce of Cartography: Making and Marketing Maps in Eighteenth-Century France and England*, Chicago: University of Chicago Press, 2005.

25  See N. Broc, *La géographie des philosophes: géographes et voyageurs français au XVIIIe. siècle*, Paris: Éditions Ophrys, 1975.

26  C. Demeulenaere-Douyère and D.J. Sturdy (eds), *L'Enquête du Régent, 1716–1718: sciences, techniques et politique dans la France pré-industrielle*, Turnhout, Belgium: Brepols, 2008.

27  Archives Nationales O1/62 fo. 183 v. 184.

28  On the politics of royal portraiture, see S. Schama, "The Domestication of Majesty: Royal Family Portraiture," *Journal of Interdisciplinary History* 17, 1986, 155–83.

29  T. Conley, *The Self-Made Map: Cartographic Writing in Early Modern France*, Minneapolis: University of Minnesota Press, 1996; and, more generally, S. Greenblatt, *Renaissance Self-Fashioning from More to Shakespeare*, Chicago: University of Chicago Press, 1980.

30  British Library G.974; J.T. Payne and H. Foss, *Bibliotheca Grenvilliana; Part the Second, Completing the Catalogue of the Library Bequeathed to the British Museum by the late Right Hon. Thomas Grenville*, London: British Museum, 1848, pp. 294–5.

31  Bibliothèque Nationale de France Res. G.2972; Anon., *Catalogue des livres de la bibliothèque de feue Madame la Marquise de Pompadour, Dame du Palais de la Reine*, Paris: J.-T. Hérissant, 1765, p. 46.

32  M. Sponberg Pedley, *The Commerce of Cartography*, p. 6.

33  C. Jones, *Madame de Pompadour: Images of a Mistress*, London/New Haven, National Gallery/Yale University Press, 2002, pp. 132–4.

34  The Pompadour copy was loaned by the Bibliothèque Nationale de France to the US Library of Congress for a 1995 exhibition entitled *Creating French Culture: Treasures of the Bibliothèque Nationale de France*. HTTP: <www.loc.gov/exhibitions/bnf>.

35  On this theme, see R. Chartier, "From Texts to Manners – A Concept and its Books: *Civilité* between Aristocratic Distinction and Popular Appropriation," in *The Cultural Uses of Print in Early Modern France*, Princeton, NJ: Princeton University Press, 1988, pp. 71–109.

# 11

# DARWINIAN LANDSCAPES

*David Livingstone*

Syms Covington was a teenager when he signed on as *HMS Beagle's* fiddler and odd-job man in 1832. Eight years later he emigrated to New South Wales at the age of around 24 and became postmaster in the settlement of Pambula. For just over six years he also served as Charles Darwin's manservant. Alongside the fledging naturalist, Covington performed the role of collector, hunter and taxidermist. The "unacknowledged shadow behind every triumph" – according to Janet Browne[1] – and the ever-present accessory to his geographical and scientific travels, it is hardly surprising that this virtually invisible Darwinian adjunct would sooner or later catch the novelist's imagination. At the tail end of Roger McDonald's novel, *Mr Darwin's Shooter*, Syms Covington is given the last word. It's staged in 1860, just shortly before his death, and just after he'd read Darwin's new book, *The Origin of Species*. His thoughts wander back to his *Beagle* adventures and in his mind's eye he catches a glimpse of his former master on a chilly, sunlit December morning: "He saw Darwin on his knees, and there was no difference between prayer and pulling a worm from the grass."[2]

Darwin was a lover of nature. His account of the *Beagle* expedition is replete with glorious landscape descriptions showing the emotional depth of his love of the natural world. And these descriptions were sculpted in dialogue with Alexander von Humboldt's writings, which were a constant traveling companion shaping his encounter with nature at every turn. Fully aware of the ways in which landscape vision was framed by what he called "preconceived ideas," Darwin confessed that "mine were taken from the vivid descriptions in the *Personal Narrative* of Humboldt, which far exceed in merit anything else which I have read."[3] Besides this, Darwin also called upon Humboldt's supporting testimony on subjects as diverse as on polished syenitic rocks in the Orinoco, on atmospheric conditions in the tropics, on crocodile hibernation, on possible connections between earthquakes and weather conditions, on miasmas in the torrid zone.

A couple of passages from the two ends of his *Journal of Researches* may be taken as emblematic of Darwin's emotional involvement with nature. Take his entry for 29 February 1832 at Salvador, Brazil, just the day after he set foot on the continent for the first time:

> Delight ... is a weak term to express the feelings of a naturalist who, for the first time, has wandered by himself in a Brazilian forest. The elegance of the grasses, the novelty of the parasitical plants, the beauty of the flowers, the glossy green of the foliage, but above all the general luxuriance of the vegetation, filled me with admiration. A most paradoxical mixture of sound and silence pervades the shady parts of the wood. The noise from the insects is so loud, that it may be heard even in a vessel anchored several hundred yards from the shore; yet within the recesses of the forest a universal silence appears to reign. To a person fond of natural history, such a day as this brings with it a deeper pleasure than he can ever hope to experience again.[4]

By April he was so entranced by the forests around Rio de Janeiro that at a height of five or six hundred feet when "the landscape attains its most brilliant tint" he confessed that to the naturalist "every form, every shade, so completely surpasses in magnificence all that the European has ever beheld in his own country, that he knows not how to express his feelings."[5] At the other end of both the voyage and the narrative, during the early days of August 1836, he recorded his thoughts on tropical scenery:

> When quietly walking along the shady pathways, and admiring each successive view, I wished to find language to express my ideas. Epithet after epithet was found too weak to convey to those who have not visited the intertropical regions the sensation of delight which the mind experiences. I have said that the plants in a hothouse fail to communicate a just idea of the vegetation, yet I must recur to it. The land is one great wild, untidy, luxuriant hothouse, made by Nature for herself ... How great would be the desire in every admirer of nature to behold, if such were possible, the scenery of another planet! ... In my last walk I stopped again and again to gaze on these beauties, and endeavoured to fix in my mind for ever an impression which at the time I knew sooner or later must fail. The form of the orange-tree, the cocoa-nut, the palm, the mango, the tree-fern, the banana, will remain clear and separate; but the thousand beauties which unite these into one perfect scene must fade away; yet they will leave, like a tale heard in childhood, a picture full of indistinct, but most beautiful figures.[6]

Such evocations, however, were not simply the outpouring of emotional responses to a sequence of ecstatic moments. Darwin was fully conscious of what might be called the analytics of landscape phenomenology.[7] As Jonathan Smith has reminded us, landscape – for Darwin – "could become monotonous without scientific understanding of its different features."[8] Thus he did pause to ponder the relations between the individual components of a landscape and its holistic impression. As he put it at the end of his journal, landscape pleasure critically depended on "an acquaintance with the individual parts of each view." He went on:

> I am strongly induced to believe that, as in music, the person who under-stands every note will … more thoroughly enjoy the whole, so he who exam-ines each part of a fine view, may also thoroughly comprehend the full and combined effect.[9]

Indeed there are suggestions that Darwin was already working toward a scientific explanation of landscape appreciation. Contemplating the capacity of natural scenery to evoke strong emotion, he mused that the pleasure of the great outdoors was "a relic of an instinctive passion;" the joys of "living in the open air, with the sky for a roof and the ground for a table" was nothing other than "the savage returning to his wild and native habits."[10]

These speculations are doubly significant. First, they connect with the ongoing problematic of the relationship between science and aesthetics, about what Driver and Martins call "the balance between holistic and analytical views of nature, and about the prospect of reconciling sedentary scholarship with observation in the field."[11] Second, they reinforce the importance of Humboldt for Darwin's outlook. As he put it in April 1832 in a letter to Henslow,

> Here I first saw a tropical forest in all its sublime grandeur – nothing but the reality can give any idea how wonderful, how magnificent the scene is.… I formerly admired Humboldt, I now almost adore him; he alone gives any notion of the feelings that are raised in the mind of first entering the Tropics.[12]

Even if Humboldt's prose fell short of the immediate impact of tropical experience, Darwin could do no other than look with Humboldtian eyes. Back in February of that same year he had scrawled in this diary:

> Humboldt's glorious descriptions are and will forever be unparalleled: but even he with his dark blue skies and the rare union of poetry with science which he so strongly displays when writing on tropical scenery, with all this falls short of the truth ….The mind is a chaos of delight … I am at present fit only to read Humboldt; he like another sun illumines everything I behold.[13]

And yet, for all that, there were significant differences between the two travelers. For one thing, while Humboldt's accounts foreground the embodied experience of tropical travel to the extent that his body became a surrogate scientific instrument for registering environmental conditions, there is a signal absence of this kind of embodied encounter in Darwin's narrative. This, I might add, is an altogether ironic circumstance in view of Darwin's later obsession with his bodily frailties and with projecting an image (a *strategic* image if I read Janet Browne aright) of himself as an invalid.

## Darwinian landscapes

As we might already have inferred, Darwin had a profound interest in geography. As his *Journal of Researches* makes clear, he saw himself – like Alexander von Humboldt – as a scientific traveler. Of course the scope of Darwin's geographical interests was not circumscribed by the riches of Humboldtian aesthetic evocations. His theory of coral atolls, for example, was originally sketched out on board the *Beagle* and was presented to the Geological Society of London in 1837. Matters of geographical distribution were critical to his thinking on evolution by natural selection too, and he devoted two lengthy chapters to the subject in *On the Origin of Species*. Figuring out why neighboring regions with remarkably similar climatic conditions should have different biogeographies, the role played by barriers to the migration of organisms, and the close affinity between different species in contiguous geographical regions all intrigued him. Inheritance, modification, natural selection and to some degree direct environmental influence provided him with the answers he sought. Explanations were to be found here rather than in the idea of "single centres of supposed creation" as naturalists like Agassiz alleged. As he put it in a single summative phrase, "all the grand leading facts of geographical distribution are explicable on the theory of migration, together with subsequent modification and the multiplication of new forms."[14] To a considerable degree, I think, it would be fair to say that Darwin's was a fundamentally bio-geographical theory.

But I want to go a little farther and suggest that the landscapes that came within the range of Darwin's vision were constitutively important in the production of his theory of evolution by natural selection. Of course I don't mean to suggest that geography *determined* Darwinian theory in any straightforward causal fashion; but I do think that the kinds of landscape he experienced shaped the way his understanding of evolution was sculpted. Three in particular seem to me to be specially important.

The first is *tropical* landscape. Darwin was entranced by the tropical world and his depictions of the landscapes through which he traveled are saturated with images of tropical abundance. The number of "minute and obscurely-coloured beetles" was "exceedingly great;" the variety of species of jumping spider appeared to be "almost infinite;" partridges were visible in "great numbers;" the birds were "extremely abundant."[15] Images of tightly wedged-in ecological niches in conditions of hyper-fecundity readily come to mind. Writing to a friend in August 1832 on the luxuriance of the Brazilian vegetation, he mused:

> ...it was realising the visions in the Arabian nights – The brilliancy of the Scenery throws one into a delirium of delight and a Beetle hunter is not likely soon to awaken from it, when whichever way he turns fresh treasures meet his eye.[16]

The organic abundance was so overwhelming that he felt that describing it to an untraveled European was like trying to convey the experience of color to a blind man.[17] For Wallace too, the claimed co-discoverer of natural selection, it was his experience in Borneo that was crucial. In his 1891 *Natural Selection and Tropical Nature*, Wallace noted that "[a]nimal life is, on the whole, far more abundant and more varied within the tropics than in any other part of the globe ....While successive glacial periods have devastated the temperate zones, and destroyed most of the larger and more specialized forms which during more favorable epochs had been developed, the equatorial lands must always have remained thronged with life." It was in the equatorial latitudes that "a comparatively continuous and unchecked development of organic forms" took place, and thus here that "evolution has had a fair chance." As he concluded:

> The equatorial regions are then, as regards their past and present life history, a more ancient world than that represented by the temperate zones, a world in which the laws which have governed the progressive development of life have operated with comparatively little check for countless ages, and have resulted in those infinitely varied and beautiful forms – those wonderful eccentricities of structure, of function, and of instinct – that rich variety of colour, and that nicely balanced harmony of relations – which delight and astonish us in the animal productions of all tropical countries.[18]

In the light of these geographical exposures we might have reason to suspect that Darwinism, in some critical sense, might be thought of as a *tropical* theory, and I shall return to this conjecture presently.

Second, the *Beagle* voyage, which Darwin himself confessed was "by far the most important event" of his life, afforded him unparalleled opportunity to experience *island environments*.[19] Apart from anything else it offered him the chance of comparing one island ecosystem with another, and his notes constantly show him seeking to dig out the underlying *vera causa* beneath surface variations. Islands were thus critically significant in the formulation of several components of his thinking. It was his first-hand observations of coral islands, for example, that led him to the theory that fringing reefs, barrier reefs and what he called "lagoon islands" – true atolls – were actually a series of one form developing into another as a consequence of subsidence. Indeed there is some suggestion that it was in this hastily composed manuscript on "Coral Islands," written on board ship sometime during December 1835, that we catch sight of his "first substantial embrace of the notion of gradualism."[20] But islands were vital for Darwinism in other ways too. What impressed him most, I think, was their relative paucity of species compared with equal continental land areas. Whether it was Madeira, St Helena, Bermuda, the Falklands or the Galapagos, the same story of the absence of biodiversity reasserted itself. It became clear to him too that, for example in the Galapagos, peculiar species were limited to their own islands. This led him to ponder more and more the relationship between organisms and their environments, and the role of migration, barriers and isolation in the processes of speciation. In the Keeling islands, for example, he stressed the way in which the large claws or pincers of some of the hermit crabs were "most beautifully adapted" to their environments.[21] And in one of his notebooks – Notebook B – he speculated that "animals, on separate islands, ought to become different if kept long enough apart, with slightly different circumstances" though he later wavered on the point and merely noted in the *Origin* that geographical isolation was "an important element in the modification of species through natural selection."[22] Nevertheless, his experience of island biogeography was critical. As he observed, after noting the "scarcity of kinds" that characterized oceanic islands, the "most striking and important fact for us in regard to the inhabitants of islands is their affinity to those of the nearest mainland without being actually the same species."[23] If continental tropical landscapes furnished him with conditions for witnessing hyper-fecundity and competition, islands showed that, by dispersal and species modification, evolution could take place. Patrick Armstrong sums up the evolutionary significance that islands had for Darwin rather nicely:

> If all life is derived from a very limited number of original forms, the life-forms found on remote islands must have made the journey thither, and will be distinctive ... In the theory of independent creation there is no reason why there should be significant differences in the forms the Creator placed on islands and those on continents.[24]

A third influential geography in the evolution of Darwinism, of course, was the *landscape of Down House* in Kent. It was here, after his return from the *Beagle* voyage and a brief sojourn in London, that he lived for forty years (1842–1882) and wrote *On the Origin of Species, The Descent of Man* and his other major works. The micro-geography of Down House and the surrounding topography of the Down landscape were critical to the development of his thinking.[25] The house, for example, became the central node in a world-wide network of correspondence and intellectual exchange. Over 13,500 letters remain from his years there. This, of course, reflects the fact that at Down Darwin became something of a recluse. His five or six years in London after the *Beagle* expedition had placed him at the center of a metropolitan set of scientific workers in which he played a prominent role, and in dialogue with whom he developed some of his early ideas. He himself was a prominent fixture in what Martin Rudwick calls the "social topography" of London science at the time and it was here that he received an "indispensable *training* as a theorist in science" – geology in particular.[26] At Down it was different. Here he occupied a decidedly more private space, interacting with the world largely (but not exclusively) through the written word. In that sense it was rather like a return to the social world of the *Beagle* where he had developed his science in relative isolation. And it was here that his most private thoughts on the question of the origin of species, on human evolution and on the development of mind moved into the public realm of publication.

The grounds of the house were also vital. They were used as experimental gardens where he tested a host of hunches and hypotheses. Experiments on cultivated and wild plants, and exotic varieties from Kew were carried out here. Great House Meadow, the fifteen-acre plot of grassland on which Darwin's house stood, was used as the site of a long-term experiment on earthworms and their working of the soil. In a letter to Hooker in June 1857, he commented on related meadow experiments:

> My observations, though on so infinitely a small scale, on the struggle for existence begin to make me a little clearer how the fight goes on: out of sixteen kinds of seen sown in my meadow, fifteen have germinated, but now they are perishing at such a rate that I doubt whether more than one will flower. Here we have choking, which has taken place likewise on great scale with plants not seedlings in a bit of my lawn allowed to grow up.[27]

The meadow too provided him with opportunities to witness what, in the *Origin*, he called "the web of complex relations" using observations on the relations between red clover, field mice, cats and bumble bees to figure out patterns of plant fertilization.[28] The interdependence of the variables was something that he gleaned from experiments in the meadow. At Great Pucklands Meadow, a thirteen-acre pasture close by, he first tested his ideas about the role of divergence in evolutionary history. There he surveyed all the plant species so as to determine the area's biodiversity. The results confirmed its remarkable richness and bore out his suspicion that the more diversified plants and animals were, the more niches they were thereby enabled to occupy. Comparing a plot sown with a single species to one with several distinct genera showed that the latter supported a much greater number of plants. This was critical to evolution for without variation, as he conceded, "natural selection can effect nothing."[29] The rural landscape in which Down House and its environs was located was thus exceptionally well suited to Darwin's needs. It "contained a diverse flora and fauna in a range of semi-natural and man-managed habitats;" its various habitat types were easily accessible to a field naturalist like Darwin who did not enjoy good health; and its diverse patterns of management – from stability to change – allowed him "in situations of both continuity and flux ... to explore the dynamics of ecological processes."[30]

## Different landscapes, different Darwinisms?

The degree to which the genesis and development of Darwin's theories were conditioned by the particular landscapes he experienced is an intriguing question. And there has been, as I have suggested, some speculation that his theory of natural selection might – in some fundamental sense – be the product of the tropical world. Some sustenance for such a conjecture may be gleaned from the experience of the Russian naturalists who congregated at the St. Petersburg Society of Natural History, and who conducted their researches in the Siberian wilderness and on the Russian steppes. They saw things differently and constructed a modified Darwinism in keeping – at least in part – with the environments with which they were familiar. For them, working in conditions where nature displayed no plenitude, no super-abundance, no swarming life-forms, the vocabulary of overpopulation and struggle between species just did not seem right. Accordingly a tradition of evolutionary zoology emphasizing cooperation developed there, in which the Malthusian components of natural selection were systematically expunged. In his 1879 essay "On the Law of Mutual Aid," for example, Karl Kessler condemned "the cruel, so-called law of the struggle for existence" and insisted that it was "more important than the law of the struggle for existence."[31] During his own fieldwork in the Crimea and Aralo-Caspian region he had had opportunity to witness the survival value of reciprocated care and cooperation among bees, beetles, spiders, reptiles and a host of other creatures. Kessler's reading of Darwin did not remain an isolated textual event. It inaugurated a reading history that steered later Russian engagements with the text. In fact a number of Kessler's associates – such as Alexander Brandt, Mikhail Filippov, Vladimir Bekhterev and Modest Bogdanov – constituted what Daniel Todes describes as the "Russian Mutual Aid Tradition."

Building on these contributions, the Russian émigré and anarchist geographer Peter Kropotkin published his famous work on *Mutual Aid: A Factor in Evolution* in 1902. Kropotkin too had carried out research in Siberia and received the gold medal of the Imperial Russian Geographical Society for his work there. As he summarized his position in a letter to Marie Goldsmith: "*We see a great deal of mutual aid*, where Darwin and Wallace see *only struggle*."[32] This was Darwinism with its Malthusian teeth extracted. Just as the teeming tropics never left Darwin's mind, the wastelands of Russia's high latitudes remained with Kropotkin. What grounded Kropotkin's enthusiasm for this Malthus-denying version of Darwinism was his own first-hand experience of the Siberian wilderness where, in the Vitim region, he engaged in zoo-geographical inquiries with Polyakoff. "We saw plenty of adaptations for struggling very often in common, against the diverse circumstances of climate," he recalled,

> ...and Polyakoff wrote many a good page upon the mutual dependency of carnivores, ruminants, and rodents in their geographical distribution; we witnessed numbers of facts of mutual support, especially during the migrations of birds and ruminants .... The same impression appears in the work of most Russian zoologists, and it probably explains why Kessler's ideas were so welcomed by the Russian Darwinists, while like ideas are not in vogue amidst the followers of Darwin in Western Europe.[33]

Daniel Todes neatly sums up the contrast:

> Darwin and Wallace shared two experiences: a sympathetic reading of Malthus's *Essay on the Principle of Population*, and an important field experience in tropical rain forests. Most Russian evolutionists shared two experiences that were roughly the opposite to these: an aversion to Malthus and life on a vast continental plain.[34]

If a different landscape shaped a different Darwinism in the Russian wastelands, what about the situation in other cruel Arctic environments? The Canadian north provides a useful comparator. While there was a signal absence of response to Darwin among Canadian practitioners of geology, during the early 1860s several botanists broached the subject.[35] The dominant motif in their endeavors was the fundamental significance of struggle against the vicissitudes of a harsh landscape. But, as in Russia, it was not struggle between species; instead it was struggle against an unyielding Precambrian shield. Success required inherited modification and the idea of environmentally induced adaptation and acclimatization was resorted to so as to account for vegetational patterns. Such circumstances encouraged the agricultural reformer William McDougall to suggest in 1854 that the Lamarckian principle of the inheritance of acquired characteristics provided a viable explanation. Yet, in contrast to Russia, Darwinian language *was* embraced. George Lawson, the Scottish-trained botanist at Queen's University, Kingston, for example, found J.D. Hooker's Darwinian account of Arctic biogeography compelling and George Dawson, son of the eminent anti-Darwinian geologist J.W. Dawson, happily resorted to Darwinian vocabulary in his anthropological studies in 1878 of the Haida people of the Queen Charlotte Islands.[36]

Still, some finessing of Darwinism was nonetheless required to make it do northern work, as a careful reading of Hooker's 1862 account of Arctic plant geography reveals. To explain Arctic floral distributions, Hooker urged that climate change had pushed plants farther south where they multiplied and branched off into new varieties through selective struggle with native species. Inter-specific struggle, plainly, was not a high latitude phenomenon. Only when plants moved south did they find themselves in competition with other species. Here they were exposed to the forces of natural selection and the modifications induced there was the explanation for the variety of plant types observable in the Arctic when warmer conditions permitted a recolonizing of the Arctic zones. The paucity of Greenland vegetation provided Hooker with a kind of hypothesis control on account of its anomalous situation. Here southward expansion was impossible and therefore many species were exterminated. Survivors were

> ...confined to the southern portion of the peninsula, and not being there brought into competition with other types, there would be no struggle for life amongst their progeny, and consequently no selection of better-adapted varieties. On the return of heat, these survivors would simply travel northward, unaccompanied by the plants of any other country.[37]

The general principle was that natural selection in Arctic plants was dependent on out-migration, invasion of new territory, competition with indigenous species, and recolonization.

Struggle between species, it seems, was not a feature of either the Asian or American Arctic. But that did not mean they reacted in the same way to Darwin's proposals. Russian naturalists could not discern inter-specific struggle in their zoological observations and thereby justified their outright rejection of it; Canadian botanists, following Hooker, called upon it as an explanatory mechanism that came into play during periods when climate change induced southern extension.

## The landscapes of Darwinism

Different renderings of Darwin's theory, then, cannot be reduced simply to landscape particularity, though that undoubtedly had a role to play both in its construction and circulation. In the Russian case, the response to Darwinian-style struggle was also moulded by earlier textual encounters with Malthus's theory of population. On both the political left and right in Russia, his atomistic conception of society had already been castigated, mostly since the 1840s, as a cold, soulless and mechanistic product of English political economy. Malthus may have rationalized poverty and inequity in England; but his commentators were certain that his theory would not apply in a harmonious Russia. It ran foul of Russian visions of a cohesive society which would jeopardize the cherished peasant commune. In Canada the relative absence of response to Darwinism in the early days sprang in part from an ingrained Baconianism that prioritized collecting and classification over theoretical speculation.[38] The absence of public controversy as indifference turned into advocacy had much to do with the ways in which religious leaders found it possible to incorporate evolutionary thinking into their progressivist conceptions of historical change. The inspiration they derived from Scottish intellectuals at the time who found it possible to harmonize evolution and theology, moreover, meant that the split that occurred elsewhere between modernists and anti-modernists was conspicuous only by its absence.[39] In addition its later adoption was also bound up with romantic nationalist notions of "the north as a source of liberty, physical strength, and hardiness of spirit."[40] Darwin delivered a scientific framework which could empower the nation's resolve to overcome its harsh environment and mould a race fitted for survival. As Zeller puts it, "inhabitants of northern lands somehow acquired the mental and physical hardiness that destined them to thrive there. Biogeographical theories anthropomorphized northern forms ... that successfully 'invaded' southerly lands and moved in as 'denizens.'"[41]

All of this suggests that there is a geography of Darwinian encounters shaped by political, social, economic and a host of other particularities. Elsewhere I have directed attention to what I call the "geographies of reading" and "speech spaces" in an attempt to come to grips with the different ways Darwinism has been constructed in different venues.[42] For in thinking about the nature of scientific culture there is much to be said in favor of Jim Secord's judgment that we should "shift our focus and think about knowledge-making as a form of communicative action."[43] By the same token an analysis of the geographies of Darwinism that ignores the salience of natural landscapes in favor of resolutely culturalist readings risks missing a basic truth: that the natural world has a role to play in the construction of scientific knowledge about it.

## Notes

1  J. Browne, *Charles Darwin: Voyaging*, Princeton, NJ: Princeton University Press, 1995, p. 229. See also A. Desmond and J. Moore, *Darwin*, London: Michael Joseph, 1991.

2  R. MacDonald, *Mr Darwin's Shooter*, London: Anchor, 1990, p. 410.

3  C. Darwin, *Journal of Researches into the Natural History & Geology of the Countries Visited During the Voyage Round the World of H.M.S. "Beagle" Under the Command of Captain FitzRoy, R.N.*, London: John Murray, new edition, 1890, p. 534.

4  Darwin, *Journal of Researches*, pp. 11, 12.

5  Darwin, *Journal of Researches*, pp. 33–4.

6  Darwin, *Journal of Researches*, p. 527.

7  Within Geography, a Darwinian perspective on landscape aesthetics surfaced clearly in the writings of Jay Appleton. According to Appleton, "pleasurable sensations in the experience of landscape [were connected] to environmental conditions favourable to biological survival." And further: if "...a landscape 'component' appears beautiful, its beauty ... derives from the contribution which it seems, *actually or symbolically*, to be capable of making to our chances of biological survival in the environment of which both we and it form a part." (See J. Appleton, *The Experience of Landscape*, Cambridge: John Wiley, 1975, pp. vii, 243.)

8  J. Smith, *Charles Darwin and Victorian Visual Culture*, Cambridge: Cambridge University Press, 2006, p. 280.

9  Darwin, *Journal of Researches*, p. 534.

10  Darwin, *Journal of Researches*, p. 536. See the discussion in J. Paradis, "Darwin and Landscape," *Annals of the New York Academy of Sciences* 360, 1981, 85–109.

11  F. Driver and L. Martins, "Views and Visions of the Tropical World," in Felix Driver and Luciana Martins (eds), *Tropical Visions in an Age of Empire*, Chicago: University of Chicago Press, 2005, p. 6.

12  F. Darwin (ed.), *The Life and Letters of Charles Darwin*, London: Murray, 1887, vol. 1, p. 237.

13  R.D. Keynes (ed.), *Charles Darwin's Beagle Diary*, Cambridge: Cambridge University Press, 2001, p. 42.

14  C. Darwin, *On the Origin of Species*, London: John Murray, 1906, p. 562.

15  Darwin, *Journal of Researches*, pp. 35, 36, 46, 54.

16  Letter, Darwin to Frederick Watkins, 18 August 1832, in F.H. Burkhardt and S. Smith (eds), *The Correspondence of Charles Darwin. Vol. 1: 1821–1836*, Cambridge: Cambridge University Press, 1985, p. 260.

17  See the discussion in L. Martins, "A Naturalist's Vision of the Tropics: Charles Darwin and the Brazilian Landscape," *Singapore Journal of Tropical Geography* 21, 2000, 19–33.

18  A.R. Wallace, *Natural Selection and Tropical Nature*, London: Macmillan, 1878, pp. 121, 122, 123.

19  F. Darwin, *The Life and Letters of Charles Darwin: Volume 1*, London, John Murray, 1888, p. 61.

20  P. Armstrong, *Darwin's Other Islands*, London: Continuum, 2004, p. 151.

21  Quoted in P. Armstrong, *Under the Blue Vault of Heaven: A Study of Charles Darwin's Sojourn in the Cocos (Keeling) Islands*, Indian Ocean Centre for Peace Studies and Geography Dept., University of Western Australia, 1991, p. 67.

22  See A. Kay, "Darwin's Biogeography and the Oceanic Islands of the Central Pacific, 1859–1909," in R. MacLeod and P.F. Rehbock, *Darwin's Laboratory: Evolutionary Theory and Natural History in the Pacific*, Honolulu: University of Hawai'i Press, 1994, pp. 49–69, at p. 53.

23  Darwin, *Origin of Species*, Ch. 12.

24  Armstrong, *Darwin's Other Islands*, p. 18.

25  What follows draws on the very useful work, *Darwin at Downe*, which was prepared under the oversight of Randal Keynes and the Steering Group of the World Heritage

Team as the Nomination Document for the proposal that Darwin's home and surrounding landscape be named as a World Heritage Site.

26 M.J.S. Rudwick, "Charles Darwin in London: The Integration of Public and Private Science," *Isis* 73, 1982, 186–206.

27 C. Darwin to Hooker, June 1857.

28 Darwin, *Origin of Species*, p. 90.

29 Darwin, *Origin of Species*, p. 137.

30 World Heritage Team, *Darwin at Downe*, p. 38.

31 Quoted in D.P. Todes, *Darwin without Malthus: the Struggle for Existence in Russian Evolutionary Thought*, Oxford: Oxford University Press, 1989, pp. 110–11.

32 Quoted in Todes, *Darwin without Malthus*, p. 104.

33 P. Kropotkin, *Mutual Aid*, London: Penguin, 1939, pp. 26–7.

34 Todes, *Darwin without Malthus*, p. 169.

35 See C. Berger, *Science, God, and Nature in Victorian Canada*, Toronto: University of Toronto Press, 1983.

36 These figures are discussed in S. Zeller, "Environment, Culture, and the Reception of Darwin in Canada, 1859–1909," in R.L. Numbers and J. Stenhouse (eds), *Disseminating Darwinism: The Role of Place, Race, Religion, and Gender*, Cambridge: Cambridge University Press, 1999, pp. 91–122.

37 J.D. Hooker, "Outlines of the Distribution of Arctic Plants," *Transactions of the Linnean Society of London* 23, 1862, 251–348.

38 Berger, *Science, God, and Nature*.

39 See M. Gauvreau, *The Evangelical Century: College and Creed in English Canada from the Great Revival to the Great Depression*, Montreal & Kingston: McGill-Queen's University Press, 1991, Ch. 4.

40 Zeller, "Environment, Culture, and the Reception of Darwin," p. 99.

41 S. Zeller, "Classical Codes: Biogeographical Assessments of Environment in Victorian Canada," *Journal of Historical Geography* 24, 1998, 20–35, on p. 27.

42 D.N. Livingstone, "Science, Text and Space: Thoughts on the Geography of Reading," *Transactions of the Institute of British Geographers* 35, 2005, 391–401; and "Science, Speech and Space: Scientific Knowledge and the Spaces of Rhetoric," *History of the Human Sciences* 20, 2007, 71–98.

43 J.A. Secord, "Knowledge in Transit," *Isis* 95, 2004, 661.

# 12

# TRAVEL AND THE DOMINATION OF SPACE IN THE EUROPEAN IMAGINATION

*Anthony Pagden*

All peoples, everywhere, perhaps, have at some point in their history conceived of a radical distinction between themselves as members of a particular ethnic group and the rest of humanity. Many, indeed, possess no terms which can adequately render the concept of "the human." As Claude Lévi-Strauss once observed, "a great number of primitive tribes simply refer to themselves by the term for 'men' in their language, showing that in their eyes an essential characteristic of man disappears outside the limits of the group."[1] This does not, of course, mean that such people cannot distinguish adequately between their neighbors and say giraffes. It merely means that they feel no obligation to extend any of the reciprocity, or the recognition, which determines life within the kin group to those outside it. The Greeks, although they did have a word, *anthropos*, which described both themselves and all human others, were no exception to this rule. In *The Statesman*, Plato made his protagonist, a "Stranger from Elea" (a town in southern Italy), complain of the Athenians:

> In this country they separate the Hellenic races from the rest as one, and to all the other races, which are countless in number and have no relation in blood or language to one another, they give the single name 'barbarian;' then because of this single name they think it a single species.[2]

The belief that the Greeks looked upon all "others" as barbarians, babblers, peoples who could no speak Greek, and who were therefore, by implication at least, possibly also devoid of true reason has been evoked down the centuries by even their most fervent admirers.[3] It had been, lamented Immanuel Kant, "a perfect source that contributed to the decline of their states."[4] As with most cultural distinctions, however, this one was never quite so simple, nor so stark, as this might suggest. For one thing, when the Greeks spoke of "barbarians," they more often than not thought of the Persians. And although the trouser-wearing Persians may have seemed, to most Greeks, soft, soggy, effeminate and over-refined, they were hardly bereft of speech or true reason. The word was also frequently applied to the Egyptians, whom the Greeks greatly respected and from whom they had drawn much of their knowledge of mathematics.[5] Neither are Plato's comments all that they might seem. The Stranger is denouncing a certain kind of chauvinism – racism – what you will. But he is doing so in order to illustrate precisely what Plato takes to be a false dichotomy. Peoples cannot be so divided, he goes on to say, any more than numbers can be divided into 2,000 and all others. Human beings, in Plato's view, should be divided into their only true dichotomies: that is male and female.

Although there was amongst the Greeks – as indeed amongst all peoples – a certain suspicion of outsiders, they had also, always been peoples on the move, *poluplanês* – "extreme travellers." One of the founding myths for all the Greek peoples, the *Odyssey*, is, after all, a poem about travel, about movement. And there were other, better-documented travelers than Odysseus: Pythagoras (sixth century BCE), for instance, who traveled from his native Samos to Egypt and Crete before settling in Croton in southern Italy, or the first of the Greek geographers, Hecateus of Miletus, who visited Egypt in the following century; and the globe-trotting "father of history," Herodotus, who had traveled as far up the Nile as Elephantine, to Babylon and Carthage, to Cyrene in Libya, and who had stood on the banks of the Dnieper in southern Russia. And it was the great Athenian sage and law-maker Solon who is said to have been the first to have made the connection – which was to have a long history in European thought – between travel and wisdom.[6] If there is something resembling a science of geography in the ancient world it is to the Greeks from Hecateus to Ptolemy in the first century CE that we owe its invention.

This knowledge of a wider world diminished the importance of race or tribe, of *genos* as a means of distinguishing between the us and the them. "She [Athena]" enthused the fifth-century Athenian orator Isocrates, "has brought it about that the name 'Hellenes,' suggests no longer a race, but an intelligence."[7] And a century later, Zeno of Citium, the founder of the Stoic School, could (supposedly) declare that:

> We should all live not in cities and demes, each distinguished by separate rules of justice, but should regard all men as fellow demesmen and fellow citizens; and that there should be one life and order as of a single flock feeding together on a common pasture.[8]

Zeno's sentiments, however, were, as we shall see, not quite as ecumenical as they might at first appear. Such cosmopolitanism was – and largely remains – an aristocratic luxury, the privilege of what the Stoics called "the wise." Yet, the desire to know – and to incorporate – "others" if only as members of the same inhabited universe, the same *oikos*, as ourselves, is as much a part of Greek, and subsequently European culture as the desire to "other," to distance and to alienate those we do not know or immediately understand. As Denis Cosgrove rightly observed, "despite constituting different nations – some yet to be redeemed – the population of the *oikoumene* constituted *humanitas*."[9]

This is not, of course, to suggest that the Greek – or indeed any other world – was a wholly, or even largely, relativistic one. Recognizing the presence of "others" and their diversity does not mean recognizing anything like an equality between "them" and "us." For even if the Greek/Barbarian distinction cannot be made to hold in the ancient world in the way it has often been described, it remains the case that the Greeks did assume a dichotomy between the wise and the non-wise, between – to use an anachronism – the civilized and the non-civilized. Not all Greeks looked upon "barbarians" they had encountered in their travels as non-beings. But true "barbarians" might nevertheless exist.

Perhaps the best-known of the non-Greek, non-men were the Cyclops who eat a number of Odysseus's crew before he manages to escape. They are cannibals – the supreme form of non-humanity. But they are also described as those who had no agriculture, lived in caves and, crucially, knew nothing of navigation.

Navigation, the desire to break the bounds, the *periodos* and *peirata*, by which the accidents of birth have confined us, was thus from as early as the composition of the *Odyssey* (probably sometime in the second half of the eighth century BCE) considered to be one of the determining features of the "human." In that curious collection of third-century (CE) texts known as the *Corpus Hermeticum*, supposedly the writings of the magus Hermes Trismegistus whose wisdom was believed to pre-date even that of Moses, Hermes is shown at work imprisoning the demiurges in human bodies as a punishment for their attempt to rival the creativity of the Gods. Even as he does so, the figure of Sarcasm (*Momos*) appears to congratulate him. "It is a courageous thing you have done to have created man," he mocks,

> ...this being with curious eyes and a bragging tongue. For he will push his designing thoughts even to the limits of the earth. [These men] will extend their audacious busy hands even to the edge of the sea. They will cut down the forests, and will drive them [as ships] over the seas from bank to bank, all the way to those lands that are furthest away.[10]

The desire to move, to travel, "even to the limits of the earth," this, together with the ability to transform nature to meet human needs, constituted not merely dominion over the world, but a form of knowledge of it.

In 1532, the German theologian Simon Grynaeus, surveying what by then had come to be recognized as the greatest navigational feats in European history: Columbus's discovery of America in 1492, Vasco da Gama's rounding of the Cape of Good Hope in 1497 and the circumnavigation of the globe under Ferdinand Magellan in 1519–22, praised these new travelers for having recovered by their action the dominion over the natural world once enjoyed by Adam.[11] Traveling into uncharted space had become, in the modern age in which Grynaeus was conscious of living, a manner of overcoming the cognitive damage inflicted by the Fall. The traveler, together with the geographer, the mathematician and the astronomer were men whose mission was comparable, both in its nobility and the distrust it aroused in the ignorant, to the early Christians, who had similarly abandoned the settled known-world in order to seek and spread the word of God.

For Grynaeus, the new navigators shared an identity with the desert fathers, even with Christ and his Disciples themselves. Nearly a century later, the tirelessly punning English geographer Samuel Purchas similarly appropriated the legend of the expulsion from Eden as the source for the transformation of man's condition from the stationary to the migratory. For mankind, he wrote, "preferring the Creature to the Creator, and therefore is justly turned out of Paradise to wander, a Pilgrime over the world." This act of divine retribution transformed for Purchas all human history – including the story of Christ's passion – "the greatest of all peregrinations," from God to man and back again – into a narrative of human movement. In this narrative, too, the itinerant Evangelists whose task it was to spread the word throughout the world, who, like Christ himself, had no settled place, whose geography was eschatological rather than real, became the source of all foundation, human and divine.[12]

For Grynaeus and Purchas, this reappropriation of the powers of Adam, the simultaneous ability to know and to master the world, becomes the definition of "modernity." The ancients had possessed a laudable curiosity about the limits of the known world. But fearful of the dangers of *oceanic* navigation, they had, as had their medieval successors, remained trapped behind the Pillars of Hercules. It had been this for this reason, claimed the great Flemish cartographer and geographer Abraham Ortelius in his *Theatrum Orbis Terrarum* of 1570, that it had been the moderns and not they who had discovered America.[13]

Oceanic travel and the discoveries by Europeans of the farther reaches of the world, discoveries which had brought them into contact with peoples whose very existence had been ignored by the ancients, became a clear indication of the ultimate superiority of the modern over the ancient world. This modernity has been given many beginnings. But from the mid sixteenth century on, most historians would point to three moments which could be said to have resulted in the transformation of the understanding of the world: they were the discovery of America (and for them it was precisely a *discovery*, a "revealing to the gaze"[14]), the new sea-route to India, the invention of gunpowder, and of the printing press. Each of these (except initially gunpowder – but I shall come back to that) had two properties. Each had made Europeans more mobile and they had made them better able to communicate with one another, and increasingly with those whom they encountered in the worlds beyond their own.

Since these achievements had been limited to, and broadly shared by, not only the "moderns," but also those whom Samuel Purchas described as "we in the West," they became a collective means of self-presentation not merely Portuguese or Spanish or later English and French but, in some broader sense, European. In 1559 the French *savant* Louis Le Roy makes "the voice of our common mother Europe" declare:

> I who in the past hundred years have made so many discoveries, even things unknown to the ancients – new seas, new lands, new species of men: with Spanish help I have found and conquered what amounts to a New World.[15]

From the early sixteenth to the late eighteenth century this message was displayed in images of the four continents which appeared in the most unlikely places as a reminder of both the newly acquired vision of a vastly enlarged world and Europe's triumph over so much of it, a triumph which only the sciences and the arts had made possible. Take one striking, but representative example. On the ceiling of the stairway hall of the Trappenhaus, the residence of the Prince Bishops of Wurzburg, a princely family in no way associated with transoceanic navigation, the great eighteenth-century Venetian artist Giambattista Tiepolo depicted in lavish detail each of the four continents. These are so arranged that no matter where the viewer stands, *Asia, Africa* and *America* can be seen only in relation to *Europe*. The allegorical figure of *Asia* is shown seated on an elephant, *Africa* on a camel and *America* on a crocodile – menacing languid and amphibious. Only *Europe* sits on a throne instead of an animal, and only Europe is surrounded not by the flora and fauna of the continent she presents but by what its peoples have themselves created, by the attributes of the arts, of music and painting, of science and of technology.[16]

Most prominent beneath Europe's throne, however, is a cannon. The presence of gunpowder in the list of significant modern achievements ties, as does Grynaeus's association of navigation with the recovery of *dominium*, travel, and the knowledge it provided – indeed all human technology – to possession. The association is, like so many of our assumptions, an ancient one. The oldest, most enduring exemplar of this is the mythological figure of Alexander the Great. Already by the time Arrian in the first century CE had written his history of Alexander's life he had become not only a great general, but also a figure possessed of an insatiable desire for knowledge and an incorrigible urge to travel. Alexander had been Aristotle's pupil, and it was widely believed that it was for him that Aristotle had written what is, in effect, not only the first treatise on politics but also one of the earliest studies of astronomy (the so-called *De Caelo*). In the Middle Ages, Alexander became a voracious legendary figure whose desire to enslave the entire world is matched only by his ambition to know all its secrets and visit all its parts, for which purposes he went in search of the hidden sources of the Nile, invented a diving bell to reach the floor of the ocean, and was carried upwards to Heaven in a great basket drawn by griffins. This Alexander, like the figure of Odysseus (Ulysses) whom Dante meets in Hell (and who is given many of Alexander's attributes), tries to sail beyond the Pillars of Hercules and dreams of conquering the western sun.[17]

But, of course, this imaginary Alexander was not only a traveler and a seer, but he was also and foremost a conqueror. In the end it is empire which brings – or forces – peoples together. The seemingly ecumenical sentiments of Zeno of Citium, which I quoted earlier, if they were ever said at all, survive only in a passage from the Greek philosopher and biographer Plutarch's *Life of Alexander*. And for Plutarch it is Alexander the empire builder, "one sent by the gods to be the conciliator and arbitrator of the Universe," who embodies Zeno's universalizing vision. This Alexander who, "using force of arms against those whom he failed to bring together by reason … united peoples of the most varied origin," who having, "bade them [humankind] all consider as their fatherland the whole inhabited earth …. should be regarded as a very great philosopher."[18] Centuries later, Montesquieu dedicated an entire chapter to Alexander's deeds in the *Spirit of the Laws*. Alexander, he said, following Plutarch, had resisted those, notably Aristotle, who had urged him to treat the Greeks as masters and the Persians as slaves. Because he "thought only of uniting the two nations, of wiping out the distinction between conquerors and vanquished," his had been a conquest of reconciliation in which "the old traditions and everything that recorded the glory or the vanity of these peoples [the Persians]" had been studiously preserved, so that Alexander himself "conquered only to be the monarch of each nation and the first citizen of each town."[19]

However improbable it might seem, it has been an enduring image. In 1926 the English jurist and historian W. W. Tarn said of Alexander that he had

> ...lifted the civilized world out of one groove and set it in another. He started a new epoch; nothing could again be as it had been ... Particularism was replaced by the idea of the "inhabited world," the common possession of civilized men.[20]

Tarn was writing shortly after the creation of the League of Nations, on a rising tide of hope that the enmities which had led to the First World War would soon be replaced by a perpetual universal peace. For Tarn, Alexander had helped to make that idea a possibility.

The mythic Alexander provided an enduring metaphor for the link between three familiar but also problematical European discourses: scientific curiosity, travel (or communication) and political *dominium*. There is a moment in the sixteenth century when these converged in a critical way. In 1539 the Spanish theologian Francisco de Vitoria delivered a now celebrated lecture at the University of Salamanca with the title "On the American Indians." He began with a question: "By what right (*ius*) were the barbarians subjected to Spanish rule?"[21] This question carried with it a very large number of implications. It had of course been asked before, and answered in a number of conscience-saving ways. Vitoria's approach, however, was different from his predecessors in one crucial respect. All humans, civilized or barbarian, Christian or non-Christian, he argued, possess by nature what he called, "the right of natural partnership and communication." This described a complex set of claims divided into five propositions. In principle, however, it was an allusion to the ancient right of hospitality, which Vitoria transformed from a Greek custom into a right under natural law. "Amongst all nations," he wrote, "it is considered inhuman to treat travelers badly without some special cause, humane and dutiful to behave hospitably to strangers."

"In the beginning of the world," he continued,

> ...when all things were held in common, everyone was allowed to visit and travel through any land he wished. This right was clearly not taken away by the division of property; it was never the intention of nations to prevent men's free mutual intercourse with one another by its division.

The right to hospitality, and in particular the right to assistance in moments of danger, is, of course, based upon the supposition that there exists a common human identity. You have no need to be hospitable to real barbarians and they – if the behavior of the Cyclops is anything to go by – are unlikely to behave hospitably to you.

"Nature," wrote Vitoria, "has decreed a certain kinship between men ... Man is not a 'wolf to his fellow men' – *homo homini lupus* – as Ovid says, but a fellow." This in turn brings with it an obligation to friendship, for "amity between men is part of the natural law." Vitoria's point is that a right to travel peacefully and to be granted hospitality is precisely a right which the creation of civil society, and of individual sovereign states, cannot erase. It was, he insists, "never the intention of nations to prevent men's free mutual intercourse with one another." On these grounds the Indians have no right to deny the Spaniards free and peaceful access to their lands. The same principle applies everywhere in the world. For it would "not be lawful for the French to prevent the Spaniards from traveling to or even living in France and vice versa."[22] It was this law which ultimately made understanding between peoples possible, for although there might exist many different kinds of peoples with many different degrees of civilization, "the whole world ... is in a sense a commonwealth."[23]

Vitoria's "right of natural partnership and communication" was to have a long history. It passes through Hugo Grotius and John Selden and Paolo Sarpi's disputes over whether or not the ocean could be subject to property rights, through Christian Wolff's notion of a world state, what he called the *civitas maxima*, whose purpose was "the promotion of the common good by its combined powers," through what the Swiss diplomat Emeric de Vattel in 1758 called the "ties of the universal society which nature has established among men," and which were based "solely on the quality of mankind" *qua* species – until it finally comes to rest in Immanuel Kant's *Perpetual Peace, a Philosophical Sketch* of 1795.[24] In this highly influential text Kant sketched out the basic conditions of what he calls the cosmopolitan right, the *ius cosmopoliticum*, a right which is restricted precisely to the "conditions of universal hospitality." This is, as it had been for Vitoria, the ancient right of all persons to be allowed free access to any part of the world. All citizens thus have the right "to try to establish community with all and, to this end, to *visit* all regions of the world." This right would, he hoped, make it possible

> ...for [strangers] to enter into relations with the native inhabitants. In this way, continents distant from each other can enter into peaceful mutual relations which may eventually be regulated by public laws, thus bringing the human race nearer and nearer to a cosmopolitan existence.

Once free access to all for the purpose of communication has been established across the globe, a state will have been reached in which: "The peoples of the earth have thus entered in varying degrees into a universal community, and it has developed to the point where a violation of rights in one part of the world is felt everywhere."[25] Only within such a "confederation of peoples" will it be possible for mankind to be at once both a citizen of a nation and a full member of the society of the citizens of the world. "This," he added, "is the most sublime idea which a man may conceive of his destiny."

Kant was reformulating the older international order of competing European states and their overseas empires as a single cosmopolitan federation. But he was also, as he says again and again, offering the image of a higher political order which, in time, all societies come to acquire – "an archetype, in order to bring the legal constitution of mankind nearer to its greatest possible perfection."[26]

Could the human urge to communicate and the seemingly insatiable desire, as Momos had put it, "to push his designing thoughts even to the limits of the earth" be satisfied without also satisfying the equally human desire to dominate and possess? Could the human race really be made as one without it also falling under the control of one group, one people and one person? Kant thought that it could. Not now, not soon, perhaps, but at some moment in future time. "For no one," he wrote, "can or ought to decide what the highest degree may be at which mankind may have to stop progressing, and hence how wide a gap may still of necessity remain between the idea and its execution."[27]

Kant's vision has, of course, never been realized. The wise have never been wise enough and the signs, as it turned out, pointed in other, more sinister directions. By the mid nineteenth century any such alliance of states had become unimaginable, and, with the advent of a new nationalism in Europe, cosmopolitanism itself all but vanished, even as an inspiration. Humans were now, as Kant's former pupil Johann Gottfried von Herder insisted, divided into peoples, not only "by languages, inclinations and characters," but even by "woods and mountains, seas and deserts, rivers and climates." Geography for Herder was the determining feature of human existence. Cultures were shaped by natural environments and as all were different, all were incommensurable.[28] Any kind of world-order could exist only by transforming, or eliminating, anything remotely "other." Today such determinism seems, rightly, suspect. But Kantian cosmopolitanism, although it has recently enjoyed something of a revival, remains overshadowed by the claim that all forms of universalism, no matter how benign they might seem, end up where, not Kant, but Plutarch's Alexander had left them: in a world of uniformity and harmony made possibly only by the assertive power of one group over all others.

## Notes

1 *The Elementary Structures of Kinship*, trans. J. H. Bell, London: Eyre and Spottiswoode, 1968, p. 46.

2 *Statesman*, 262d.

3 According to the first-century BC geographer Strabo, the word *barbaros* is an onomatopoeia, describing someone who, instead of true speech, merely says "bar bar" (*Geographia*, 14.2.27–8). Thucydides writing in the fifth century BC says of Homer that "he did not use the term [*barbaros*] because the Hellenes had not yet been distinguished under one name as opposed to them" (*History*, 1.3).

4 I. Kant, "Kant on the Metaphysics of Morals: Vigilantius's lecture notes," in P. Heath and J.B. Schneewind (eds), *Lectures on Ethics*, Cambridge: Cambridge University Press, 1997, p. 406.

5 H.C. Baldry, *The Unity of Mankind in Greek Thought*, Cambridge: Cambridge University Press, 1965, pp. 20–4.

6 F. Hartogh, *Mémoires d'Ulysse. Récits sur la frontière en Grèce ancienne*, Paris, 1996, pp. 12–13.

7 *Panegyricus*, 50.

8 As recorded by Plutarch, *On the Fortune of Alexander*, 326b.

9 "Globalism and Tolerance in Early Modern Geography," *Annals of the Association of American Geographers* 93 (4), 2003, pp. 852–70.

10 A.J. Festugière and A.D. Knock (eds), *Corpus Hermeticum*, Paris: Les Belles Lettres, 1954, IV Fr. 23, pp. 14–16.

11 *Novus orbis regionum ac insularum veteribus incognitarum* (Basle, 1532), "Epistola nuncupatoria," ff. 92r–93r, and see A. Pagden, "La Découverte de l'Amérique et la transformation du temps et de l'espace en Europe," *Revue de synthèse* 129, 2008, 1–16.

12 *Purchas his Pilgrimes*, 5 vols, London, 1625, I, pp. 49–50.

13 J. Romm, "New World and '*novos orbes*': Seneca in the Renaissance Debate over Ancient Knowledge of the Americas," in *The Classical Tradition and the Americas. Vol. I: European Images of the Americas and the Classical Tradition*, Part I, Berlin; New York: Walter de Gruyter, 1994, p. 105.

14 "Discovery" derives from the late ecclesiastical Latin term, *disco-operire* which means literally to "take the lid off," or "reveal to the gaze."

15 *De la vicissitude ou variété des choses en l'univers*, Paris, 1579, ff. 98v–99v.

16 See S. Alpers and M. Baxandall, *Tiepolo and the Pictorial Imagination*, New Haven and London: Yale University Press, 1994, p. 154; "*Asia, Africa* and *America* are depicted in their relation to *Europe. Europe* is the rubric, the initial code."

17 *Inferno*, Canto, 28, 94–120. For Cicero, too, Ulysses is a seeker after truth, *De finibus* 5.18.49.

18 *The Fortunes of Alexander*, 326b.

19 *L'Esprit des lois*, X, 14, in Roger Caillois (ed.), *Oeuvres complètes*, Paris: Bibliothèque de la Pléiade, 1951, 2 vols. II, pp. 389–90. Plutarch is also the source of the story, alluded to here, that Aristotle had told his pupil to treat only Greeks as human beings and to look upon all the other peoples he conquered as either animals or plants. This advice, Plutarch says, Alexander wisely ignored for had he accepted his mentor's council, he would have "filled his kingdoms with exiles and clandestine rebellions." *On the Fortunes of Alexander*, 326b.

20 W.W. Tarn, *Alexander the Great*, Cambridge: Cambridge University Press, 1948, I, pp. 145–8.

21 In A. Pagden and J. Lawrance (eds), *Political Writings*, Cambridge: Cambridge University Press, 1991, p. 233.

22 Pagden and Lawrance (eds), *Political Writings*, p. 278.

23 "On Civil Power," in Pagden and Lawrance (eds), *Political Writings*, 40.

24 C. Wolff, *Jus gentium methodo scientifica pertractatum*, Oxford: The Clarendon Press, 1934, 2 vols. Vol. 2 (translation), 11 (*Prolegomena # 8*). E. de Vattel, *Le Droit de gens, et les devoirs des citoyens, ou principes de la loi naturelle*, Nimes, 1793, 2 vols. I, 149–50.

25 "Perpetual Peace: A Philosophical Sketch," in Hans Reiss (ed.), *Political Writings*, Cambridge: Cambridge University Press, 1991, pp. 106–8.

26 *Critique of Pure Reason*, AA IV, 20–1.

27 "Reflexionen zur Rechtsphilosophie," in *Kants gesammelte Schriften*, Herausgegeben von der königlich preussischen Akademie der Wissenschaften, Berlin: G. Reimer, 1902–97, vol. IX, p. 609.

28 See, A. Pagden, "Die Auslöschung der Differenz Der Kolonialismus und die Ursprünge des Nationalismus bei Diderot und Herder," in S. Conrad and S. Randeria (eds), *Jenseits des Eurozentrismus. Postkoloniale Perspektiven in den Geschichts- und Kulturwissenschaften*, Campus: Frankfurt/New York, 2002, pp. 116–47.

# 13

# THE GOOD INHERIT THE EARTH

*Yi-Fu Tuan*

Iniquities and injustices in the world are legion. For thousands of years, since the first appearance of complex societies, rich and poor, powerful and powerless, exploiters and exploited have lived side by side. Biology further contributes to human inequality: some are born with genes that make them strong and beautiful, others weak and plain, or worse, susceptible to crippling diseases. Awareness of these differences in the human condition makes the more conscientious among us despair, for whereas we can, through great effort, make adjustments, shocking inequalities remain. Religion may console us by saying that although all too often bad people prosper and good people suffer, nevertheless, on God's ledger, the good do or will inherit the earth.

Is this gross wishful thinking? Perhaps. Still, I believe that there is something to the idea, and I will show why by drawing attention to the inequalities, arguing that though they are unconscionably large they are not quite as large as we think. They loom so large for two reasons. First is our tendency to focus on external environments rather than on the ways people experience them. Second is our neglect to weigh certain compensatory factors that diminish the inequality. Moral pathology, for example, can so distort the experience of the rich and powerful that they do not truly have what they legally own. The experience of good people is, by definition, the least so distorted. If people can have only what they clearly see and love, then the good are the only ones who have. Appearance to the contrary, they inherit the earth.

## Natural and built environments

This, then, is a brief preview of my line of argument. What follows will be an elaboration, beginning with the environments. A striking fact about the natural environments humans occupy is their extraordinary range: everything from frozen Arctic, barren desert, and high mountains to indented seashore, tropical forest, and islands of great floral and faunal wealth. What this means to human occupiers is that the resources available to them vary greatly, depending on the location. Like other students of the earth, I have simply accepted this fact, without further questioning, until through some strange quirk of the mind I find myself indulging in a fantasy. In the fantasy, natural environments are estates open to settlement. An agent is giving a conducted tour to prospective settlers. The fantasy makes me ask: would anyone choose the frozen Arctic over a tropical island, the desert over woodland? Such choice, needless to say, never arose. What happened was that through serpentine twists of history and fate, some people ended up living in rich natural environments and others in poor ones. In today's world, many thoughtful individuals are concerned with fairness and injustice, yet they do not see these sharp differences in resource availability as unfair or unjust. They raise no hue and cry because the locals themselves appear to be perfectly content. The fact is: wherever people have settled down, they find ways to savor life's small satisfactions. As one day follows another uneventfully, they have no reason to survey the horizon for greener pastures.

Built environments also show great variation and there, too, envy of what neighbors have is far from common. In the Congo rainforest the shelters of hunter-gatherers are of the simplest construction, yet they do not in the least want to move into the substantial houses and villages of their neighbors – the Bantu farmers. Nor, for that matter, do the Bantu farmers want to move into a modern city.[1] The desire for better accommodation emerges only in layered societies in which rich and poor share certain values and no insurmountable barrier prevents those living in one layer from moving up to the next level.

The possibility of moving up points to an important difference between natural and built environments, namely, the idea of an upper limit. In natural environments, upper limit in the luxuriance of floral and faunal life is set by nature and rises, if at all, very slowly in the course of biological evolution. By contrast, upper limit in the luxuriance of built environments is set by human desire and skill, which appear to have no bounds. It is in built environments, then, that we are prompted to ask: how high will people want to go? In all known civilizations, the answer is maniacally high. Consider aristocratic England in the eighteenth century.[2] English aristocrats, like the elites of other rich societies, had no notion of the human scale – of restrictions dictated by the mere fact of possessing a body. One common extravagance was, and still is, in the number of residences or seats the elite thought necessary to their comfort and status. In the period we are considering, a duke had ten seats, an earl nine, and a baron eight. These residences boasted many amenities and art treasures that enhanced the owner's status, if not necessarily his quality of life. Given, however, almost unlimited wealth and means, whimsy inevitably also played a role. One English castle, for example, was equipped with twenty pianos that, apparently, nobody played. A duke owned 365 pairs of shoes; an earl's castle had 365 rooms.[3]

Facts like these make the question: "Who inherit the earth?" fatuous, for how can the answer be other than that the rich and powerful do? Even so, a different answer is plausible if we do not focus on externalities that are easily inventoried, but rather on what I have described earlier as the daily satisfactions of life, for which I will henceforth use the word "experience." A shift from environment to experience narrows the inequalities. Needless to say, huge gaps remain if the desperately poor and sick are included in my account. The only reason I have for leaving them out is my own inadequacy. Still, even with this glaring omission, I believe it worthwhile to explore how the quality of life between elite and ordinary folk is significantly narrowed when the focus is on experience. But, first, what do I mean by that word?

## Experience

"Experience" I take to mean all the ways that humans perceive and understand reality through their senses and mind. It is passive when the senses act like mechanical instruments, registering impacts from the environment. If this were all, people who live in a rich environment can be expected to record a higher level of satisfaction than those who live in a dull one. But experience is also activity. The senses do not merely record but they engage, compose, and explore. Absent these acts of reaching out, even the richest environment can do little for its occupant.

The senses may be divided into proximate and distant. Proximate senses are those of taste, touch, and smell; distant senses are those of hearing and sight. Proximate senses yield a diffuse, unstructured reality close to the body that is charged with emotion; distant senses yield a composed world that is less emotional, more coolly aesthetic, and intellectual. All humans start their life in proximate reality. Infants all over the world know what it is like to slumber in the warm hollow of their mother's arm. Since their eyes can see only a very short distance – about a foot or so – it hardly matters whether their larger environment is a hovel or a mansion, Arctic coast or tropical forest.[4] The fact that the range of experience is much the same for all healthy infants means that their pleasures and satisfactions are also much the same. Worldwide equality in living standard, if it exists, exists in the first month or so of life.

As infants become toddlers, and then prepubescent children, details in perceived reality begin to differ. At the same time, with the added years, children acquire an overall gain, which is perceptual enlargement. They come to know beyond the intimacies of place to open space that carries hints of adventure and mystery. Almost all communities offer their young both place and space, though the details in them vary greatly. In developed countries, children have, for intimacy, mother's lap or its substitute, the teddy bear; and, for adventure, mysterious nooks and corners in the basement, attic, or backyard. Houses in poorer neighborhoods are not equipped with attic or basement. Yet children know well the difference between protective home turf and challenging street. The culture of hunter-gatherers is quite unlike that of the developed world; their shelters are lean-tos, not houses. Nevertheless, their children are able to enjoy the two poles of existence: cooking fire and its inviting odors, and fields where wild animals roam. Given the two poles of place and space, the young do not need separate playground and adult-made toys to thrive. Indeed, without them, they may be challenged to put more of their imagination to work and so transform pebble into turtle, stick into spear, and unkempt backyard into pristine wilderness. Imagination, in other words, can make up for certain deficiencies in the material environment. It can be a powerful equalizer.

## Imagination and culture

Having said this, I must quickly qualify it by adding that for imagination to truly take flight it does need some cultural support. Children who have never heard stories of adventure in the wilds are unlikely to see the backyard of their house as an inviting wilderness. Just how culture affects the imagination is a complicated matter, which I cannot take up here, except to say that an important difference exists between children and adults. Culture's impact on children is direct and immediate. On adults, it is all too often diverted or muddied by extraneous social considerations. Let's say that I have a well-stocked library. It will not do much for my mind if the books are selected and put on display largely to impress my colleagues. On the other hand, if I do read them, I will benefit in ways unavailable to children, and this is so even if the reading material is quite simple. I can illustrate my point with a single line taken from a story. The story is set in Wisconsin in winter. The line reads: "She opens the door and sunlight falls flat on the floor like a penitent." A child doesn't have the necessary cultural background to make much of these words. I do. Reading them, I see the sun as the Bible's Prodigal Son who, having lavished his riches on alien latitudes, returns home to Wisconsin and me.

Here is another illustration, one that derives its emotional tone from nature rather than religion. Wordsworth, in his old age, scribbled the following lines in a little girl's notebook:

> Small service is true service while it lasts:
> Of humblest Friends, bright Creature! scorn not one:
> The Daisy, by the shadow that it casts,
> Protects the lingering dewdrop from the sun.

The language, again, is the simplest, and the idea that even small service is true service is easy enough for a child to understand. The child and I both benefit, but not quite at the same level. When I read the poem I am aware, as the child is not, of the vast difference in size between sun and earth, as well as the astronomical distance that separates them, facts that make the daisy – already known to me as a common flower – even more negligible. And yet, there it is, gallantly protecting its still weaker sister, the dewdrop. That is the power of the poem. It enables me to see, as it will one day enable the child to see, significance – even grandeur – in the insignificant. It adds for me, as it will one day add for the child, a moral dimension to the aesthetic.

## Proximate senses and luxury

Experience is enriched by mind and language. I have emphasized visual perception thus far. Experience is also kinesthetic, tactile, or olfactory sensation stimulated largely and even solely by the environment. Let the environment be a great building, say, Amiens Cathedral in northern France. When one enters the huge, well-proportioned space, one feels lifted up, afloat, the weight of years falling off, the nimble lightness of youth regained.[5] Knowledge of Christian doctrine and symbolism is not a precondition for this feeling. I choose the cathedral because few have commented on its power to act on the human body and spirit even in the absence of cultural preparedness. Another reason is that, whereas monumental buildings have historically catered to the privileged and contributed to inequality, cathedrals are an outstanding exception. All have access – rich and poor, old and young, healthy and lame. In the Western world, cathedrals were and are the only architectural luxury that everyone can enjoy.

I use the word "luxury" somewhat hesitantly, for it suggests goods and materials that pamper the flesh, and makes me think more of the bedroom than of the cathedral. Let me, then, turn to the bedroom, a space designed to indulge the body. What is it like to sleep in one that is tastefully voluptuous? Society photographer Cecil Beaton gives a hint. After spending a night in the bedroom of a Rothschild house, he has this to say: "To defile, even by lying against ... the exquisitely embroidered linen pillows, is the greatest luxury." Luxury, however, is not just fine cloth and precious stone; even more it is thoughtfulness. Beaton notes that next to the bed are baskets with sharpened pencils, a half-bottle of whisky, with a half-bottle of Perrier water and ice in case one needs a refresher. On the rim of the bath there is a frilled lamp by which one can read, and a row of Floris scents, violet, gardenia, etc. The linen is impeccable. Every detail in the bedroom is perfect in its finish.[6]

Luxury implies wealth. Wealth excludes and deepens inequality. But is it quite correct to equate luxury with wealth? The poet Rupert Brooke reminds us that even people of modest means can luxuriate in

> ...white plates and cups, the strong crust of friendly bread, and many-tasting food; the cool kindliness of sheets that soon smooth away trouble, the rough male kiss of blankets, the benison of hot water, the good smell of old clothes; and other such.
>
> *(The Great Lover)*

Brooke also mentions "hair's fragrance" and "the comfortable smell of friendly fingers," which take us to the human body as the source of luxury. In its older sense, luxury meant the erotic and nothing but the erotic.[7] Accepting this meaning frees us from fretting about material inequality, for, though only the rich can sink into plump linen pillows, even the indigent can sink into the warm flesh of a beloved. Certainly, all infants enjoy that kind of luxury as they hold and fondle their mother's breast.

## Quality, not quantity

Luxury is quality rather than quantity. Let's say that I have a pair of hand-crafted shoes. I slip my feet into their soft interior with a contented sigh. That an English duke has 365 pairs arouses in me more disbelief and amusement than envy. In any case, no sane person today approves of mindless consumption. What if the goods are not clothes and chinaware but works that elevate the mind, such as books and paintings? Our attitude toward them is somewhat different. For one, access to cultural goods matters more than legal possession. For another, to the extent that one does own them, there is less onus on quantity. Books, in particular, appear to be exempt. A scholar who builds up a large personal library is not likely to be accused of greed. Still, there are sharp limits to the benefit one can derive from cultural goods. Suppose I crave knowledge, and I live in Madison, Wisconsin. The Madison campus of the University of Wisconsin has 7.3 million books, compared with the 29 million in the Library of Congress. Should I feel deprived? Irrationally I do, even though at two books a week, in fifty years I will have read only 5,000, a pitifully small number compared with what is available. Worse is that I can recall the content of fewer than fifty in any detail. Fewer still, perhaps only a dozen, have helped to make me into the sort of person I am.

Paintings are another cultural resource. The University of Wisconsin has the Chazon Museum and the city of Madison has a Contemporary Arts Center, but the collections there are very modest compared with what Yale galleries and museums have to offer. Again, should I feel deprived? Or should I ask myself just how will I be enriched by a great collection when I visit art galleries more to provide ammunition for cultural one-upmanship than for aesthetic enlightenment? The austere philosopher Ludwig Wittgenstein would be appalled by the waste of treasure on philistines like me. He admonished a student:

> If you must go to … an exhibition, there is only one way to do it. Walk into a room, select one picture that attracts you, look at it for as long as you want to, then come away and don't look at anything else. If you try to see everything you will see nothing.[8]

Thus far I speak as a consumer. What if I have talent enough to make, create? There, too, I question quantity's overriding importance. A basic resource of culture is language. As a writer, I use words. The English language at present contains some 600,000 words. Elizabethan English is thought to have had 150,000, yet ample even for Shakespeare, who used − astonishingly − more than 30,000. But did he really need even that many to achieve what he did? The King James Bible, a work that matched Shakespeare's in scope and depth, used only 6,000. Shakespeare's French rival Racine needed a mere 2,000 to ensure his immortality.[9]

## The costs of ambition

People differ in motivation and drive. That, too, may be considered unfair if only because the motivated and ambitious possess one incontrovertible advantage: they know what to do with their lives. Even if the goal at the end turns out to be illusory, traveling down the path can still be pleasing. One goal that is not illusory and common to all is knowledge. It is not illusory because knowledge is demonstrably practical: we all need it to survive. However, pursuing it passionately and without regard to practicality has undesirable consequences, and this is true whether the knowledge pursued is of the inner self or of external reality. Relentless introspection of the inner self in the hope of grasping its essence is a will-o'-the-wisp, an undertaking more likely to end in bewilderment than enlightenment. As for external reality, studying it can become an obsession that costs heavily in money, physical fitness, family cohesion, and emotional well-being. Charles Darwin is a case in point. He noted that up to the age of thirty, poetry, painting, music, and landscapes all gave him intense pleasure. Years of relentless work changed him so that he became, as he put it, "a machine for grinding laws out of large collection of facts, leading to the atrophy of that part of the brain on which the higher tastes depend." Darwin believed that the loss of these tastes not only impoverished his emotional life but also damaged his intellect and moral character.[10]

Darwin is a genius and his contribution to knowledge stellar. What if the scholar or scientist is neither? And what if he is driven more by the desire for recognition and worldly success than by the yearning for new knowledge? Such a person could feel regret late in life. I have in mind the Cambridge classicist, F.L. Lucas, even though he is not the best example, for he was honest enough to suspect that he took the wrong road. Taking the wrong road meant that he spent the most productive years of his life poring over second-rate works in the interest of research, publication, and career advancement, when he could have spent them studying and teaching the masterpieces, with greater benefit to himself and to his students.[11]

I have no reason to envy F.L. Lucas, but do I envy Charles Darwin? No. In a way, I am even glad I don't have his gift, for, without it and the obligation it entails, I am free to benefit from the knowledge he opens up and am still able to enjoy nature and the arts. Moreover, the admiration I feel for him is itself a rewarding experience. The root of the word "admire" is the Latin for miracle (*mirari*) – that which catches one by surprise and arouses wonder. Imagine a world without admiration, without *mirari*!

## Admiring other people's excellence

Envy all too often stands in the way of admiration. In one area, however, it seldom does and that is physical prowess, perhaps because it is either quantifiable, as in speed running, or is just too evident to be denied, as in gymnastic and balletic excellence. Appreciation of mental and moral achievements is more qualified, less generous. True, some people quickly see value in another's work; others do so slowly; still others reject that which is presented to them without quite knowing why. In the extreme case, lack of appreciation is traceable to a physiological defect. Color-blindness makes one immune to the chromatic appeal of paintings, tone-deafness to music. The most cultivated connoisseurs of art may be thus handicapped. Vladimir Nabokov, for example, is tone-deaf. At a concert he cannot follow the sequence of sounds for more than a few minutes, and so finds himself admiring the "reflection of hands in [the] lacquered wood" of a fiddle.[12]

More baffling and far more common is rejection or indifference without clear cause. People, otherwise receptive to culture, may dislike, say, Peking opera, post-modern architecture, or the novels of Dostoevsky. Indifference toward the aesthetics of a work is of no great consequence. When the indifference or hostility is toward the moral stance of someone like Isaiah, Buddha, Jesus, Gandhi, or Dostoevsky, one has to ask, is it culturally conditioned or are some people just morally tone-deaf? The prophet Isaiah thunders against people who hear yet do not understand, look yet do not see – in other words, who are morally tone-deaf and blind. Jesus, who quotes Isaiah, has more reason than most to wonder at human obtuseness (Mark 4: 10–12). For here he is, the Son of God in person, voicing sublime truths. Yet few recognize who he is or understand what he says. All too often, then, it is not the environment that is at fault. The problem lies rather in us – in a dullness of perception that is so out of line with our natural power as to seem willed.

## The constraints of time

In the last thirty years or so, time shortage is a common excuse for our failure to look at the wonders of life and creation. We fail to look because more than ever we are loaded with specific projects, each of which has to be completed by a certain day or hour. Time is cut into units. Each has a ceiling or cap. Nightfall is perhaps the most commonly recognized cap. But the one that haunts us most is the one that marks the end of a life-span. The life-span of the rich is no longer than that of ordinary folk. Unlike ordinary folk, however, the rich are keenly aware of the chasm between their wealth in money and their pittance in hours and days, a consequence of which is their willingness to do without certain pleasures. Take a modern tycoon. He calculates that since he makes an extra $100,000 for his company every hour he is on the job, it is absurd for him to use that hour in the art gallery, the admission fee to which is trifling. It is much more in line with his position to buy a painting for a million dollars and put it in the bank vault as an investment. And, again, rather than go to the symphony and have his soul filled with glorious music, the tycoon may feel that his time is more rationally spent gaining prestige through sending a large check to the arts foundation of which he is the chairman. As for the beauties of nature, they cost even less to visit and take even longer to enjoy. And so our tycoon stays in his air-conditioned office, sits in a swivel chair next to potted plants, and writes another large check, this time made out to the Nature Conservancy.[13]

There is something heroic, even saintly, about such a life. A man with less money and subject to less time constraint can lead a life that is self-indulgent by comparison. I, for one, willingly spend a couple of hours in the Museum of Modern Art, an evening listening to Mahler's interminable ninth symphony, and, possibly, even a long weekend in the wilderness.

## Sin and its handicaps

I have noted that experience, imagination, and time are all equalizers: in different ways, they compensate for deficiencies in one's external circumstance. Sin is the fourth and most powerful equalizer: it so blights its victims that, for all their natural and worldly advantage, they cannot know, as good people do, the true rewards of life. What, then, is sin? What are the moral failings that come under this old-fashioned word? According to one dictionary, sin is the "disorientation in our relation to God [that] causes immediate confusion in our judgments concerning what is good and right, resulting in disorientation in all our relationships: to ourselves, our bodies, one another, and the natural world."[14] All have sinned, though not equally. The powerful and rich more readily succumb if only because more temptations come their way. So disabled, the lords of the earth have, yet have not: *have* in all things that can be counted, yet have not in what matters, which include the undistorted vision, the happiness of service, and the chaste felicities of the passing hour.

Tradition speaks of seven deadly sins – pride, covetousness, lust, anger, gluttony, envy, and sloth. Each impoverishes life in its own way. Pride is Lucifer's sin, and considered by Christianity to be the most deadly. What it does to Lucifer is to make him see himself as God's rival, the source of all value and creativity. Pride in the Luciferian sense is, however, quite rare. Few people see themselves as God's rival. Quite the opposite. Western civilization would seem to be the exception. In classical antiquity, already, Protagoras could coin the boastful phrase "man is the measure." The ancients differ from we moderns, however, in two respects. One: they considered pride a virtue: pride so elevated a man's sense of self that it forbade him all petty or dishonorable action. Christian teaching makes us less willing to call pride a virtue, yet because we recognize its power for good, we commend it under the name of self-esteem. The second difference is that, despite Protagoras, the ancients did not extend the rule – "man the measure" – beyond the human world to the cosmos. Placed against the cosmos, the ancients recognized their own insignificance. For this reason, they could find inspiration in a night sky filled with stars. This remained largely true during subsequent expansions of human power in the Renaissance, in the age of great explorations, and in the age of great scientific discoveries. Only in the late twentieth century did a radical shift occur. In the academies, a style of thinking called post-modernist or critical theorist so expands the imperium of human language that all, including the forces of nature and the cosmos itself, come under it. There is nothing outside verbal constructs, themselves subservient to social forces, to command respect and awe. The pride of post-modernists is become truly Luciferian. The result is a certain languor. How can post-modernists be starry-eyed when there is no knowledge, only knowingness, no gain in truth, only gain in smartness?

Covetousness, the next sin on the list, is perhaps less deadly than pride if only because it implies a lack and not a vaunting self-sufficiency. One covets other people's possessions in the belief that having them will enhance one's own life. The desire, unworthy in itself, has the further disadvantage of being an illusion. Possessions do not have the power to enhance life in any real, tangible sense unless one has the knowledge, the time, and a state of mind untainted by guilt, to enjoy them. Otherwise, they clutter space and collect dust, their satisfaction reduced to the tawdry prestige of mere ownership.

Next is lust. Although its old meaning of pleasure and even playfulness is innocent enough, the meaning it has now is uncontrollable passion and naked power-play. A lustful man is one who puts his own sensual intoxication ahead of any concern for the other. This is not only morally wrong, it is also a feeble surrogate for true erotic passion in which one self merges with the other, and for a moment all egoism disappears. Such a state is heavenly while it lasts. But it cannot, as the myths of Tristan and Isolde, Romeo and Juliet remind us. In the imagination, erotic passion at its most intense culminates in a soaring Liebestod. What a contrast with Casanova's serial couplings and conquests that become increasingly monotonous and boring as they drag on. In the end, they are good only for boasting.

The next two sins are anger and gluttony, different emotions and states of being, of course, but they have this in common. Both are turbulent, barely controllable feelings in the self and, at the same time, targeted at something outside the self. Their difference lies in that whereas anger seeks release by exiting from the self and destroying the other, gluttony futilely seeks relief by incorporating the other – food – into the self. In Western countries both have declined sharply in modern times. One cause is a higher standard of living; another is a change in perception. Through the ages anger stood for masculinity and power. From the eighteenth century onward, however, it was to seem not strength but weakness, not a man in control but a child in tantrums.[15] As for gluttony, fewer famines made it less exigent; improved table manners made it less acceptable.

Envy, like covetousness, is pathetic, not a feeling that I, for one, would care to admit, for whereas I may brag about my pride and anger, deeming the one proper self-esteem and the other an outburst of righteousness, I can hardly admit to being envious or covetous without drawing attention to a deficiency in me. And what is it that others have and I don't? Wealth? Power? Status? Reputation? Family happiness? Any one of them can help boost my sense of self-worth, but to yearn for it is not only to admit a lack, it is also to make all too evident the absence of any core value in the depth of my being that leaves me vulnerable to other people's judgment and whim. Moreover, unlike the other sins, envy is especially deadly in that recognizing it is not necessarily liberating. On the contrary, dwelling on the feeling can be addictive, making me see what others have as simultaneously desirable in their exaggerated glamour and hateful in their unavailability.

Sloth, the last of the seven sins, is both a moral and an intellectual failing. It shows ingratitude for existence; it wastes the gifts of senses and mind that, properly used, reveal to us "the many splendored" works of creation; it denies our nature as active beings, capable of evil but also of good.

## Knowing and enjoying other humans

If sins disable, goodness enables. Absent the weight and entanglements of envy, covetousness, anger, gluttony, and such like, a man or woman is free to enjoy the encompassing world and, especially, one element in it – other humans. Other humans matter because, practically, they are indispensable to an individual person's survival. They matter, intrinsically, because they are by far the most highly evolved, the most complex, beings on earth. They are the earth's one true wonder. Recognition of this fact means that so long as a place is peopled, it cannot be dull, without interest.

How are humans a wonder? No mother will think that I exaggerate. To the mother, her baby encapsulates the universe: its toenails quasi-mineral, its warm body animal, its tantrums devilish, its smile and efforts at speech human, its touching attempts to help angelic. A mother sees this way because she sees truly, through the eyes of love. Skeptics say that the mother is biased. Yet science – that is to say, empirical knowledge – largely agrees with her. What is a human being to science? I will translate its technical language into the more poetic words and cadence of a layman, as follows. Humans are made of stardust and to stardust they will return. For the briefest time, they are resurrected from dead matter to become living creatures that are an amalgam of the sordid and the sublime: at one extreme, bone and flesh, viscid fluids and putrid wastes, urges and passions, that point to their cousinhood with beasts of swamp and jungle, except for a moral rottenness that makes them seem creatures from hell; and at the other extreme, speech, creativity, and an ability to rise to moral heights that make them seem only a little lower than the angels.

## The power of the social

I am leading up to the question: how is it that we do not – or hardly ever – look upon our fellow humans this way? How is it that we show a keener appreciation of handbags in a shop window than of the saleswoman, a dog tied to a lamp post than a mother pushing a stroller? Various reasons account for this unbalanced attentiveness. One is practical: if we give humans the attention they deserve we would be incapacitated for action. Precisely because handbags and dogs do not merit our full attention, we can attend to them as time and circumstance allow, and still get on with our lives. Another reason, also practical, is a human being's conscious awareness that society has commanding power over an individual. This conscious awareness appears at a certain stage in life – commonly, around age seven. Before it is an idyllic period when the child is not only physically attractive but has a moral quality that is called innocence. To the child herself, the world is good – nurturing and cozy, yet full of surprises. After age seven, a change comes about that tempts one to use the word "fall." The child's world is still good, but it has shed a little of its dewy freshness, its fairy-dust enchantment. She herself is less exuberantly creative, making up fewer stories and using fewer and less colorful metaphors in her speech.[16]

What is happening is the child's growing need to be accepted by her peers – her awareness of society. To be understood and accepted, she must use words that others also use. She must, above all, restrain her fantasies. The result is a diminishment in linguistic flair and, with it, a faint graying of her world. Of course, group membership has compelling advantages. It is, after all, only in the group, among one's fellows, that one can develop such social virtues as regard for others, cooperation, generosity, and self-sacrifice. However, there is a downside to this sensitivity. In a group that is large and stratified, the maturing child and, later, the grown woman will feel a certain helplessness – a certain dependence on other people's favors – that produce moral pathologies such as flattering those in power, seeking status and material advantage at another's expense, moves in the game of life that require one to ignore or exploit those who don't count. Even the social virtues do not go far. They may encourage one to notice and be helpful toward one's fellows, but only within the bounds of convention: people remain not quite real and are easily eclipsed by simpler and more easily appreciated objects such as the handbag in the shop window.

## The Kingdom of God

Children, as Jesus repeatedly said, are the most likely candidates for the Kingdom of God. Some adults have a childlike quality and so they too are candidates. To them, the earth in which even the most common objects hum with vitality and meaning is already a shadow of the Kingdom. However, the Kingdom is not and cannot be static; it cannot be just a playground for the perpetually childlike. There has to be room for growth. And growth does occur. Good can become better. The steps may be so gradual that we are hardly aware of them. Yet that is what happens. We take such a step when a daisy is become, for us, more than just a pretty flower, and a patch of sunlight on the floor more than just a harbinger of warmth. In the moral-aesthetic realm, growth essentially means an ever deepening appreciation of the other – in particular, the most difficult of others, the other human being. For this to happen, we have to acknowledge our moral failings and, at the same time, be confident of our ability to change.

Is change possible? It is, if only because change is embedded in language, and that amounts to saying that it is embedded in our nature. A common feature of language is the metaphor, which is the substitution of one image or idea by another. We use metaphors to shift images and thereby our understanding of the world all the time, though – more often than not – subconsciously. Why, then, not use them consciously for a moral end? Take my twinge of envy for my friend's marital bliss. That envy fades when I see it as a remarkable fact in the world and, as such, it can be likened to something altogether delightful, say, a summer's breeze. As for my envy of a colleague's talent for truth, it fades when I see it as light that dispels darkness, a wonderment in nature for which I should feel only gratitude.

## Good scientist and good person: what they share

What I have just said must seem naive to people steeped in the ways of the world, whom I will call "sophisticates." They are wary of goodness; even mentioning it can provoke a pitying smile. As for metaphors, sophisticates are more likely to use them to tear down than to build up. Scientists also eschew the metaphor; as scientists, they do not seek added emotion-tinted meaning and resonance in objects and events.

On the other hand, unlike sophisticates, they build; and if they criticize and tear down, it is only to prepare the ground for building. What they build are frameworks and theories that fit the known facts as fully and as elegantly as possible. To sophisticates, scientists can seem naive, a little unrealistic despite their proven ability to harness the energies of nature. Their dedication to what they do makes them seem children at play. They are indeed at play, only it is serious play, their reward being not the game itself, as it is with some sophisticates, but rather that which lies revealed at its successful conclusion – the kernel beauty and mystery of the universe.

I began with the child, then the good person who is like a child, then the dedicated scientist. They have in common openness and a feeling that things "out there" are worthy of attention. Another trait they share is attending to what most people neglect. The child may do so from ignorance. It is otherwise with the exceptional scientist and the good man or woman. The exceptional scientist lights on evidence that run-of-the-mill practitioners either do not see or see but deem insignificant and so miss out on the chance for new ideas and paradigms to emerge. The man or woman of exceptional goodness is the same. Recall the Samaritan. He notices a wounded man by the roadside and stays to help, thereby opening the window to a broader horizon of charity, unlike the other passers-by who choose not to see or see but decide not to help. Responding creatively to what others neglect is equally the mark of genius and of charity.[17]

## Goodness – an unearned gift?

I have given hints of what good people are like, including the indirect method of providing a sketch of their opposites – the less good and the bad. I have chosen hints and indirection because true goodness and the truly good life are elusive; perhaps only poets can capture their texture and essence. Even more elusive is the origin of true goodness. Is it an outcome of religious faith and its broader culture? Is it the product of a benign social system? Do philosophical wisdom and empirical knowledge play a part? I assume that they all do. Again and again, I have tried to show that even words can make a difference. Still, at the end of the day, I remain skeptical as to the power of words, culture, and environment to produce something so rare.[18] More and more I lean toward the belief that luck matters, that some people are just born under the right conjugation of stars. If so, this makes the uneven allotment of goodness in people the greatest inequality and injustice. Of course, the good suffer hardship and pain like everybody else; in fact, given their sensitivity and lack of protective armor, they are more likely to suffer than most. And they may die young if only because they give unstintingly of themselves. Nevertheless, despite afflictions and in even a short life, they touch the hem of God. The notion that the good inherit the earth, far from being a fantasy, is axiomatic. And so I end on an ironic note: all my attempts to argue that inequality is not quite as large as we believe founders on one irreducible inequality – the gift of goodness.

## Notes

1 C.M. Turnbull, "The Mbuti Pygmies of the Congo," in J.L. Gibbs (ed.), *Peoples of Africa*, New York: Holt, Rinehart & Winston, 1965.
2 P. Laslett, *The World We Have Lost*, New York: Scribner's, 1971, p. 65.
3 R. Perrott, *The Aristocrats: A Portrait of Britain's Nobility and Their Way of Life Today*, London: Weidenfeld and Nicolson, 1968, p. 202.
4 D.M. Maurer, "Newborn Babies See Better Than You Think," *Psychology Today*, October 1976, 87.
5 D. Blum, "Walking to the Pavilion," *New Yorker*, August 30, 1990, 51.
6 C. Beaton, *The Parting Years: Diaries, 1963–74*, London: Weidenfeld and Nicolson, 1978, p. 5.
7 L. Trilling, *Beyond Culture*, New York: Harvest/HBJ Book, 1965, p. 56.
8 M. O'C. Drury, "Conversations with Wittgenstein," in R. Rhees (ed.), *Recollections of Wittgenstein*, Oxford: Oxford University Press, 1984, p. 118.
9 G. Steiner, *Language and Silence*, London: Faber, 1967, pp. 43–4.
10 *Charles Darwin's Autobiography*, New York: Collier Books, 1961, pp. 69–70.
11 F.L. Lucas, *The Greatest Problem and Other Essays,* London: Cassell, 1960, p. 173.
12 V. Nabokov, *Strong Opinions*, New York: Vintage Books, 1990, p. 35.
13 S.B. Linder, *The Harried Leisure Class*, New York: Columbia University Press, 1970.
14 *The Oxford Companion to Christian Thought*, Oxford: Oxford University Press, 2000, pp. 665–6.
15 C. Zisowitz Stearns and P.N. Stearns, *Anger: The Struggle for Emotional Control in America's History*, Chicago: University of Chicago Press, 1986.
16 H. Gardner, *Art, Mind, and Brain: A Cognitive Approach to Creativity*, New York: Basic Books, 1982, p. 94.
17 W.H. Auden, *The Prolific and the Devourer*, Hopewell: The Ecco Press, 1994, p. 41.
18 Y.F. Tuan, *Human Goodness*, Madison: University of Wisconsin Press, 2008.

# PART III
# Representing

# 14

# PUTTING PABLO NERUDA'S
# *ALTURAS DE MACCHU PICCHU*
# IN ITS PLACES

*Jim Cocola*

Pablo Neruda's *Canto general* (1948) is an epic of various interrelated parts, among which the most prominent has often stood alone: *Alturas de Macchu Picchu* (1945). Though it reads as a work of startling immediacy, nearly two years elapsed between Neruda's visit to the heights of Machu Picchu, in the fall of 1943, and the writing of *Alturas* itself, in the late summer of 1945.[1] The interval proved crucial for the famously peripatetic Neruda, and for this reason *Alturas* might be better understood by putting it in its places, treating it as a poem not only of Peru but also as a poem by one who returned to Chile from Mexico via Peru, and who subsequently assumed a particularly fraught position in Chile with respect to Peru.

Neruda was a prolific and many-sided poet, and there are several ways of reading him productively. In this case, rather than relying strictly on the dense textual fabric of *Alturas*, or construing it as the work of a unified voice speaking forth from a single location, I turn to the contextual framework of *Alturas* as it developed over a much larger span of space and time. Using a number of geographical and historical particulars, I will aim to fashion a spatial-temporal explication of this famously opaque set piece, a poem which presents itself as a paean to the American continent generally, even as it keys itself more specifically to the artifacts of one of America's most prominent civilizations.

In literary terms, *Alturas* can be classified under the rubric of the prospect poem. Set at an elevation, fixed both on places below and places beyond, the prospect poem is as old as Parnassus itself. As a modern poem concerned with the broadest of American prospects, Neruda's *Canto general* stands among estimable company: Robinson Jeffers's *Such Counsels You Gave To Me and Other Poems* (1937), written from the coast ranges of Monterey County, California; William Carlos Williams's *Paterson* (1946–58), imagined largely from Mount Garrett, in Paterson, New Jersey; Lorine Niedecker's *North Central* (1968), fashioned in part from Wintergreen

Ridge, in Door County, Wisconsin; and Charles Olson's *The Maximus Poems* (1960–75), which redounded from the kames of Cape Ann, in Gloucester, Massachusetts.

Neruda's chosen prospect proved an especially vertiginous one: set in the Andes, at over 13,000 feet, and addressed to an entire hemisphere, spanning nearly 13,000 miles from north to south. But while *Alturas* is an exemplary prospect poem in many respects, the circumstances of its composition run more strongly to the desert plateau of the Atacama than to the alpine peaks of the Andes. For not only did Neruda's visit to Machu Picchu and Peru come en route from Mexico back to his natal Chile, but it was there in Chile, upon his return, that he ran a successful campaign for the national legislature, winning a 1945 election to represent the northern mining provinces of Tarapacá and Antofagasta. Though most familiar with the more verdant landscapes of southern Chile, Neruda suddenly found himself as the representative of a desert landscape replete with the lucrative copper and nitrate reserves that Chile had wrenched from Bolivia and Peru in the late nineteenth century.

Thus, during the nearly two-year interval between Machu Picchu and *Macchu Picchu* Neruda's life changed dramatically; and yet, to posterity, the distance between the two dates has all but collapsed. Nevertheless, the interval between his experience of the place and this execution of the poem remains crucial: whereas it was Citizen Neruda (lately Ambassador Neruda) who visited Machu Picchu in 1943, it was Senator Neruda who completed *Alturas* in the late summer of 1945, writing from his coastal retreat at Isla Negra. In the meantime, Neruda's attention had shifted decisively from the Andes to the Atacama, and accordingly from the Peruvian past to the Chilean present. In this respect, just as criticism fails in failing to separate the poem from the mountain it represents, so too does criticism fail in failing to return the poem to the desertic dialectic it emerged from,[2] for it was there in Tarapacá and Antofagasta that Neruda's most immediate political concerns were most intensely focused when he set out to write a poem ostensibly preoccupied with a mountain in the Peruvian Andes.[3]

In its original conception, the set of poems that came to preoccupy Neruda between 1938 and 1950 was called *Canto general de Chile*: a work he later extended, reformulated and incorporated into the longer work that became known as *Canto general*.[4] Critical consensus has identified *Alturas* at the pivot of "Neruda's new poetic expression, the shift from a purely nationalistic (i.e., Chilean) to a broader Latin American worldview,"[5] in which he upped the ante from merely national to more decidedly hemispheric yawp. Of his journey to Machu Picchu, Neruda later claimed that "aquella visita cambió la perspectiva" [that visit changed the perspective] he had previously held toward his burgeoning epic, pushing it beyond a merely Chilean work and in the direction of a "canto general americano," so that ultimately, "unir a nuestro continente, descubrirlo, construirlo, recobrarlo" [to unite our continent, to discover it, to construct it, to recover it.][6] But while *Alturas* aspires toward American and socialist commitments that are both broad-ranging and deeply rooted, it is difficult to value *Alturas* as a specifically Peruvian or Quechuan manifesto given its concomitant status as a Chilean proclamation.[7]

The return to Chile and things Chilean was one that Neruda made throughout life, from early work as a consul to political exile in middle-age and later travels as an internationally recognized man-of-letters and elder statesman. At the same time, Neruda was as internationally minded and well traveled as any poet of the twentieth century. Though clearly writing from what one critic has styled as "a tradition of poets obsessed with writing the song of their place,"[8] it's often unclear which place Neruda's songs are directed most decidedly toward. Writing from Chile about Peru, and in writing about Peru writing about Chile: such contortions of place came naturally for the junior senator, who had spent a long political career abroad in the lead-up to his election. In Asia he had been preoccupied with Chile, in France he was worried about Spain, in Mexico he was concerned yet again with Chile, and yet upon returning to Chile he began writing more generally about America, and in due course about the Soviet Union. Over time Neruda came to rely most of all on an aesthetic of dislocation by which absence, rather than presence, provided him with poetics.[9] Becoming more cosmopolitan, Neruda also came to a better sense of what he was, and of what he was not: of European descent, but not of Europe; of American birth, but not of America.

In these realizations, Neruda's double consciousness regarding colonialism began to emerge. For if Neruda, insofar as Spanish, inherited the role of colonizer (*vis-à-vis* the *indígenas*) he was also, insofar as Chilean, placed in the role of the colonized (*vis-à-vis* the controlling interests of the *anglos* and *norteamericanos*). This dual position was replicated during his years as a consul in the colonial ports of Rangoon, Colombo, Batavia, and Singapore, where Neruda's status was subaltern with respect to the European colonials but hegemonic with respect to the Asian colonials. To the extent that he sympathized more closely with the colonized than with the colonizer in Asia, Neruda gradually disaffiliated himself from earlier ideological identifications with his Spanish heritage, more closely aligning himself with the colonized *indígenas* of the Americas. Thus, in *Canto general*, and in *Alturas* specifically, Neruda moved to employ the earlier indigenous struggle against European imperialism as a cipher for the more recent Chilean struggle against the imperialism of *anglos* and *norteamericanos*.[10]

Still, in essentials, as far as Chile itself was concerned, Neruda remained an identified member of the colonizing establishment, even as he began to speak out against certain established patterns of colonization. Octavio Paz, for one, censured Neruda in 1943 as a potential turncoat with a tendency "a aceptar cada seis meses, banquetes y homenajes de ... una jauría que adula su resentimiento" [to accept every six months, banquets and honors of ... a pack that flatters his resentment.][11] Though Neruda came to consider himself as a writer from what he termed "tierras literarias semicoloniales" [semicolonial literary territories], it is difficult to discern whether such "agrupaciones semicoloniales" [semicolonial groupings] should be taken primarily as semi-colonizing or semi-colonized. Neruda, for his part, claimed that "en un país en que persisten todos los rasgos del colonialismo ... todo intento de exaltación nacional es un proceso de rebeldia anticolonial" [in a country in which all of the characteristics of colonialism persist ... all attempts at nationalist exaltation are a process of anticolonial rebellion.][12]

Here we arrive at the vexed fact of what John Felstiner termed Neruda's "nationalist persuasion."[13] Though retaining the identity of colonizer by heritage, Neruda simultaneously attempted to speak on behalf of the colonized through an anti-colonialist ideology.[14] The nationalism of the colonizer may be tantamount to colonialism, and the nationalism of the colonized may be an expression of anti-colonialism, but what to make of the figure whose double consciousness remains poised between (and acts at once as and on behalf of) both colonizer and colonized? Is it possible, when reading Neruda, to distinguish the nationalism that speaks as colonizer from the nationalism that speaks as colonized?

To complicate matters further, Neruda's thorny position with regard to the question of colonization became even more overdetermined upon his senatorial election, for he served as the representative of Chile's most tenuously held region: a desert landscape replete with the lucrative copper and nitrate reserves that had been wrenched from Bolivia and Peru during the War of the Pacific.[15] This rather perversely named war, fought between 1879 and 1884 over mining rights in the Atacama Desert, saw a decisive Chilean victory, thanks in large part to superior naval resources and as a result of British collaboration. In the aftermath, Chile extended its northern border by 600 miles, thus creating the material conditions for much of its future prosperity. Meanwhile, Bolivia became landlocked (through the Chilean annexation of Antofagasta), Peru lost significant and valuable territory on its southern coast (including the regions of Arica, Tacna, and Tarapacá) and both nations, in defeat, entered an extended period of economic catastrophe whose ripple effects persist to the present day.

It is therefore somewhat remarkable that in the voluminous criticism on *Alturas*, only a handful of writers so much as mention the conflict, generally relegating it to a footnote if discussing it at all. Felstiner is one of the few who even allude to the War of the Pacific in discussing Neruda's cultural politics. In a footnote, he explains it to be "worth remembering that Chile and Peru were old enemies, from the Inca incursions through the Independence period to the War of the Pacific (1879–83) and after."[16] But far more than a mere footnote, this historical antagonism between Chile and Peru must be emphasized as a critical aspect in the analysis – or, as Gaston Bachelard might describe it, the topoanalysis[17] – of *Alturas*, an anti-colonial Chilean poem about a Peruvian landmark written by a colonizing authority purporting to represent Tarapacá, a contested region commandeered by Chile from Peru some sixty years earlier.

If, in terms of its narrative setting, *Alturas* aspires to the condition of an internationalist poem, preoccupied with both the generalized indigenous past and the generalized revolutionary future, its compositional setting simultaneously reveals it to be a nationalist poem, all but oblivious to the particulars of the indigenous present, and more closely engaged with the intensely local details of an intensely nationalistic Chilean revolution whose economic well-being has been predicated to a large extent upon the continued suppression of Peruvian claims to the Atacama.

Thus, while from a distance *Alturas* may read as the ecstatic conversion experience of a provincial political ingenue addressing the whole of America in radical tones, upon closer inspection, it stands as a testament of political anxiety, produced by a cosmopolitan government insider, outflanked on both the right and the left, and thus unable to speak for the contested ground of a regional constituency that he was finally incapable of representing. When Neruda invokes the Inca through phrases such as "ya no sois" [you no longer are] and "cuanto fuistes cayó" [what you were fell away] (*Alturas*.vii.32), repeating "No volverás ... No volverás" [You will not return ... You will not return] (*Alturas*.xii.66), he speaks not so much from historical necessity as from a specific political imperative than looks to elide and to overwrite the indigenous realities of the Peruvian present.[18]

Thus, to the extent that Chile held a colonizing role with respect to Peru, and to the extent that Neruda came to serve as a pivotal functionary within that colonizing authority, his double consciousness found itself doubled over into a very complicated tangle indeed. In essentials, as a Chilean, Neruda remained an identified member of the colonizing establishment, even as he began to speak out against certain established patterns of colonization. Here again we return to the strange – and characteristically American – paradox by which, though in deed and fact a colonizer, Neruda attempts to speak on behalf of the colonized, through an anti-colonialist ideology. Faced with this irreconcilable divide Neruda's rhetoric shifts from invocation toward insistence, thus exposing his compromised position.

"Sube a nacer conmigo, hermano," [Rise up to be born with me, brother] (*Alturas*.xii.66): so Neruda enjoins the reader in a moment of socialist rhetoric at the beginning of the final section of *Alturas*. Yet his litany of directives in this concluding movement are couched throughout in the imperative tense: "Mostradme vuestra sangre y vuestro surco" [You must show me your blood and your furrow]; "señaladme la piedra en que caísteis" [You must point out to me the rock on which you stumbled]; "Contadme todo" [You must tell me everything]; "Dadme la lucha" [You must give me the struggle] (*Alturas*.xii.66–70).[19] With such language, Neruda tacitly assumes the imperial prerogative by which all commands are issued, even as he makes an apparently anti-imperial gesture of solidarity with the *indígenas*.

Such was Neruda's tone in poetry toward the Peruvian and American past; meanwhile in prose he was somewhat more unassuming in his address to the Peruvian and American present. Though highly laudatory in his description of Peru, referring to it as the "matriz de América" [womb of America], one also notices a slight glimmer of guilt buttressing Neruda's praise, if not quite an outright plea for forgiveness:

> Americanos del Perú, si he tocado con mis manos australes vuestra corteza y he abierto la fruta sagrada de vuestra fraternidad, no penséis que os dejo sin que también mi corazón se acerque a vuestro estado y a vuestra magnitud actuales. Perdonadme, entonces, que, como Americano esencial, meta la mano en vuestro silencio.

> [Americans of Peru, if I have touched your crust with my austral hands, and if I have opened the sacred fruit of your brotherhood, do not think that I leave you without also steeling my heart to your present state and status. You must forgive me, therefore, that, like an essential American, I insert a hand into your silence.][20]

"Perdonadme" [You must forgive me]: again we see Neruda's plea expressed as a demand, via the imperative tense. Yet in that demand comes a sort of obligatory *mea culpa* that could stand as a précis of the unspoken motives which rest behind *Alturas*, gesturing as it does toward a sense of accountability that nonetheless occludes a fuller accounting of Chile's relationship to Peru.

As the senator representing the formerly Peruvian Tarapacá, Neruda's recourse to an Incan trope in *Alturas* functioned as a re-inscription of Chile's enduring colonial authority over Peru, casting a new spell on Peru's national icon, and rewriting the history of the defeated in the victor's terms. Moreover, Neruda's choice to valorize Incans in place of Araucanians of his native Chile further reinforced a sense of identity with colonizer over colonized, for among the *indígenas* themselves, the Incans betrayed colonialist ambitions, while the Araucanians – even prior to European colonization – were fiercely anti-colonialist.

By his own admission, neither Machu Picchu nor Tarapacá was a region in which Neruda felt entirely comfortable. Thus, when he exulted of his visit to Machu Picchu that "en aquellas alturas difíciles, entre aquellas ruinas gloriosas y dispersas ... me sentí chileno, peruano, americano" [in those difficult heights, among those glorious and dispersed ruins, I felt Chilean, Peruvian, American] he seems to have overstated his case. As he subsequently admitted, in enumerating those places that he had not understood, "tampoco entendí bien las resecas colinas del Perú misterioso y metálico" [nor did I well understand the parched and mysterious hills of metallic Peru] for "yo soy un patriota poético, un nacionalista de las gredas de Chile" [I am a patriotic poet, a nationalist of the Chilean white clay.][21] And yet, the Atacaman landscape was as alien to Neruda's Patagonian roots as Machu Picchu itself, if not more so. As Neruda explained elsewhere of Tarapacá and Antofagasta,

> The mere act of facing that lunar desert was a turning point in my life. Representing those men in parliament – their isolation, their titanic land – was also a difficult task. The naked earth, without a single plant, is an immense, elusive enigma. In the forests, alongside rivers, everything speaks to a man. The desert, on the other hand, is uncommunicative. I couldn't understand its language: that is, its silence.[22]

What Neruda understood best of all was not the Atacama, but rather, "the other end of the republic," for he "was born in a green country with huge, thickly wooded forests ... [and] had a childhood filled with rain and snow."[23]

Although in that green country the Araucanians had once thrived, by the twentieth century they had been pacified and in large part confined to reservations. To Neruda, this conflict fell outside of the bounds of colonialism altogether; he preferred to construe the conflict between the "semi-savage" *indígenas* and the *latinoamericanos* as an ongoing "civil war,"[24] thus denying the territorial question which rendered the conflict between peoples rather than a conflict among a people. Moving on to *indígena* land, and moving on to Peruvian land, as a *latinoamericano* and a Chilean, Neruda had certain ghosts to exorcise in writing *Alturas*, whether he acknowledged them or not.

In the naïve reading of Neruda, *Alturas* in particular and *Canto general* as a whole are styled as pinnacles of poetic achievement produced by a prophet and a seer of the unified American field. Yet, beyond this accepted hemispheric valance, when seen through a more precise geographical frame, *Alturas* simultaneously functions as Neruda's exculpatory attempt to validate his own particular position as a usurper of Incan and Peruvian heritage. As such, it is also possible to read *Alturas* as a masked elegy for one of Neruda's most heralded contemporaries, the Peruvian poet César Vallejo, who died in exile in Paris in 1938. Though Neruda knew few Peruvians, and fewer *indígenas*, he did know, in Vallejo, one who was – in part by birth and in part by heritage – everything that Neruda was not. Vallejo stood as subaltern to Neruda in myriad respects: mestizo to his creole, *serrano* to his cosmopolite, Peruvian to his Chilean, and Trotskyist to his Stalinist. Like Neruda, Vallejo had staked an early claim to poetic mastery in the Americas: a claim that was cut short by his premature death in 1938. At this very moment, Neruda was poised on the verge of *Canto general de Chile*, which, via *Alturas*, would lead to the fuller realization of *Canto general* and to recognition as the unofficial poet laureate of Latin America.

Eulogizing Vallejo in the *Aurora* of Santiago de Chile on August 1, 1938, Neruda described him as "el espectro americano, – indoamericano como vosotros preferís decir –, un espectro de nuestra martirizada América, un espectro maduro en libertad y en la pasión" [the American specter, – Native American as you prefer to say, – a specter of our martyred America, a specter mature in liberty and in passion]. But more than a specter, Neruda saw Vallejo in the landscape itself, explaining that "tenías algo de mina, de socavón lunar, algo tercamente profundo ... Eras interior y grande, como un gran palacio de piedra subterránea, con mucho silencio mineral, con mucha esencia de tiempo y de espacie" [you had something of the mine, the lunar cavern, something stubbornly profound ... You were interior and grand, like a great underground palace of stone, with much stony silence, with much of the essence of time and space].[25]

But if Vallejo was the essence of America, how could Neruda himself hope to become the poet of America? Ethnically, geographically, and ideologically, the two were so close, and yet so irredeemably far apart. Ultimately, Neruda and Vallejo were always poised on career paths at cross-purposes, despite Neruda's later protests to the contrary.[26] While both were attuned to Spanish heritage, to the American Pacific, and to Marxism, the gulf between them was profound. Whereas Vallejo died in exile, Neruda parlayed his way into unrivaled prominence, thus assuming as the self-proclaimed poet of the Americas a position he may have known to be better reserved for Vallejo.

As it happened, it was scarcely five years after Vallejo's death that Neruda ascended the heights of Machu Picchu. In fact, the ascension occurred on the first of the *dias de los muertos*, the three-day holiday observed throughout Latin America between October 31 and November 2. At Machu Picchu, meditating upon the plight of the *indígena*, surrounded by the plight of the Peruvian, and later in *Alturas*, meditating on his own situation as a Chilean and an American, Neruda most likely had Vallejo in mind to some greater or lesser extent. Thus if *Alturas* reads as the "mineralización de un vivir casi objetal" [mineralization of a near object-living],[27] the *minerίalidad*[28] of the poem simultaneously references the *minerίalidad* whereby Neruda subconsciously commemorated Vallejo and the *minerίalidad* that so divided their nations, their nation's fortunes, and, by extension, their own fortunes.

Neruda's use of the *minerίalidad* metaphor runs through the twelve sections of Alturas.[29] In the first section, Neruda describes "aceros convertidos/al silencio del ácido" [steels converted/to the silence of acid] and speaks of "la paz sulfúrica" [a sleep of sulfur] and "la gastada primavera humana" [our exhausted human spring] (*Alturas*.i.2). In the second section, he refers to "el metal palpitante" [the pulsing metal] (*Alturas*.ii.6) and in the seventh, to "las rocas talaradas" [the perforated rocks] (*Alturas*.vii.32). The ninth section is especially replete with *minerίalidad*, including several references to "piedra" [stone] as well as to a "Lámpara de granito" [Granite lamp], a "Serpiente mineral" [Mineral snake], an "Inmóvil catarata de turquesa" [Still turquoise cataract], a "Burbuja mineral" [Mineral bubble] and a "luna de cuarzo" [moon of quartz] (*Alturas*.ix.46–50).[30] There is a piece of Vallejo in each of these moments, for, as Volodia Teitelboim explained, "al ojo nerudiano, Vallejo tenía algo de mina, de socavón lunar" [to the Nerudian eye, Vallejo had something of the mine, of the lunar cavern].[31]

Beyond *minerίalidad* are a series of larger gestures regarding the *indígena* vis-à-vis Vallejo: "Qué era el hombre" [What was man] (*Alturas*.ii.8) he asks in section two, and in section ten, "el hombre, dónde estuvo" [man, where was he?] (*Alturas*.x.56). While these are questions of an *ubi sunt* character, interrogating the larger genocide of the *indígena*, they also suggest a concern for the personification of that larger trauma as exemplified by Vallejo's death. This may account further for Neruda's repeated use of stone imagery, in a nod to Vallejo's famous mortality poem "Piedra Negra Sobre Una Piedra Blanca."[32] In this way, Machu Picchu itself comes to stand as a symbol for Vallejo's life, implicitly described in section seven as "una vida de piedra después tantas vidas" [a life of stone after so many lives] (*Alturas*.vii.34).[33] Like his Incan forebears, Vallejo too has since departed, becoming like them "Piedra en la piedra" [Stone within stone] (*Alturas*.x.56).

With the Peruvian context more fully in view, *Alturas* can be construed as a stone within the stone of Vallejo's *oeuvre*, or, perhaps, as a stone upon that stone. The litany of directives in the final section of *Alturas* is thus in a certain sense addressed not only to the *indígena* dead, but also more specifically to the dead poet Vallejo, of whom Neruda ultimately insists: "Hablad por mis palabras y mi sangre" [You must speak through my speech, and through my blood] (*Alturas*.xii.70): a demand that goes unanswered, as the poem leaves off in haunting silence. As Nathaniel Tarn concluded, "Vallejo ... must have always haunted Pablo Neruda, the Castilian of Chile, as is clear from Neruda's efforts to identify with the Indian masses" ("Latin" 34). Basque critic Juan Larrea takes Tarn's thesis a step farther, going so far as to claim that Vallejo was in fact the true source of *Alturas*, and that

Neruda "no es el poeta de Machupicchu," *Alturas* having been based on a 1935 manuscript attributed to Vallejo titled "Los Incas Redivivos."[34]

Whether directly derivative or no, *Alturas* has never been able to escape the specter of Vallejo, which has long loomed large over Neruda's personal, poetic, and political commitments and trajectories. In his *Memoirs*, Neruda had already come to lament the situation[35] by which "César Vallejo's ghost, César Vallejo's absence, César Vallejo's poetry, have been thrown into the fight against me and my poetry."[36] Even Neruda's advocates, among them Rodríguez Monegal, have acknowledged that "muchos críticos marxistas coincidirán con Larrea en sus objeciones concretas a la poesía política a Neruda" [many Marxist critics will agree with Larrea in his concrete objections to Neruda's political poetry]. Despite the concession, Rodríguez Monegal refused to believe that "la poesía hispanoamerica no tolerase la existencia de dos grandes poetas" [Latin American poetry cannot tolerate the existence of these two great poets].[37]

If Neruda largely occluded Vallejo's achievement, Vallejo's influence on Neruda, while largely subterranean, remained and remains enduring. For just as *europeos* displaced *indígenas* as the primary political authority in South America, so did Neruda come to displace Vallejo as the primary poetic authority on the continent. And if Neruda's literary usurpation was a small detail within a much larger dynamic of colonization for which he was not culpable, his election to the Chilean senate at the very nexus of that colonization implicated him within that larger dynamic, thus triggering the attempted expiation of *Alturas*: by no means an obvious aspect of the poem, but perhaps its overriding *raison d'être*.

While there is certainly space for Neruda and Vallejo in the American canon, a topoanalytic critique of their respective poetics demonstrates that the tension between them does not spring merely from personality or prosody, but also from a legacy of cultural and political fractures that is ultimately geographical in nature. *Alturas* is without question a poem about Machu Picchu, but behind the scenes of Neruda's professed call for solidarity rests a much more complex story, embedded in a complex matrix of places ranging from the United Kingdom and the United States to the disputed provinces of Antofagasta and Tarapacá and to various other sites along the Pacific Rim. In the final analysis, if the task of literary criticism is to put a poet such as Neruda in his place, then that task can be well served by recourse to geography. Here and elsewhere, putting Neruda's poetry in its various places might allow us to position him more precisely within the larger framework of literary history.

## Notes

1 That Neruda composed *Alturas de Macchu Picchu* in August and September of 1945 is by now well established, yet the poem itself remains sufficiently shrouded in mystery (and yoked to his 1943 visit to Machu Picchu) that many mistakenly attribute it to an earlier period. By adding an extra 'c' to Machu for the *Macchu* of his title, Neruda encoded the trace of a difference between the poem and the place it purports to evoke. Even so, or, perhaps, in consequence, the place is commonly misspelled as "Macchu Picchu," and the poem is commonly misspelled as *Alturas de Machu Picchu*. For a recent

example, see Erik Camayd-Freixas's *Alturas de Machu Picchu* and the Modern Revival of Pre-Columbian Cultural Artifacts," *Revista de Estudios Hispánicos* 36:2 (May 2002): 277–90.

2  Of Neruda's critics to date, Enrico Mario Santi has come nearest to crediting this viewpoint. Though he advanced a reading predicated on the biblical resonances of *Canto general*, he also acknowledged the possibility of "a purely biographical reading" that "stems rather from Neruda's experiences during his tenure as senator from the northern provinces of Antofagasta and Tarapaca." E.M. Santi, "Canto General: The Politics of the Book," *Symposium*, 1978, vol.32, pp. 254–75.

3  Neruda later acknowledged that, relative to his Machu Picchu visit, "escribí mucho tiempo más tarde este poema de Macchu Picchu ... en la Isla Negra, frente al mar," as qtd. in Santi's *Pablo Neruda: The Poetics of Prophecy*, Ithaca: Cornell University Press, 1982, p. 119. From Isla Negra, in evoking Machu Picchu, the competing claims of the littoral and the montane must have been mediated by that third landscape which then had the foremost claim to his attention: the desertic spaces that were home to his electorate in Antofagasta and Tarapacá.

4  While it has long been known as *Canto general*, the shift from a national to an international poem seems to have been a vague and incomplete one. As late as 1947, Neruda ambiguously described his poetic errand as the attempt "a cantar de nuevo ensimismándome en la profundidad de mi tierra y en sus más secretas raíces" [to sing anew becoming absorbed in the profundity of my land and my most secret roots]. The following year, in his 1948 "Yo Acuso" address to the Chilean senate, he referred to this new reverie-in-progress as "el vasto poema titulado *Canto general de Chile*" [the vast poem titled *Canto general de Chile*] Qtd. in Neruda's *Para nacer he nacido,* Barcelona: Seix Barral, 1978, pp. 310, 338, in English as *Passions and Impressions*, trans. Margaret Sayers Peden, New York: Farrar, Straus and Giroux, 1983.

5  H. Méndez-Ramírez, *Neruda's Ekphrastic Experience: Mural Art and Canto General*, Lewisburg: Bucknell University Press, 1999, p. 14.

6  In Emir Rodríguez Monegal's *Neruda: el viajero inmovil*, 1966, Buenos Aires: Losada, pp. 156, 185, the former as qtd. in Sonja Karsen, "Neruda's *Canto General* in Historical Context," *Symposium* 1978, vol. 32, pp. 220–35, 222.

7  Moreover, if it is a socialist manifesto at all, it is so within a specifically Stalinist (and Chilean, and Nerudian) context, rather than from the Trotskyist perspective that Neruda had long since forsworn.

8  G. Brotherson, *Latin American Poetry: Origins and Presence*, Cambridge: Cambridge University Press, 1975, p. 3.

9  Santí, who wrote of *Canto general* that "the experience of exile" rests "at the core of the poem's allegory" (*Prophecy* 173), could in fact have been talking about any number of Neruda's works.

10  Recalling his journey to Machu Picchu, Neruda remarked:

> pensé muchas veces a partir de mi visita al Cuzco. Pensé en el antiguo hombre americano. Vi sus antiguas luchas enlazadas con las luchas actuales ... Allí comenzo a germinar mi idea de un Canto General americano.... Ahora veía a América entera desde las alturas de Macchu Picchu [I thought often of my visit to Cuzco. I thought of ancient American man. I saw his old fights connected with present ones ... There I began to germinate my idea of an American Canto General ... Now I saw the whole of America from the heights of Macchu Picchu].
>
> (Neruda, *Prophecy* 118)

Once again, note Neruda's elision of distance between his journey to Machu Picchu and the composition of *Alturas*, as though the two were one and the same.

11  Méndez-Ramírez, "Neruda's Ekphrastic Experience," p. 17.

12  P. Neruda, *Para nacer he nacido*, Barcelona: Seix Barral, 1978, pp. 255, 393.

13  See Felstiner's *Translating Neruda*, Stanford: Stanford University Press, 1980, p. 41.

14 Terry DeHay argues of Neruda that an "absolute, manichean attitude underlies the structure of his moral system," and yet Neruda's insistent manicheaism collapses in the midst of anything resembling a close scrutiny of the American scene. See DeHay's "Pablo Neruda's *Canto General*: Revisioning the Apocalypse," in D. Bevan (ed.), *Literature and the Bible*, Amsterdam: Rodopi, 1993, 47–59, 50.

15 For more on the War of the Pacific see the chapters on Bolivian, Chilean, and Peruvian history in volumes 4, 5 and 8 of *The Cambridge History of Latin America*, ed. Leslie Bethel, Cambridge: Cambridge University Press, 1984. See also V.G. Kiernan, "Foreign Interests in the War of the Pacific," *The Hispanic American Historical Review* 35.1, 1955, 14–36; J.R. Brown, "Nitrate Crises, Combinations, and the Chilean Government in the Nitrate Age," *The Hispanic American Historical Review* 43.2, 1963, 230–46; W.F. Staer, "Chile during the First Months of the War of the Pacific," *Journal of Latin American Studies* 5.1, 1973, 133–58; H. Bonilla, "The War of the Pacific and the National and Colonial Problem in Peru," *Past and Present*, 1978, vol. 81, 92–118; and L. Ortega, "Nitrates, Chilean Entrepreneurs and the Origins of the War of the Pacific," *Journal of Latin American Studies* 16.2, 1984, 337–80.

16 Felstiner, "Translating," p. 145.

17 Bachelard uses the term "topo-analyse," [topoanalysis] in *The Poetics of Space*, trans. M. Jolas, New York: Orion Press, 1964, p. xxxii.

18 References to Neruda's *Alturas*, which will occur parenthetically, are from the bilingual edition of *The Heights of Macchu Picchu*, trans. N. Tarn, New York: Farrar, Straus and Giroux, 1966. Neruda's larger body of work can be found in *Obras Completas*, 4th edn, Buenos Aires: Losada, 1973. See also a complete English version of *Canto General*, trans. J. Schmitt, Berkeley: University of California Press, 1991.

19 In these instances I have modified Tarn's translation to convey the imperative tense in English. Whereas "you must ..." is understood in the original Spanish of *Alturas*, it should be spelled out explicitly in an English rendering for the political valences outlined above.

20 Neruda, "Nacer," pp. 168–9.

21 Neruda, "Nacer," pp. 403–4.

22 First in Neruda's *Confieso que ha vivido*, Barcelona: Seix Barral, 1974, in English as *Memoirs*, trans. H. St. Martin, New York: Farrar, Straus and Giroux, 1977, p. 167. Neruda, who likened the experience of "coming into those lowlands, facing those stretches of sand" to "visiting the moon," acknowledged that "this region that looks like an empty planet holds my country's great wealth" (167), though he did not, here or elsewhere, elaborate on the historical sources of that wealth.

23 Ibid.

24 In R. Guibert, "Pablo Neruda: The Art of Poetry XIV," *Paris Review*, 1971, vol. 51, 149–75, 165.

25 Neruda, "Nacer," p. 79. Neruda would revisit these same themes in his in his "Oda a César Vallejo," from *Odas Elementales* of 1954, which, to Giuseppe Bellini, found Neruda "representado en él al símbolo de su raza" [representing in him the symbol of his race] such that "Vallejo es, para Neruda, la esencia de su raza y de su mundo" [Vallejo is, for Neruda, the essence of his race and of his world]. From this conjecture Bellini proceeded to speculate that Vallejo "posiblemente influida en Neruda por el deslumbramiento provocado en él por las *Alturas de Macchu Picchu*, visión luminosa en la que se expresa la sustancia misma de una América permanente, que en Vallejo tiene su poeta" [possibly influenced Neruda by the glare that he provoked in him in *Alturas*, a luminous vision in which he expressed the same substance of a permanent America that has in Vallejo its poet]. See Bellini's "Vallejo-Neruda: Divergencias y Convergencias," *Cuadernos Hispanoamericanos*, 1988, *vols. 454–5*, 27–37, 28–9.

26 In his *Memoirs*, Neruda described Vallejo as "my dear friend, my good comrade," reminiscing of "our friendship, which was never interrupted by time or distance" (284). Yet, at another point, Neruda also acknowledged their differences, speaking of a "heated conversation" in which Neruda and Vallejo went to lengths to rehabilitate a fractured

friendship (131–2). This we can presume to be among the last, if not the very last, of the meetings between the two poets.

27 According to Juan Loveluck, in his "Alturas de Macchu Picchu': Cantos I–V", *Revista Iberoamericana*, 1973, vol. 39, 175–88.

28 In the spirit of Vallejo, I've coined a nonce word here, compounding *minería and realidad* to suggest the degree to which the mining industry shaped the reality of the America that Neruda and Vallejo lived in.

29 *Mineríalidad* is, in fact, one of the larger tropes within *Canto general*. Most obviously, there is the poem "Minerales" (*Canto* 1.5), which gestures toward the deep history of colonization while suggesting more indirectly the imperial greed of modern times.

30 In the poem itself, these references are so ethereal and so remote from the *mineríalidad* behind them that they are easily glossed over. Consider Felstiner's claim that "a few of Neruda's epithets could even be scrambled mischievously in translation without much damage: *Serpiente andina, frente de amaranto*, could come out "Serpent of amaranth, Andean brow," and go unnoticed; *Serpiente mineral, rosa de piedra*, might as well be "Serpent of stone, mineral rose'" (181). Yet the actual Andean cordillera is serpentine, not brow-like, and its minerals have served as the serpentine tempter prompting America's fall from grace.

31 V. Teitelboim, *Neruda*, 1991, 4th edn, Santiago de Chile: Ediciones Bat, p. 234.

32 First published in Vallejo's posthumous *Poemas humanos* (1939); rep. in Vallejo's *The Complete Poetry: A Bilingual Edition*, 2007, trans. C. Eshleman (ed.), Berkeley: University of California Press, pp. 380–1.

33 Neruda once again associated Vallejo with stone in his "Oda a César Vallejo," which opens "A piedra en tu rostro,/Vallejo," [To stone in your face, Vallejo] and closes with Neruda hoping "que un dia/te verás en el centro/de tu patria,/insurrecto,/viviente,/cristal de tu cristal, fuego en tu fuego,/rayo de piedra púrpura" [that one day/you will see in the center/of your country/rebel/living/crystal of your crystal, fire of your fire/ray of your purple stone] *Obras completas*, vol. 2, 4th edn, Buenos Aires: Losada, pp. 195–8.

34 See Larrea's *Del surrealismo a Machupicchu*, 1967, Guaymas, Mexico: Joaquín Moritz, p.163. In support of his argument, Larrea pointed out that Vallejo "caracterizaba en 1935 a "las alturas" de Machupicchu en términos que escandalizarán sin duda al autor del poema titulado precisamente de ese modo once años después" [characterized in 1935 "las alturas" of Machu Picchu in terms that will indeed scandalize without doubt the author of that poem titled in precisely that fashion eleven years later"] (216–17). Quoting briefly from Vallejo's text, Larrea points out that Vallejo muses upon the "barriadas, escalinatas, poternas, torreones" [quarters, stairways, posterns, towers] and explains that "'una gran parte de Machupicchu ha sido labrada en la peña viva de los Andes. El resto ha sido enclavado y construido piedra a piedra ...'" [A great part of Machu Picchu has been worked into the living rock of the Andes. The rest has been nailed and constructed stone by stone ...] (217). If Vallejo's language here does converge quite significantly with Neruda's, Larrea seems to exaggerate for effect in concluding that while "todo lo que en Neruda es, principalmente, producto de un oportunismo de superfice en provecho de su glorificanción personal, y supeditado, por ello, a la maquinación política, era en Vallejo impusión auténtica, desinteresada y sacrificada de punta a cabo, negadora de sí" [everything in Neruda is, principally, the product of superficial opportunism in the benefit of his personal glorification, and for that reason, toward a political machination, whereas there was in Vallejo an authentic impulse, a disinterested sacrifice and a denial of self from beginning to end"] (221).

35 Neruda was not blind to Larrea's role in setting him against Vallejo. For Neruda's vindictive dismissal of Larrea's views, see his "Oda a Juan Tarrea" (*Obras Completas*, vol. 2, 4th edn, Buenos Aires: Losada, pp. 342–7), which begins with the defensive "Sí, conoce la América,/Tarrea./La conoce," [Yes, I know America, Tarrea. I know her] and continues to retaliate from there.

36 *Memoirs*, trans. H. St. Martin, New York: Farrar, Straus and Giroux, 1977, p. 284.

37 E. Rodríguez Monegal, *Neruda: el viajero inmovil*, 1966, Buenos Aires: Losada, p. 140.

# 15

## GREAT BALLS OF FIRE

### Envisioning the brilliant meteor of 1783

*Stephen Daniels*

This chapter brings together two fields of inquiry, landscape representation and the geographies of knowledge. It is focused on the representation of a particular event, a brilliant meteor which passed over Britain one evening in the summer of 1783, an event which lasted less than a minute but which had, through its depiction and description in a range of published work, a more longer-lasting and wider-ranging, reverberative, effect on the many people who didn't witness the event. This chapter is concerned with the framing of the event in terms of both contemporary scientific theory and of landscape aesthetics, and how the recording of the event in two views from different places in Britain involved composite forms of image making, including diagrams and verbal recollections as well as poetic and cartographic codes of landscape taste. In contrast to the traumatized views of some ordinary people who experienced the sight of the meteor at the time, the pictures examined in this chapter represented the meteor as an uplifting spectacle, and an opportunity to display a culturally enlightened view of a physically luminous event.

### Beacons in the sky

The skies of Europe exhibited strange and disturbing effects in the summer of 1783, lightening storms and fireballs, earthquakes and volcanoes, and a persistent dry fog through which the sun shone a lurid red, vegetation withering in the noxious air and people suffering an epidemic of fatal fevers.[1] Writing that summer from the eastern English village of Olney, Huntingdonshire, the poet William Cowper, like many evangelicals, saw a fiery firmament that heralded the end of the world:

> Fires from beneath, and meteors from above,
> Portentious, unexampled, unexplained.
> Have kindled beacons in the skies....[2]

The season forms a set piece in Gilbert White's 1789 *Natural History and Antiquities of Selbourne*, placing this cleric's Hampshire parish in a wider world.

> The summer of the year 1783 was an amazing and portentious one, full of horrible phaenomena; for besides the alarming meteors and tremendous thunder storms that affrighted and distressed the different counties of this kingdom, the peculiar *haze*, or smoky fog, that prevailed for many weeks in this island, and in every part of *Europe*, even beyond its limits, was a most extraordinary appearance, unlike anything known within the memory of man ...The country people began to look with a superstitious awe at the red, louring aspect of the sun; and indeed there was reason for the most enlightened person to be apprehensive; for, all the while, *Calabria* and part of the isle of *Sicily*, were torn and convulsed with earthquakes; and about that juncture a *volcano* sprung out of the sea on the coast of Norway.[3]

A little after nine in the evening, on Monday, 18 August 1783, a brilliant, explosive fireball streaked across the length of Britain. It was another twenty years before advanced opinion accepted that such fireballs were extra-terrestial in origin, matter burning up on entering the earth's atmosphere. At the time they were seen as originating, in various ways, within the earth's atmosphere, through various forms of energy disturbance, conjectures encouraged by the current weather patterns. Breaking up into a band of smaller fiery globes, the fireball lasted about thirty seconds. As the evening was sultry many people were outside and the fireball was frequently and widely observed and reported, from many parts of Britain, in correspondence to newspapers, magazines and scholarly journals. From most places its elevation and direction could be recalled in terms of its relation to rooftops, treetops, steeples and the horizon. Indeed it gave the landscape a clarity it had lacked that hazy summer. The course of the fireball was soon established, emerging into view from over the Shetland Isles and passing southwards over Scotland and eastern England, before fading from view over northern France (Figure 15.1).[4]

Readers nationwide were invited to collect reports from their locality and, in the process, sift reliable observations from uneducated impressions. A letter from the village of Kirkella near Hull to the *London Chronicle*, from the area where the fireball appeared to explode, noted that "hundreds of people here give wonderful and extravagant accounts", including his frightened cowman who likened the fireball to a flaming sword, while "an intelligent farmer in Holderness" managed to follow its course carefully from his farmyard, counting the seconds and calibrating its direction. The correspondent noted a high level of conjecture among his peers, some in the sulphurous air associating the fireball with the "heat of this summer" and the "exhalations and vapours that must have been discharged by the convulsions of the earth," another basing his explanation on a visit some years before to a scientific demonstration in a local country house Burton Constable, the "lambent stream of vapours [from the fireball] exactly resembling in colour sparks taken from the conduct of an electrical machine."[5] The published descriptions give a generally

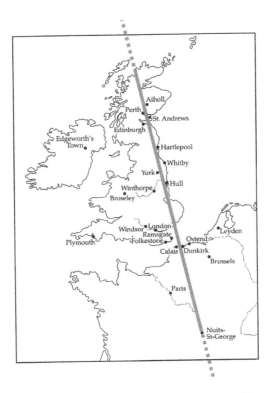

**FIGURE 15.1**   Track of the Meteor of 18 August 1783 and Main Sites of Observation. Drawn by The Cartographic Unit, University of Nottingham.

positive view of the spectacle, the hazy atmosphere dramatically breaking to reveal a radiant spectacle which illuminated the landscape.[6] "It could not I think have astonished or terrified any other than the ignorant part of the beholders," noted a correspondent from Chelsea to the *Morning Herald and Daily Advertizer*, "It was the most pleasing and beautiful phenomenon ever seen."[7] A letter from Greenwich to the *Gentleman's Magazine* went further:

> The balls were first tinted by a pure bright light, then followed by a tender yellow, mixed with azure, red, green &c, which with a coalition of bolder tints, and reflection from the other balls, gave the most beautiful rotundity and variation of colours that the human eye could possible be delighted with.[8]

Charting the track of the fireball along a north–south meridian coordinated a national network of local observers, drawing in reports from many parts of Scotland and England. As a collective act of observation, reasoned and refined, and represented through the modern media, the reporting of the fireball recalled that of the Great Storm of 1703, which was also designed to counter vulgar superstition and quell rising anxieties whipped up by religious enthusiasts.[9] It mobilized the local army of amateur weather spotters, including clerics, squires, schoolmasters, physicians and retired military men, who seized every opportunity to send in their reports of remarkable events to the Royal Society.[10] Fireballs, like star haloes, northern lights, thunderstorms, earth tremors and waterspouts, were classed as meteoric, that is fleeting, spectacular, extraordinary and airborne. This tradition of meteoric reportage was consciously local and topographical, often proudly provincial. Accounts of meteors, and their effects, give highly detailed descriptions and depictions of places, where precisely they passed and the trees, rooms, yards and gardens they struck, forming landmark episodes in narratives of a parish, estate or county.[11] In the case of the fireball of 1783, such local events were plotted in terms of a wider world.

While there are scores of written accounts of the meteor, there are just a handful of visual images, some diagrams for journal articles, and two published views, both issued in October 1783. One is a view of the meteor over the Nottinghamshire village of Winthorpe made by a schoolmaster, one Henry Robinson (Figure 15.2a); the other is a print by leading, London-based, professional landscape artists Paul and Thomas Sandby (Figure 15.2b), of the fireball seen from the terrace of Windsor Castle. Meteors were not uncommon in eighteenth-century artworks and visual imagery, usually as portentous signs in literary, biblical or mythological scenes, or in satires to signify passing fame or celebrity.[12] What distinguishes Robinson's and the Sandbys' views is their observational precision and scientific conjecture, if they are in their way emblematic, radiating cultural and political symbolism.

**FIGURE 15.2a**   Henry Robinson, *An Accurate Representation of the Meteor which was seen on Aug 18, 1783* (1783). Mezzotint. 188 × 251 mm. Department of Prints and Drawings, British Museum.

**FIGURE 15.2b**  Paul Sandby, with Thomas Sandby. *The Meteor of Aug. 18th, 1783, as it appeared from the NE Terrace, at Windsor Castle* (1783). Aquatint. 345 × 498 mm. Department of Prints and Drawings, British Museum.

## The view from Winthorpe

The letterpress to Robinson's print gives a narrative exposition of the view, setting out various sources of authority.

> An accurate Representation of the Meteor which was seen on Augt 18th 1783 __ At first it appeared as one Ball of Fire, but in a few Seconds, broke into many small ones. Its course was from N.W. to S.E. __This extraordinary Phaenomenon was of that species of Meteor which the great Phisiologist Dr. Woodward and others call the Draco Volans or Flying Dragon __The above View was taken at Winthorpe near Newark upon Trent, by Henry Robinson, Schoolmaster __and published by him as the Act directs, 14 Octr 1783 __This Plate is inscribed to Roger Pocklington Esq. by his much obliged humble Servant, Henry Robinson.

The print is dedicated to "Roger Pocklington Esq." who owned the lordship of Winthorpe, a village of around 200 inhabitants, two miles from Newark. Pocklington came from a local yeoman family, who had once owned property in the village; now a wealthy partner in Newark's main bank, he purchased the Winthorpe Hall estate in 1765 when the owner went bankrupt. Pocklington completed the mansion, constructed or renovated village buildings and made exchanges of land, confirmed by the 1778 enclosure award, to consolidate his estate, lay out a new pleasure ground, with conspicuous garden buildings, temples and follies, plant woods and improve farms.[13] Winthorpe Hall was a smart, showy modernized estate, of a kind which ringed many commercial towns at this time, and Pocklington was a conspicuous figure in the region's culture of improvement. Among his connections was Sir Joseph Banks, owner of the Revesby estate just 35 miles from Winthorpe and Britain's most prominent scientist and patron of exploration, president of the Royal Society. Banks cultivated many such connections in the country, patronizing local natural historians as well as improving squires, as a way of building a network of intelligence, if Pocklington seems to have been a more personal contact than many. He sent Banks trees for planting and also a 17-year-old plough boy named Claughton for his talent as an artist, who lodged in Banks' London house in Soho Square, probably as one of the engravers for Banks worked on natural history.[14] Patron of Robinson's living as a schoolmaster, Pocklington may have paid for the meteor print to be issued, perhaps as a speculative investment to market such a topical subject regionally, or more widely, to raise the cultural profile of this land-scape and put his village on the map.

The text emphasizes that the fireball was more than a local marvel; rather a partic-ular, regional recollection of a national event, observed at a village located "near Newark on Trent," a place widely known as a key coaching town on the Great North Road from England to Scotland, as well a river port on a major national waterway, the river Trent which flows due north here. The meteor's track is coordinated by the cardinal compass points marked around the borders of the view, the nucleus appearing due north at the center of the picture, at the moment it broke into several balls of fire over this region of England. While the image may be synoptic, informed by various published reports, its status as an "accurate representation" is as a first-hand eyewitness report, as represented by the figure to the right, whose experience, or rather perform-ance of that experience, forms part of the event and its scenic status. The figure beholds the meteor, arms raised, striking a pose of awestruck spectatorship. This presumably represents Robinson taking a riverside walk that sultry summer evening, but the cultural identity of the figure is larger: "Henry Robinson, Schoolmaster." The historical record reveals little about the real Henry Robinson, other than he seems to have delegated much of the day–to-day teaching at Winthorpe to an assistant.[15] What the view projects is the cultural reputation of schoolmasterly know-ledge, enshrined in the authority figure of Oliver Goldsmith's popular poem *The Deserted Village*, the Village Schoolmaster who amazed local people with the scope and performance of his learning, from oratory to land surveying and weather forecasting.

The print's text identifies the fireball as "the Draco Volans or Flying Dragon," an antiquated name for the form of the meteor which appeared in some popular texts, including almanacs of country lore like the many versions of *The Shepherd's Calendar*. Along with astrological prophecy, land measuring and bell ringing, such almanacs offered a practical, experiential form of country wisdom, with the figure of the shepherd watching his flock by night, as an authority figure for field observation, witnessing strange and portentous events but also through his experience able to discriminate regularities from curiosities, and translate them into proverbial fore-casting, including the shepherd's delight at red skies at night. Such field knowledge was contrasted with the learning of urban natural philosophers, in their studies pedantically poring over their books and globes or tucked up in bed. This was not

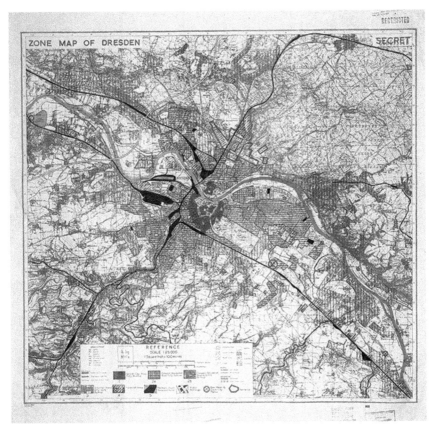

**PLATE I** RAF zone map of Dresden, Copyright British Library Board (Maps MOD GSGS 4399).

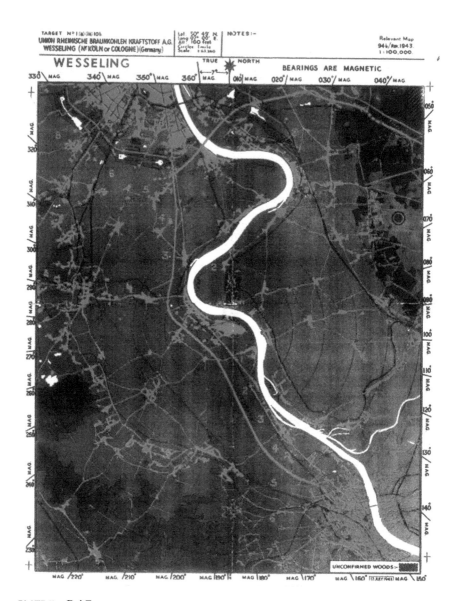

**PLATE II** RAF target map.

**PLATE III** *Dresden*, elin o'Hara slavick, 1945.

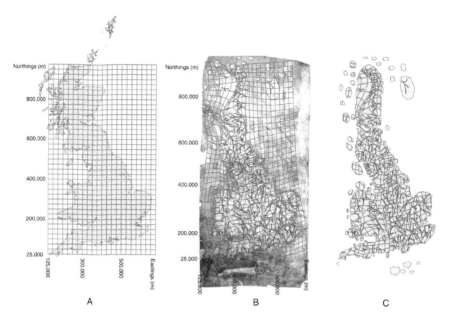

**PLATE IV** (a) Britain mapped in National Grid coordinate space, (b) Gough Map distortion grid derived using Euclidean transformation, (c) Gough Map distortion grid clipped coastline. Units are meters. (Courtesy of the Bodleian Library, University of Oxford. MS. Gough Gen Top. 16)

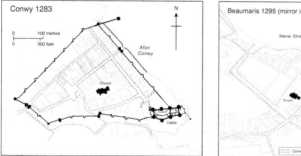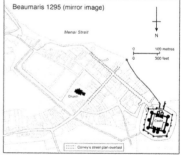

**PLATE V** New towns of Edward I compared; the plan of Conwy (left), and the plan of Beaumaris overlaid with Conwy's (inverted) street-layout (shown in pink) (right). Copyright Keith D. Lilley.

**PLATE VI** *Le Sahara,* by Gustave Guillaumet, 1867. Courtesy of Art Resource, New York.

**PLATE VII** *Au Pays de la soif,* by Eugène Fromentin, 1869. Courtesy of the Royal Museum of Fine Arts of Belgium, Brussels.

PLATE VIII   "Icons stare back"; they invite *katányxē*, penetration, transformation – here, now. Icon of the Panaghia Theosképastē with *támata* (ex-votos), Holy Monastery Panaghias Theoskepástou, Sochos, Greece. Photograph by the author.

PLATE IX   Sacred space is configured and enacted through performance: in between the gazing viewer and the gazing holy figure; in between the icon and the faithful venerating it. Mount Athos monk standing before the icon of the Mother of God, Holy Monastery of Docheiariou, Mount Athos. Photograph by Fr. Apollò Docheiarite.

PLATE X   The "wrapping effect" of icons' inverse perspective is magnified on frescoes and mosaics. Fresco of the enthroned Mother of God flanked by the archangels painted in the *katholikón* of Docheiariou Monastery, Mount Athos, 1568. Photograph by Fr. Apollò Dochiearite.

PLATE XI   The "transfigured" and "transfiguring" landscape. Enclosed in a mandorla of light, the transfigured Christ on Mt. Tabor becomes the focus of the scene and the entire wall. Mount Tabor is transfigured into the mountains of the Old Testament's revelations: on the left Mount Sinai (topped by Moses) and on the right Mount Horeb (by Elijah). In the upper part of the scene the poses of the two prophets mirror each other, as they engage a timeless dialogue with a motionless Christ. On the lower part of the icon, time continues to flow. The three apostles below are set against flowering terrain indicative of movement and depicted at different stages of their ascent. Below the scene of the Transfiguration are other scenes from the New Testament connected to the manifestation of Christ's divine nature and to the last part of His ministry: from the left, the Resurrection of Lazarus, the entrance into Jerusalem, Maria pouring ointment on Jesus' head. *Katholikón* of Docheiariou Monastery, Mount Athos, 1568. Photograph by Fr. Apollò Dochiearite.

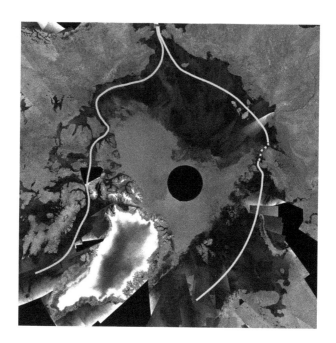

**PLATE XII** Envisat ASAR mosaic of the Arctic Ocean for early September 2007, clearly showing the most direct route of the Northwest Passage open (orange line) and the Northeast passage only partially blocked (blue line). The dark grey color represents the ice-free areas, while green represents areas with sea ice. Credits: ESA

**PLATE XIII** Greenland, 2008: Photograph by Kathryn Yusoff.

so much the virtue of folklore, or customary practice, as of classical agrarian knowledge, the property of schoolmasters, parsons, gentlemen farmers and improving squires.[16] In eighteenth-century almanacs the evidence for events attending fireballs, such as exhalations and sickness, included quotation from Virgil. Progressive almanacs tended to be skeptical of prognostications, and traded the figure of the smocked shepherd for one dressed improbably in frock-coat and hose, like the figure in Robinson's print, standing for an educated form of country knowledge.

"The great Phisiologist Dr Woodward" cited in the text is John Woodward, author of *The Natural History of the Earth*. First published in 1697, and lastly in 1726, over half a century before Robinson's view, Woodward's work was old-fashioned but still highly influential in popular publications, in best-selling text books and magazine essays, including until late in the century the entry on meteorology in *Encyclopedia Britannica*. Woodward maintained that aerial meteors, of all kinds, fireballs, lightening, aurora borealis and falling stars were, like volcanoes and earthquakes (with which he reckoned they were reciprocally connected), caused by inflammable effluvia rising from the earth. This explanation, which goes back to classical antiquity, was given a biblical significance, a great subterranean internal fire presumed to form a volatile mix with the submerged waters of the Mosaic Flood. A sustained period of hot summer days was thought to draw up thermal exhalations from bodies of water and set off fireballs at a high altitude, as they more frequently set off lightening storms at lower levels. There was little advanced empirical evidence for this theory, for no available instrument could measure the relevant signs, in particular the concentrations of sulphurous matter, but it did find warrant in certain correlations and associations, and bodily apprehensions, strange sights and bad smells; moreover, according to one popular text a fireball might be replicated in miniature at home by dropping gunpowder on a red hot poker.[17] Reporting the meteor, and collating other observations, the local press endorsed this explanation, *The Nottingham Journal* reckoning the cause was "inflammable air, for which a great abundance is discharged by stagnant pools ... a flash of lightning set fire to this accumulated inflammable air."[18]

Robinson's print is part of the fashion for spectacular nocturnal scenes at this time, in a variety of artworks, such as scenes of volcanoes, burning cottages and scientific demonstrations, including figures exhibiting various forms of spectatorship, notably in works by Joseph Wright of Derby.[19] The medium of Robinson's print, mezzotint, a skilled, relatively expensive procedure enabling half-tones to be produced, without hatching or stippling, demonstrates cultural ambition. The view is overlaid by genres of landscape aesthetics. The site and situation of this bend in the river are topographically precise,[20] while the combination of cottage and ruined tower is a standard picturesque design. Archival, map and field evidence does not confirm the existence of a cottage at this site at this time, although ones like it existed elsewhere in the village.[21] As a pictorial motif, brilliantly lit, it signifies a vernacular domesticity derived from Dutch-style landscape art set in lowland river scenery.[22] The silhouetted church tower, ruined and overgrown, is a clearer case of invention, a framing device and poetic motif which characterized nocturnal scenes in art and literature with philosophical reflections, especially on passing time.[23]

## The view from Windsor

Paul and Thomas Sandby made their name depicting Britain as a newly unified, culturally improved, national landscape, incorporating the scenery of Scotland in the wake of the military survey designed to pacify the country after the failure of the 1745 rebellion and pioneering the taste for Welsh scenery. The southern pivot of their north by north-west vision of Britain was the royal fortress of Windsor Castle, where Thomas Sandby was appointed Deputy Ranger (that is estate steward) of Windsor Great Park and Architect of the Kings Works, living there in a grace-and-favor residence. Both brothers were founder members of the Royal Academy of Arts, and Thomas Sandby was also its first Professor of Architecture, emphasizing in his lectures and designs extensive panoramic prospects.[24]

The meteor view is one of a long series of views of the Castle and its precincts made by the Sandbys over a fifty-year career, in many collaborative productions.[25] This print is a larger, more luxurious and more expensive work than Robinson's. The medium is aquatint, an inventive process named and popularized by Paul Sandby which enabled the reproduction of watercolor wash drawings with atmospheric effects. Sandby first developed aquatint in a set of prints of South Wales views based on a tour with Sir Joseph Banks, who also purchased upwards of seventy of the Sandbys' Windsor Castle views.[26] The print of the meteor is dedicated to Banks in his office as President of the Royal Society.

In many views of Windsor Castle, at various times of day, the Sandbys show the royal residence as a series of public as well as private spaces, populated by tourists, street vendors and beggars as well as guards, servants and guests. The North Terrace was a famous promenade, "the noblest walk in Europe," the guide books claimed, and one whose noble views of the Thames valley were open to many, not reserved, as in other European palaces, for the monarch and titled nobility. Sandby did a series of sometimes satirical scenes taking in various forms of conduct and misconduct, family groups promenading and taking in the prospect, encountering soldiers groping young women and men sleeping off drink, a liberal view of the male British public "preferring sex and booze to the awesome views of monarchs."[27]

At the time of the Sandbys' meteor picture, Windsor Castle and its environs were being renovated and regulated as a more private, more studious space, as a summer residence and palace of state for George III and his large family retinue. New rules were made to keep the precincts more polite, banning tradesmen and street sellers and keeping out beggars and prostitutes from the North Terrace.[28] Thomas Sandby was involved in documenting and perhaps contributing to the building and landscaping improvements of the 1780s, including proposals for the North Terrace.[29]

With Joseph Banks cultivating the monarch's patronage of science and its applications, the Castle also became a domestic center of improved knowledge. Banks secured a private royal position and pension at Windsor for William Herschel in recognition of his discovery in 1781 of a new planet, which he named the Georgian Planet in honor of the King (a name never popular on the continent where it was called, according to classical precedent, Uranus). Herschel was given a house and premises to make larger telescopes (a twenty-foot reflector in 1783) to undertake a systematic survey of the heavens, notably for nebulae and comets, in return for being summoned to the Castle with a portable telescope to give astronomical demonstrations to the royal family.[30]

The scene shown on the meteor print is assembled from a number of drawings, showing the figures and the meteor in different positions, and with varying elevations over the landscape and of depictions of the sky. The fireball illuminates a schematic, renovated-looking terrace space, the perspective lines described by the walls, from the very edge of the foreground, both establishing the vista and showing the terrace as an instrumental part of the depiction, like the balcony box of a theater. The figure group are a mixed company of four men and two women and are an integral part of the scene, watching the meteor as its moves in time lapse, in three phases, across the sky, each marked on the nucleus with a letter A, B, C.[31] The print's text explains:

> A. Its appearance soon after it emerged from a cloud, in the NW, by W. where it was first discovered. B. Its further Progress when it grew more oblong. C. When it divided & formed a long train of small luminous bodies, each having a tail; in this form it disappeared from the interposition of Cloud in E. by S.

At the time the cultural status of spectacular scientific scenes, either produced in nature or through human technology, was a major issue, particularly for their tendency to bedazzle and inflame credulous spectators.[32] Paul Sandby addressed this issue in a series of satires on the folly of fashionable balloon ascents, showing one from outside a Chelsea coffee house watched by a motley crowd including Sandby's stock figures of delusion, Scottish highlanders.[33] In contrast, the Sandbys' scene of the meteor and its spectatorship is one of refined observation and reflection.

The meteor print's text, which includes details of time, place, altitude and luminosity, is an abstract of a fuller report, read to the Royal Society in January 1784 as part of a symposium on the event. This was authored by Tiberius Cavallo, a leading authority on electricity who explains he was in the terrace at the time with Thomas Sandby, John Lockman, canon of St George's Chapel, his friend James Lind, physician to the royal household, electrical experimenter, astronomical observer and surveyor who accompanied Banks on his voyage to Iceland and western isles of Scotland, and "a few other persons." Of the group in the picture, the figure to the right in clerical clothes probably represents Canon Lockman, Cavallo the one pointing at the sky, Thomas Sandby the other figure with his back to us, James Lind in profile. Two female figures represent "a few other persons"; one or other may be Lind's wife Ann who made observations with Herschel. According to Cavallo's report, "everyone one of the company remarked some particular circumstance, the collection of all which furnished the materials for this account." It is Cavallo's sketch of the path of the meteor, later published, which the Sandbys incorporated in their print.[34] "The weather was calm, agreeably warm," the report continues "and the sky was serene." "As soon as the meteor emerged behind the cloud, its light was prodigious. Every object appeared very distinct; the whole face of the country in that beautiful prospect before the terrace being instantly illuminated." The print shows this moment, the clouds breaking (as they rarely did that hazy summer) in the fashion of heavenly scenes painted for altarpieces, church windows or palace ceilings.

In February 1784 Charles Blagden, Secretary of the Royal Society, and Banks' right-hand man, presented a long, synoptic account of the August fireball, and subsequent ones, including one in October he glimpsed from the study of Banks' London house. This may well have been to upstage another synoptic account by the Astronomer Royal Neville Maskelyne, part of the rival mathematical faction to Banks in the Royal Society, who resented his version of science, its affiliations with natural history and antiquarianism, and trust in gentlemanly testimony. In November 1783 Maskelyne initiated through the press and pamphlets a systematic nationwide "Plan" to observe meteors, which involved a strict methodological procedure and having quadrants, compasses and watches at the ready as well as paper and pencil to make a sketch. Expert observers could also enlist "observations of the most illiterate" by asking them to "trace with a stick the path of the Meteor in the Heavens," taking them to the "very spot" to "assist [their] memory" in terms of neighboring roads, houses and trees.[35] The project came to nothing but there are occasional reports on the lines of the Plan, which question the observational consensus of the August fireball.

Blagden's account was more international, based on the many reports submitted to Banks as well as reports from the European mainland, from Holland, Denmark and France. Blagden also incorporated reports of earlier fireballs to the Royal Society, throughout its history, and, where, relevant, "in writings of the ancients." The August fireball is largely British in Blagden's account, but not exclusively so. Its path is traced over the straits of Dover with reports from Calais, Ostend, Paris and Nuits, Burgundy in the same peacetime international spirit as the cross–Channel trigonometrical survey to coordinate astronomical observations between Greenwich and Paris.[36] In plotting the course of the meteor across Britain, Blagden's account overlooks many reports in the national and local press and, predictably, attributes particular authority to testimony from Fellows of the Royal Society, from the Duke of Atholl observing the meteor passing his mansion directly overhead and especially that from Windsor Castle, "because the gentlemen present, two of them [Cavallo and Lind] Fellows of the Royal Society, were remarkably well qualified for such an estimation." Blagden also enlists the Sandbys' "beautiful Drawing" for his argument. Also cited as an expert witness is William Herschel, who was observing from his house near Windsor that night, and while he missed the beginning of the meteor "must have kept it in sight long after other observers had thought it extinct" for up to 45 seconds when it was within a few degrees of the horizon, confirming "foreign accounts of its long progress to the southward." Herschel estimated the meteor's duration by reconstructing his "positions and gestures" at the time, evidence for Blagden of its velocity, "it must have passed over the whole island of Great Britain in less than half a minute, and might have reached Rome within a minute afterwards, or in seven minutes have traversed the whole diameter of the earth!" In conjecturing an explanation of the meteor, particularly its velocity and polarity, Blagden uses the observation to reject the hypothesis that fireballs were "terrestial comets" ignited by combustible vapours from the earth. "What then can meteors be?" he asks. "The only agent in nature with which we are acquainted, that seems capable of producing such phenomena, is electricity."[37]

Electricity was the dominant paradigm of the Royal Society, deployed to explain a great range of phenomena, and also a highly fashionable form of instruction, entertainment, natural theology and medical therapy, over-exalted according to skeptics as "a universal empire in the philosophical world."[38] Tiberius Cavallo's published works on electricity, including his *A Complete Treatise on Electricity* (which went into a second edition in 1782), chart a broad field of largely benign power. They maintain that electrical therapy had proven medical benefits, notably for clearing inflamed eyesight, and that electricity was responsible for various kinds of meteors now that the power had been drawn down from the aurora borealis, which occurred frequently over northern Britain throughout the eighteenth century, and fireball-like effects could be replicated by electrical machines, "darting about with great velocity, and even leaving a train behind like the common fire-balls." "We are then involved in a maze, that leaves nothing to contemplate but the inexpressible and permanent idea of admiration and wonder."[39] In his report from Windsor Castle, Cavallo observed that the meteor was heralded by "some flashes of lambent light, much like the *aurora borealis*" and Blagden uses this evidence of the northern lights to confirm and extend electrical power in polar regions. He noted the course of meteors, large ones at least, "to be constantly from or toward the north or north-west quarter of the heavens, and indeed to approach very nearly to the present magnetic meridian." It was projected that the meteor would have been observed over Iceland, which lay in the direct line of its trajectory over Britain, but no credible reports were forthcoming.[40] In a spiralling finale to his account Blagden describes "masses of electrical fluid repelled, or bursting from that great collected body of it in the north," allotting different electrical phenomena to "distinct regions" at different elevations, lightening at a low level, above the clouds falling stars, "till beyond the limits of our crepuscular atmosphere the fluid is put into motion in sufficient masses to hold a determined course, and exhibit the different appearances of what we call fire-balls; and probably at a still greater elevation above the earth, the electricity accumulates in a lighter less condensed form, to produce the wonderfully diverse streams and coruscations of the *aurora borealis*."[41]

In contrast to the mineral model for meteors, with its emphasis on earthly airs and country matters, the electrical model was predicated on high places, both in terms of latitude and altitude, and a planetary field of circulation.[42] While the aurora borealis was seen much farther south in the later eighteenth century than it is now, over large parts of England, the sphere of the aurora borealis was seen to define a northern global space, one which enlightened opinion and imperial ambition found alluring, redirecting the focus of natural and cultural history, and the advancement of knowledge, away from the heritage of southern Europe. If Britain was seen as a northern nation, it shared this cultural territory with other powers. British, French and German *savants* combined and competed with Scandinavian men of science; newly founded academies in Nordic nations claimed the aurora as their territory of investigation, as a way of raising their cultural profile, in response to sustained British and French interest which maintained that their own centers of observation, analysis and learning were better placed to explain it. For enlightened Britons, investigation of the aurora borealis was also a way of dispelling the literary associations of the northern lights as the Celtic Twilight of second-sighted Highlanders swayed by beliefs in divine omens, and firmly annexing this region to its empire of knowledge.[43] After bringing back books and manuscripts from his voyage to Iceland, along with surveys of its people and landscape, Joseph Banks had the island engraved on his visiting card. With the outbreak of war with Napoleonic France, Banks argued that Britain should annex Iceland from Denmark, for the islanders "are universally desirous of being placed under the dominion of England" and be part of that group of islands "called by the Ancients Britannica" and being "eminently fitted for the establishment of a Naval Empire."[44]

Paul and Thomas Sandby's vision of Britain as an expansive, culturally improved, nation-state had a pronounced north by north-west trajectory, incorporating the scenery of Scotland and Wales. Their view of the meteor put this picture in a more global perspective. It may be seen as part of a mapping project, a form of cultural cartography which like all surveying was mindful of the sky as well as the ground, and an enlarged view of knowledge which included the geographical program of connecting the celestial and terrestrial worlds.

## Mapping the moment

Robinson's and the Sandbys' views of the meteor of 1783 are complex and capacious images, deploying landscape as a genre to represent a sudden force of nature and figure it out. They depict and describe a fleeting, flashpoint of nature's power in terms of long-term, widely distributed conditions, not just converting the momentary into the momentous, but through their deployment of theory, an exceptional event into an expected one. The two prints chart a broad field of knowledge and imagination, scientific and artistic, popular and professional, local and cosmopolitan, here combining, there competing, the common ground of representation and its contested terrain. The respective views display contrasting atmospheric explanations of the meteor, and the significance of a range of geographical locations and networks, from villages to empires, to the pursuit of knowledge.[45] They each reveal the cultural capacity of landscape at this time as a medium of observation and speculation, experience and imagination.

Extreme events have been seen historically as heaven-sent opportunities to assess and improve knowledge of the world and its workings. Through the aperture of an instance in the historical record, they offer an opportunity for modern scholars to consider what nature means in particular times and places, to appreciate how nature is pictured and narrated.[46] Such studies of past geographies assume significance in the present as concerns about extreme environmental events, and the relations of climate and culture, again assume an urgent currency.[47]

## Acknowledgment

My thanks are due to John Bonehill, Georgina Endfield, Paul Elliott and Richard Hamblyn for their help.

## Notes

1 J. Grattan and M. Brayshay, "An Amazing and Portentous Summer: Environmental and Social Responses in Britain to the 1783 Eruption of an Iceland Volcano," *The Geographical Journal* 161, 1995, 125–34; R. Hamblyn, *Terra: Tales of the Earth*, 2009, London: Picador, pp. 65–121.
2 W. Cowper, *The Task, A Poem in Six Books*, London: J. Johnson, 1785, p. 48.
3 G. White, *The Natural History and Antiquities of Selbourne*, London: T. Bensley, 1789, p. 301–02.
4 M. Beech, "The Great Meteor of 18th August 1783," *Journal of the British Astronomical Association* 99, 1989, 130–4.
5 *London Chronicle*, 30 August 1783.
6 *Morning Herald and Daily Advertizer*, 18 August 1783; *St James's Chronicle* or the *British Evening Post*, 19 August 1783; *Whitehall Evening Post*, 23 August 1783; *London Chronicle*, 4 September 1783; *Gentleman's Magazine*, August 1783, pp. 711–12, 744, 795, 888.
7 *Morning Herald and Daily Advertizer*, 22 August 1783.
8 *Gentleman's Magazine*, August 1783, p. 744.
9 J. Golinski, *British Weather and the Climate of Enlightenment*, Chicago: Chicago University Press, 2007, pp. 41–76.
10 T.S. Feldman, "Late Enlightenment Meteorology," in T. Frangsmyr, J.L. Heilbron and R.E. Rider (eds), *The Quantifying Spirit in the 18th Century*, Berkeley: University of California Press, 1990, pp. 143–78.
11 V. Jankovic, *Reading the Skies: A Cultural History of the English Weather, 1650–1820*, Manchester: Manchester University Press, 2000.
12 R.J.M. Olsen and J.M. Pasachoff, *Fire in the Sky: Comets and Meteors, the Decisive Centuries, in British Art and Science*, Cambridge: Cambridge University Press, 1998.
13 R. Thoroton, *The Antiquities of Nottinghamshire*, Nottingham: G. Burbage, 1790, vol.1, 366–177; Village Scrap Book Volume 4, www.winthorpe.org.uk/Winthorpe-Hall (accessed August 8, 2009)
14 Roger Pocklington to Joseph Banks, letter, 12 March 1787, 3 July 1787, Yale University JSB930824/001.10787; JSB930824/001.10787, copies in Banks Archive, Nottingham Trent University. According to one London newspaper Claughton surprised "several of the first painters and engravers in town," but not it seems sufficiently to make the progress Pocklington hoped, who admitted "we in the Country are bad judges of the Performance." *Whitehall Evening Post* 24 April 1787.
15 On Robinson www.winthorpe.org.uk/the-school-in-winthorpe1, accessed August 8 2009.
16 Jankovic, *Reading the Skies*, pp. 125–42.
17 Jankovic, *Reading the Skies*, pp. 26–32; B. Martin, *The Young Gentleman and Lady's Philosophy*, vol. 1 (1781), 289–91.
18 *Nottingham Journal*, 30 August 1783.
19 S. Daniels, *Joseph Wright*, London: Tate, 1999. See also C. Klonk, "Science, Art and the Representation of the Natural World," in R. Porter (ed.), *The Cambridge History of Science: Eighteenth Century Science*, Cambridge: Cambridge University Press, 2001, 584–608.

20 N. Alfrey (ed.), *Trentside*, Nottingham: University of Nottingham, 2000.

21 Map of Winthorpe Lordship in the County of Nottingham, part of the Estate belonging to Roger Pocklington (1775), Nottinghamshire Archives Acc 5502.

22 A. Bermingham (ed.), *Sensation and Sensibility: Viewing Gainsborough's Cottage Door*, New Haven and London: Yale University Press, 2004.

23 L. Hawes, *Presences of Nature: British Landscape 1780–1830*, New Haven: Yale Center for British Art, 1980, pp. 35–9.

24 J. Bonehill and S. Daniels (eds), *Paul Sandby: Picturing Britain*, London: Royal Academy, 2009.

25 J. Roberts, *Views of Windsor: Watercolours by Thomas and Paul Sandby*, London: Merrell Holberton, 1990. The print states that it was drawn by Thomas, engraved by Paul, who also published the print from his London address, and the various preparatory drawings, from which the print is assembled, suggest that Paul may have had a hand in the figures.

26 Bonehill and Daniels, *Paul Sandby*, pp. 18–19, 26, 69–70, 178.

27 M. Craske, "Court Art Reviewed: the Sandbys' Vision of Windsor and its Environs," in Bonehill and Daniels (eds), *Paul Sandby*, 48–55.

28 J. Roberts, *Royal Landscape: The Gardens and Parks of Windsor*, New Haven and London: Yale University Press, 1997, pp. 53–61, 167; J. Martin Robinson, *Windsor Castle: A Short History*, London: Michael Joseph, 2006.

29 Bonehill and Daniels, *Paul Sandby*.

30 H.B. Carter, *Sir Joseph Banks 1743–1826*, London: British Museum, 1988, pp. 179–87.

31 Roberts, *Royal Landscape*, pp. 68–9.

32 S. Schaffer, "Natural Philosophy and Public Spectacle in the Eighteenth Century," *History of Science* 21 (1983), 1–43.

33 P. Sandby, *John Bul-loons Asses*, 1784, *Coleum Ipsum Petimus Stultitia*, 1784, *An English Balloon*, 1784, British Museum Prints and Drawings BMSat 6700,6701,6703. On-line database www.britishmuseum.org/research/search_the_collection_database (accessed August 5 2009).

34 T. Cavallo, "Description of the Meteor, Observed Aug. 18, 1783," *Philosophical Transactions of the Royal Society* 74 (1784), 108–11.

35 N. Maskelyne, *A Plan for Observing the Meteors Called Fireballs*, London, 1783.

36 S. Widmalm, "Accuracy, Rhetoric, and Technology," in Frangsmyr et al. (eds), *The Quantifying Spirit*, pp. 179–206.

37 C. Blagden, "An Account of Some Late Fiery Meteors; with Observations," *Philosophical Transactions of the Royal Society* 74 (1784), 201–32. This effectively upstaged Maskelyne's plan which was probably designed to confirm a theory of fireballs as terrestrial comets, and probably reflects the so-called "dissensions" within the Royal Society between mathematical men of science like Maskelyne and those who followed the dominant, gentlemanly Banksian model which included botany and antiquarianism. See D. Howse, *Nevil Maskelyne: The Seaman's Astronomer*, Cambridge: Cambridge University Press, 1989, p. 151.

38 P. Fara, *Sympathetic Attractions: Magnetic Practices, Beliefs and Symbolism in Eighteenth Century England*, Princeton, NJ: Princeton University Press, 1996.

39 T. Cavallo, *A Complete Treatise on Electricity*, 2nd edn, London: C. Dilly, 1782, p. 17.

40 Howse, *Nevil Maskelyne*, 149.

41 Blagden, "An Account of Some Late Fiery Meteors," pp. 231–2; on the imagination of high altitude and latitude, see D. Cosgrove and V. della Dora (eds), *High Places: Cultural Geographies of Mountains, Ice and Science*, London and New York: IB Tauris, 2009.

42 Jankovic, *Reading the Skies*, pp. 146–51.

43 P. Fara, "Northern Possession: Laying Claim to the Aurora Borealis," *History Workshop Journal* 42 (1996), 37–57.

44 A. Agnarsdóttir, *"This Wonderful Volcano of Water": Sir Joseph Banks, Explorer and Protector of Iceland, 1772–1820*, London: Hakluyt Society, 2004, pp. 21–3; J. Gascoigne, *Science in the Service of Empire: Joseph Banks, the British State and the Uses of Science in the Age of Revolution*, Cambridge: Cambridge University Press, 1998, pp. 174–6.

45  C. Withers, *Placing the Enlightenment: Thinking Geographically about the Age of Reason*, Chicago: Chicago University Press, 2007.

46  D. Lowenthal, "Geography, Experience and Imagination: Towards a Geographical Epistemology," *Annals of the Association of American Geographers* 51.3, 1961, 241–60; for a discussion of the cultural geography of destructive extreme events see J.N. Entrikin, "Place Destruction and Cultural Trauma," in I. Reed and J. Alexander (eds), *Culture, Society and Democracy: The Interpretative Approach*, Boulder and London: Paradigm, 2006, pp. 163–79.

47  S. Daniels and G. Endfield, "Narratives of Climate Change," *Journal of Historical Geography* 35 (2009), 215–22.

# 16

## READING LANDSCAPES AND TELLING STORIES

### Geography, the humanities and environmental history

*Diana K. Davis*

Landscapes are like libraries whose information is ignored by most academics.[1]

Just as reading a landscape can tell us much about its past uses, reading and taking seriously stories about landscapes over time can provide a window on human–environment relations which are at the heart of geographical inquiry. By studying environmental history, one can learn how different cultures view, interpret, and represent various environments, how they value and use them, and even how they perceive and understand the humans and non-humans who live in these environments. As Kate Showers demonstrates in *Imperial Gullies*, much harm can be done to people and to the environment when those in power misread, or perhaps worse, misrepresent landscapes and their histories. Studying environmental history thus often reveals as much about social and political negotiation and conflict over the environment as it does about environmental change. In "doing environmental history" from a geographical perspective, geographers can and do engage with the humanities in several important ways. Geographical environmental histories can also help to unravel the complex and subtle relationships between the humanities and environmental science, for example the development of potential vegetation maps, relationships influential today in ways that impact both people and the environment. An examination of forests planted where they have likely not existed since the last ice age, for instance, can reveal a great deal about cultural ideals and the history of ecological science at the same time that it may expose political, economic, or social aspirations of many different actors.

Environmental historians use many tools and sources to understand environmental change and its context. We analyze biophysical data from scientific sources to try to reconstruct histories of vegetation change including pollen cores, tree-ring studies, sedimentation studies, charcoal deposits, etc. We also need to understand what various people and cultures believed about the environment and environmental change over time and how and why they acted on those beliefs. We trace such narratives of environmental change by analyzing many different kinds of texts ranging from archival documents, letters of influential people, botanical and agricultural treatises, to forestry manuals, to name just a few. The humanities also provide a rich source of material for environmental history research, especially national and regional histories, literature (novels and stories), and the arts. Some environmental narratives, for example, may be traced back to the writings of the ancient Greeks and Romans. The visual arts provide sources of information for environmental history as well through depictions of nature that convey general cultural views of the environment and the peoples living therein. This is especially true of landscape painting. Environmental historians benefit from using the tools of the humanities to analyze these visual and textual narratives.

Many environmental historians endeavor to tell stories about others' stories of nature and environmental change over time in order to make clear the political, economic, scientific, and social ramifications of what is widely accepted by society at large as "environmental history." This is because environmental history today – that is, the most commonly accepted current story of environmental change – frequently has serious and widespread consequences since it forms the base on which myriad environment and development programs are planned. If the environmental history is misrepresented, not only people, but also the environment, often suffer(s).

The environmental history of the Maghreb (Algeria, Tunisia, and Morocco) provides a useful example of the value of studying environmental history from a geographical perspective that engages with the humanities. An analysis of the evolving account of the Maghreb's environmental changes under French colonial administration (from 1830 to about 1956) shows the profound importance of the humanities, especially the classics, histories, and the visual arts, in the creation of a colonial narrative of environmental change and also shows their usefulness in supporting and propagating the colonial narrative once it was firmly established. This examination of the French story of nature in North Africa likewise demonstrates the complex and long-lasting legacies of the humanities for contemporary questions of environmental degradation and for pressing debates over sustainability.

The most widely accepted environmental history of the Maghreb today was created during the French colonial period, derived primarily from literary sources – especially ancient Greek and Roman writings about Africa. It is a false story of environmental decline at the hands of the "invading Arab nomads" and their voracious herds. It was wielded like a weapon during the heyday of French colonialism in North Africa from the 1860s until independence to justify and facilitate imperial goals.[2]

Before the French occupation of Algeria in 1830, though, the consensus of opinion in Europe, including France, was that North Africa was a fertile and beautiful land that was not quite living up to its potential due to inept Turkish administration and the inherent laziness of the "natives." Importantly, not even the nomads were perceived as destructive in these early views. The agricultural production of North Africa was greatly admired and, in fact, France and other European countries relied on the importation of large amounts of grain from the region. This view of

North Africa as an abundantly fertile and pleasant land was drawn from readings of ancient texts by, for instance, Herodotus, Pliny the Elder, and Ptolemy, parts of whose work extolled the spectacular fertility of North Africa and described thick forests and abundant game. This literary information was bolstered by first-hand accounts by European and French travelers, emissaries, and religious figures who had visited or lived in the Maghreb during the eighteenth century.

Within just a few years of the French occupation of Algeria this environmental story of North Africa began to change to one that blamed the native North Africans, especially the nomads, for degrading the environment. The nomads were accused by the French of overgrazing, deforesting, burning, and turning Algeria into a desert. Beginning in the 1830s this story was developed and used to justify sedentarizing nomads, and appropriating forests, agricultural, and grazing lands, all of which furthered imperial goals in Algeria. This colonial environmental narrative included Tunisia and Morocco and was later carried to these territories as they were conquered, and applied in very similar ways.

The evidence used for this new, colonial environmental history of North Africa drew heavily and very selectively from the writings of medieval Arab historians including Idrissi, El Bekri, Ibn Khaldoun, and Leo the African to make the case for wanton destruction by the "invading Arab nomads," the Beni Hillal, in the eleventh century.[3] These passages describing the alleged environmental destruction of the invading Beni Hillal were contrasted with the Greek and Roman descriptions of the incredible fertility of North Africa and its thick forests. From this, the French deduced that the descendants of the Beni Hillal had continued to deforest and overgraze North Africa to the present. They incorrectly assumed that the contemporary nomads had simply continued the destructive ways of their ancestors and that the sparse forests, vast grassy plains, and desert areas visible in the landscape were the result of 800 years of continuous degradation. The paleoecological record, in fact, does not support the existence of vast, thick forests in the Maghreb for the last several thousand years and does not record the massive deforestation assumed by the French during the colonial period. Most documented deforestation actually resulted from colonial activities from the 1890s into the 1940s. Moreover, the local vegetation is highly adapted to drought, grazing, and fire and there is no evidence that the North African environment at the time of occupation was degraded in any significant way.[4]

The colonial environmental narrative, then, was constructed primarily from humanistic sources including the writings of the ancient Greeks and Romans and medieval Arab histories. The picture derived from these sources was compared with what the French thought they saw in the landscape, a poor, degraded remnant of a previously lush and fertile land. The final part of the narrative included the need for the French to restore the previous fertility of North Africa, to quite literally resurrect the granary of Rome. It became commonplace for government and popular writers to argue that it was the duty of France to reforest North Africa to bring back the rains. It became equally common to claim that since the North Africans, especially the nomads, had ruined the land, the land must instead be tended (and owned) by French and other European colonists. Thus the French narrative was used to justify the expropriation from North Africans of their lands and resources. This provided the large amounts of agricultural land needed by the settlers and later the industrial agricultural companies that came to dominate Algerian agricultural production. The narrative was also used to justify the appropriation of nearly all of the forests and much of the grazing land by the colonial state. These actions quickly impoverished and marginalized the majority of the indigenous population, a condition that remained until, and indeed after, independence.

By the 1860s, when the colonial narrative was well formed and widely used in Algeria, it began to be reflected in influential works of art and literature that were welcomed by French audiences. Many of the works of the artist Gustave Guillaumet (1840–87), for instance, were "positively received by the state as illustrative of its Algerian policies."[5] One of Guillaumet's most influential paintings was *Le Sahara* painted in 1867 (Plate VI). This large and beautiful painting depicts a faintly visible group of people with camels, likely representing nomads, leaving a desert landscape. In the foreground the parched bones of a long-dead camel are prominently displayed. The only vegetation to be seen is a few dry sticks and tufts that suggest overgrazing. Adding to the negative connotations of the painting is the image of a downed camel and rider, presumably callously left to die by the departing nomads.

Although several interpretations of this painting exist, it may be seen to illustrate the destructiveness and brutality of the nomads, an image that likely provided the French government broader public support for their Algerian policies including sedentarization of the nomads. The scene is set at sunset, strongly suggesting that the nomads' way of life must end. This painting was very well received at the Paris Salon of 1867, just a few years before the Algerian insurrection was brutally crushed by the French, paving the way for the ensuing period of hegemonic rule by the settlers in Algeria. It quickly became one of the key works in the "Orientalist" school of painting. The influence of this and similar paintings helped to propagate the colonial environmental narrative and to garner public support for colonial activities and the implementation of related laws and policies in Algeria.[6]

Travel literature was also influential in shaping French public opinion in support of French policies in colonial Algeria. A good example is found in the writings of author and artist Eugène Fromentin (1820–76), particularly his 1857 *Summer in the Sahara*. This book portrayed the Algerian landscape in ways that reflected the colonial narrative. The final sentence of the book, "I will salute with a profound regret this menacing and desolate horizon that has been so precisely named – *Land of Thirst*" also provided the title of one of Fromentin's influential paintings: *The Land of Thirst (Au Pays de la soif)*.[7] (Plate VII)

This painting shows a group of five men and a horse, all dead or dying, in a rough, stony desert, with not a single trace of vegetation. Frequently interpreted as a call for France to rescue an expiring Algeria, it also depicts the colonial environmental narrative particularly well. Fromentin was well known for his opinion that the Algerians hated trees. The landscape shown here has been so deforested and overgrazed that it can no longer support human or animal life. Although painted in 1869, only two years after the appearance of Guillaumet's *Le Sahara*, it was not available to the public until it was used as an engraving in the illustrated version of Fromentin's book in 1879.[8] Published posthumously, this illustrated volume was extremely popular and went through multiple editions over the years, the most recent being an English translation in 2004.[9] Both Fromentin's book and his painting, like Guillaumet's *Le Sahara*, portrayed the Algerian desert as a ruined landscape in ways that helped to naturalize the environmental story that was being utilized in Algeria to bring about many changes which benefited the settlers and the colonial administration.

One of the most tenacious legacies of the colonial environmental narrative of the ruin of the North African environment has been the way in which it has informed scientific endeavors during the twentieth century, both during the colonial period and since. In the early years of the French occupation of Morocco (1912–56), the French narrative, largely informed and justified with humanistic sources as we have seen, came to be incorporated into the young science of plant ecology. Through the work of men like the head of the Moroccan Forest Service, Paul Boudy (1874–1957), and the prominent ecologist Louis Emberger (1897–1969), this spurious environmental history was used to develop natural vegetation maps and calculations of deforestation in Morocco and across North Africa.

The methods used to generate these maps and statistics, however, relied on the subjective identification of what the "natural" vegetation of a given region was (or should be). The ways in which such natural vegetation was deduced relied on individual, "expert" opinions of what the dominant vegetation was in a particular zone, or region. In Morocco, for example, Emberger defined all the major vegetation zones by their trees – even in regions with few or no trees. He reasoned, based on the dominant environmental narrative that had been taught for decades, that since North Africa was so egregiously deforested the few clumps of trees seen in a few places in otherwise tree-free areas were the relics of former forests. He deduced, therefore, that these areas were formerly forested and that the "natural," or potential dominant vegetation was trees.[10] He created maps in the 1930s and 1940s that indicated this potential vegetation for Morocco and later Algeria and Tunisia. These maps, using circular logic, were then used to "prove" the fact of deforestation in North Africa.[11] That is, having been constructed using the narrative, the maps were then used to support the pre-existing French colonial environmental narrative of North Africa. They are still invoked today for similar reasons, as are Boudy's deforestation statistics for the Maghreb published in the 1940s and 1950s. Both continue to be used for the formulation of numerous environmental projects in Morocco and across the Maghreb that have questionable ecological effects and, all too often, deleterious social consequences.

This humanistically informed French environmental history was so dominant at the end of the colonial period in the Maghreb that it persisted without question for those in the post-colonial Maghreb and for most scholars and others in France. This is in large part because it was institutionalized in the apparently scientific work of Emberger's (and others') ecological treatises and maps as well as in forestry tomes like Boudy's magisterial four-volume series on forestry in North Africa.[12] It was also written in myriad agricultural tomes, histories of the region, and more recent ecological literature. Many of these writings were used in the education not only of young Maghrebis but also of young French students, thus ensuring the propagation of the narrative.

So hegemonic was this colonial story that in the 1970s a French scientist of historical vegetation change published a series of pollen cores that appeared to show significant deforestation coinciding with the eleventh-century "Arab invasion." It is very telling, though, that his samples are neither carbon-dated nor dated in any other scientific way. Rather, it seems that he inferred the dates for his pollen core analysis from the colonial narrative, concluding that "from the tenth century onwards the Arab invasion has been the indirect cause of an intensive deforestation."[13] His interpretation is perhaps understandable given the dominant force of this narrative over the last 150 years or more. His work, however, has further institutionalized the colonial narrative in a significant way because it has been cited more often by scholars of environmental history on North Africa than other scientific literature on historical vegetation change that is carbon-dated and does not show widespread massive deforestation in North Africa over the last two millennia.[14]

The frequency with which this colonial environmental history appears in contemporary national, international and NGO (non-governmental organization) environmental project documents for the Maghreb farther attests to the staying power of this inaccurate environmental history.[15] It also demonstrates the long-lasting legacies of the humanities for some contemporary environmental problems and their solutions. My analysis of the French colonial environmental history of North Africa suggests that incorporating study of the humanities into what is all too often a "scientific" approach to studying environmental change can reveal important information to improve our understanding of environmental history. It may also help to provide better understanding of some of our contemporary environmental problems. In this instance, incorporating the humanities in environmental history research helped to locate serious weaknesses not only in the standard history, but also in the ecological science itself used in the history. Such inconsistencies in the science and the history would likely otherwise have remained invisible. Taking the humanities seriously, in this case art, literature, and a variety of histories including the classics, provided significant new windows of understanding on the evolution and use of the standard stories of environmental change in the Maghreb dominant over the last 150 years. Geographers are increasingly engaging with the humanities in a variety of interesting and innovative ways, as this volume shows. In doing so, I believe we may be able to write better environmental histories and to help design environment and development projects that do less ecological and social harm in the future.

## Notes

1 K. Showers, *Imperial Gullies: Soil Erosion and Conservation in Lesotho,* Athens, Ohio: Ohio University Press, 2005, p. 1.

186Diana K. Davis

2 For details, see D.K. Davis, *Resurrecting the Granary of Rome: Environmental History and French Colonial Expansion in North Africa*, Athens, Ohio: Ohio University Press, 2007.

3 The writings of these Arab historians are actually quite complex and many of them also praised nomads in parts of their writings not cited by the French in the colonial environmental narrative. For details, see Davis, *Resurrecting the Granary*, pp. 54–7.

4 See Davis, *Resurrecting the Granary*, chapter 1 and pp. 177–86.

5 A. Delage, "Guillaumet, Gustave," in Grove Art On-Line, http://www.groveart.com.

6 For a discussion of French Orientalist landscape painting and its relation to the colonial environmental narrative in Algeria, see Davis, *Resurrecting the Granary*, pp. 68–71.

7 E. Fromentin, *Sahara et Sahel*. Illustrated edition, Paris: E. Plon, 1879, p. 188.

8 This book, *Sahara et Sahel* (see note 7) included the original *Summer in the Sahara* paired with the later *Summer in the Sahel*.

9 See E. Fromentin, *Between Sea and Sahara: An Orientalist Adventure*, London: Tauris Parke Paperbacks, 2004.

10 The extensive problems with interpreting "relict vegetation" have been discussed by numerous authors. For two particularly well-known examples, see J. Fairhead and M. Leach, *Reframing Deforestation: Global Analyses and Local Realities: Studies in West Africa*, London: Routledge, 1998; W. Cronon, *Changes in the Land: Indians, Colonists, and the Ecology of New England*, New Haven, CT: Yale University Press, 1983.

11 For details, see Davis, *Resurrecting the Granary*, Ch. 5.

12 P. Boudy, *Économie forestière nord-africaine*, vols 1–4, Paris: Larose, 1948–58.

13 M. Reille, "'Contribution pollenanalytique à l'histoire holocène de la vegetation des montagnes du Rif (Maroc septentrional)," *Supplément au Bulletin AFEQ* 1.50, 1977, 53–76. Tree-ring studies from the Rif mountains show that there was a severe drought in approximately 1100–1250 CE and it is more likely this drought that the pollen cores showed rather than the "Arab invasion." See C. Till and J. Guiot, "Reconstruction of Precipitation in Morocco since 100 A.D. Based on *Cedrus atlantica* Tree–Ring Widths," *Quaternary Research* 33.3, 1990, 337–51.

14 See, for a detailed discussion of the available paleoecological data, Davis, *Resurrecting the Granary*, pp. 9–12. A few French scientists have begun to question and even to contest the colonial environmental narrative. See the work of Anik Brun, including "Étude palynoloque des sediments marins holocènes de 5000 B.P. à l'actuel dans le golfe de Gabès," *Pollen et Spores* 25(3–4), 1983, 437–60, and "Microflores et paléovégétations en Afrique du nord depuis 30,000 ans," *Bulletin de la Société Géologique de France* 8.1, 1989, 25–33.

15 For numerous examples, see Davis, *Resurrecting the Granary*, Ch. 6.

# 17

# PARTICIPATORY HISTORICAL GEOGRAPHY?

## Shaping and failing to shape social memory at an Oklahoma monument

*Dydia DeLyser*

In December of 2004 a stranger who had discovered my research on the internet called my home phone in Louisiana to speak about what the stranger referred to as a "local history problem" in the town of Edmond, Oklahoma — a place I had never visited or even heard of. As the stranger explained it, a proposed monument to a local historical figure was going to "distort history" by commemorating an event that never occurred: the monument would show one-time area resident Nannita "Kentucky" Daisey leaping from the cowcatcher at the front of a train to claim land in Oklahoma's first predominantly white-settler land run of 1889. This, as the stranger explained it, appeared to be a locally prominent exaggeration of what had really happened, and the monument, to be built with locally raised funds and the support of the City government, would now commit this false version of the story to bronze. The stranger — a self-described "history buff" whom I'll call Pat but whose gender I will not identify — asked me if I, as a credentialed scholar, could help generate what Pat termed "publicity" about how such mythologized stories develop and get perpetuated, and if I would join Pat's campaign to set the record straight. Since the topic basically fitted with my research on the mythologizing of the American West and the ongoing creation of American social memory, I agreed to look into it. And so began some five years of what I'll call "participatory historical geography," where I, now the stranger in town myself, worked collaboratively with Pat on a project of local interest in Pat's community.[1]

As we worked together I came to trust and rely on Pat's dedicated and exhaustive research, and Pat came to trust and rely on my research — with dramatically different backgrounds, skill sets, access, networks, and job statuses, our abilities complemented one another, and we shared all of our research materials with one another. At great length, and over literally hundreds of phone and written conversations as well as multiple in-person visits, Pat and I discussed different interpretations of the materials we'd uncovered, different possible understandings of what had "really happened" and of how the story had come to be told, and different analyses of what the story meant. Though we agreed to differ on a few minor points, our discussions led us to a deeply shared understanding and interpretation of the person and the events we were researching.

Pat had no interest in writing or publishing, ardently wished to remain anonymous, and had contacted me expressly in the hopes that I would write and publish about the monument and the story behind it. So when it came time to write, I wrote but I also shared multiple drafts of my work with Pat, and Pat gave me invaluable feedback and insights into it, meticulously fact-checking every detail. Before submitting a journal-article manuscript for publication, I made sure Pat approved of how I presented the story in written form, and also carefully obtained Pat's approval of how I presented Pat in the story.

So viewed, this collaborative research fits closely within a long-accepted model of humanities scholarship: grounded in close readings of historical texts, and put forward in an interpretive mode.[2] But the work was also explicitly dedicated to community-based social change, in this case specifically to consciousness-raising about an historical issue. That makes my work with Pat fit within a model of humanities scholarship prominent at least since the 1960s, where advocacy in teaching and research has been sufficiently strong that the works of some scholars cannot be seen separately from "the hopes these scholars cherish [...] for their society and the ambition they harbor [...] for their scholarly work as a possible influence."[3]

In my case, I chose to understand our work together as "participatory research" even though the "object" of our study lay in the past. Seen in this way, though Pat was not really the "subject" of the research (as is the case in most participatory research), my interactions with Pat fulfilled the three "core principles" of participatory research in including the participants, creating co-ownership of the research process, and relying on participants' local knowledge.[4] Though Pat refused co-authorship in favor of anonymity, we engaged in the research together. My goal, even as the single author of the manuscript, was to enable Pat's knowledge and allow for its "extension, enhancement, and analysis" through our research and my writing, writing which I sought to make both empathetic and distanciated – combining my academic skill and training with Pat's local knowledge and understanding.[5]

In March of 2006 I submitted a journal-article manuscript that detailed what, to the best of our knowledge at that time, Pat and I deemed to be the actual history behind the myth portrayed by the planned monument, and that outlined just how we understood that that myth had arisen and come to be perpetuated. In October of 2006 I also wrote a Wikipedia page outlining the story and made sure that site would come up first on Google searches about the monument. The journal article, after two rounds of revisions and waiting in the publication cue, was published in January 2008.[6] But the statue (Figure 17.1) had been completed and unveiled in July 2007.

Initially that led me to believe that my project had "failed," because the scholarly work I had been enlisted to do had failed to change the statue, and had failed to

**FIGURE 17.1**   The completed statue of Nannita Daisey, '*Leaping into History*' showing her mythic jump from the cowcatcher, here with petticoats flying. Photo by "Pat."

even appear before the statue was unveiled. In this chapter, I discuss my attempt at participatory historical research, question whether or not this project should be seen to have failed, and ponder what that might mean for humanities-oriented (historical) geography.

Failure or no, it seems obvious that most participatory-action research, because of its active engagement with local communities, be contemporary. Since we who engage in historical research are most often interested in those already dead, it seems obvious too that historical work could wind up excluded from the participatory-action rubric. But, since the past is more than just a series of actual past events, but is also a fluid construct, remembered anew in each new present, historical research too can serve progressive ends, with goals similar to those of much partici-patory-action research. These typically include bringing a community-engaged, action dimension into academic work that may otherwise be seen as quietist; having a liberatory agenda seeking to empower local people and reverse discriminatory or exclusionary practices and/or policies; and engaging in collaborative work that gives voice to research participants and validates their local knowledge (rather than only the expert knowledge of the academic).[7] In what follows I describe the participatory historical research I undertook together with Pat, and show how historical geography, in this case an active narrative intervention, can fit models of participatory-action research – regardless of whether our project succeeded or failed.

I begin here by describing Nannita Daisey and her now-completed monument, along with the ways that my collaborative research with Pat sought to understand the story behind what the monument was designed to depict, and the establishment of a mythology behind Nannita Daisey's life story. Of course, though few geographers have seized opportunities to study monuments not yet completed,[8] we have long sought to understand the roles and places of monuments in both the past and the present.[9] Most often such works have detailed narratives about a monument already built as it moves through different times, examining a monument's meaning(s) for different people in different eras, in the particular place where the monument was built.[10] In this case, I (with Pat's help and support) strove to set the (to the best of our knowledge) actual story of one woman's life in the broader context of a more nuanced and multicultural telling of Oklahoma's history. In other words, I deliberately sought not only new facts, but also a new narrative context for the Daisey story and monument.

## Daisey's story

When Pat and I began our research, little was known about Nannita Daisey, beyond her alleged leap from the cowcatcher. By searching land-office, newspaper, census, and probate records as well as marriage-, birth-, and death-records in multiple cities and states, we pieced together a telegraphic biography.[11] Nannita Regina H. Daisey was born in Pennsylvania in 1855 and orphaned as a child.[12] She moved west to Saint Louis where she earned an education, eventually traveled to Kentucky, and then to the lands known as "Indian Territory" in the contemporary state of Oklahoma, to teach school. In Kentucky she had begun also working as a journalist, and she took that profession, then unusual for women, with her to the Territory.[13] Daisey did claim land near Edmond in the 1889 land run and was reported to be active in subsequent land runs as well.[14] By 1890 she had married a Scandinavian immigrant then serving as a US soldier in the Territory, Andreas E. J. Ueland Svegeborg.[15] Daisey proved up on her claim outside of Edmond, and built a small house on the property.[16] Later she moved to Chicago, where she died in 1903; after her death, her Edmond property was sold.[17]

But the lingering point of interest about her life has been her dramatic original land claim. A witness to the original event (her editor) declared that she jumped from the coach platform at the front of the first passenger car, rushed to the site of her claim, threw her cloak over a pole she planted in the ground to demonstrate "improvements" on her site, fired her revolver in the air to announce her claim, and then re-boarded the train before the final car passed her by.[18]

Three years later Daisey herself described the event: "I stood in with the engineer" (but not on the cowcatcher), jumped from the train, and "planted my stakes [to mark the claim], threw my cloak over one, then fell on my knees and discharged my revolver in the air."[19]

These details are indeed dramatic, but they are not what is commemorated in the Edmond monument. Those, Pat and I learned, emerged later, after Daisey's death, when an obituary and subsequent stories embellished the tale, claiming that she was "sitting gracefully" "on the cow-catcher," before her leap, and that, to demonstrate the improvements on her claim, she removed not her cloak but her petticoat, which she "quickly tied ... to a nearby bush."[20] And thus, we confirmed that two of the most compelling details of the story were either fake, or just not mentioned by the only known published eyewitness accounts.

Indeed, we found that the exaggerated version of the story was retold in various newspapers in subsequent years.[21] Much more influential though was its repetition in books about Oklahoma's history. In fact, without exception, every book about Oklahoma's history, Edmond's history, the history of women in Oklahoma, and the history of Oklahoma's land runs that mentions Daisey at all has repeated the exaggerated version of her story. Each book, to the degree it was referenced at all, referenced previous books that told the exaggerated version of the tale; no one returned to quote the original accounts.[22]

The embellishments I believed were telling in themselves, for they helped me to set Daisey's story in the context of the mythologizing of the American West, and of westerners, both men and women. Though some townspeople and a local newspaper had lauded the statue as a "progressive" undertaking,[23] an understanding of women's history and a more multicultural understanding of the history of the US West yields a rather different interpretation.

For decades feminist scholars have worked to include women's stories in narratives of US western history, and in so doing have brought to light tales of both ordinary and extraordinary women.[24] As such new narratives have traveled from academia to broader audiences, the more public versions have often focused on heroicized accounts of women's lives.[25] Significantly, in landscape, monuments to "women pioneers" have often sprung from the heroicized accounts.[26] Thus, I realized, the Daisey monument could be set in a broader, more multicultural context of US western historical geography.

## Oklahoma's land runs

The tale of Oklahoma's occupation by predominantly white settlers has long been told as a heroic one, focused around several "land runs" that "opened" the land to settlement by non-Indians. In fact though, those land runs occurred in an area dubbed "Indian Territory" on land once set aside in perpetuity by the US Federal Government for Native Americans, and on land where, beginning in the 1830s, the Government had forcibly resettled Native Americans from all over the US. Later, when pressure from railroad companies and prospective non-Indian settlers mounted, the Government forced the Native peoples out of various parts of that land, and, on April 22, 1889, declared a two-million-acre area called the "Unassigned Lands" open for non-Indian settlement under the Homestead Act. Prior to the "opening," parcels were laid out across the Lands, and at noon the borders were thrown open to non-Indians eager to claim land. Some 50,000 rushed to claim for free what had previously been set aside for Native Americans. So many reporters from around the country covered the land run that on the first train the very front car was reserved for the press. With such access, some sought to do more than just report, and one of those was Nannita Daisey herself, who worked her land-claiming jump into her reporting duties. Thus, with betrayal and displacement began the first of Oklahoma's now-famous mainly white-settler land runs, deeds now commemorated by the Daisey monument.[27]

Despite such grounding, the telling of dramatic and heroic tails of land claiming has been widespread, reaching then and now far beyond accounts of Daisey's claim. In other words, the reporting about Daisey, while dramatic sounding, was consistent with the reports of other events and people, particularly men, during the land runs.[28] Indeed, at one time the history of the US West was told as a triumphal westward march of white males.[29] Women (along with non-whites, and members of other minority groups) were either absent or granted lesser (or opposing) roles. For women, as feminist historians Joan Jensen and Darlis Miller observed, their portrayals commonly fell into one of four stereotypes: gentle tamers, sunbonneted helpmates, soiled doves, or hell-raisers. And these stereotypes are still common today, not for their accuracy, but for their perpetuation in narratives about the American West – in books, magazine, and newspaper articles, as well as on television and in film. The hell-raiser, according to Jensen and Miller, drew the most attention: the "super cowgirls, the Calamity Janes, who acted more like men than women and became the heroes of dime novels and wild west shows." Feminist scholarship has moved well beyond such images, but as the unfolding of Nannita Daisey's story demonstrates, in popular media, and in landscape, such stereotyped images of western women persist.[30]

But it comes as no surprise that such legendary figures as Nannita Daisey should arise, for, as scholars of the US West have pointed out, for more than one hundred years the American West has been "the most strongly *imagined* section of the country" existing as a real region but just as importantly as a mythic place. Along with their region, westerners, men and women, white people and people of color, have long been widely mythologized. It began in the nineteenth century, when many of the subjects of mythologized accounts were still living, and only later grew more intense. Some mythologized their own lives, others had it largely done for them.[31]

In the case of Nannita Daisey, from the moment that she first leapt from the coach platform her exploits were chronicled – and mythologized – by others; the dramatic reporting of her land claiming made virtually all else she did worthy of another story. Even a break-in at her home made the papers.[32] But while (white) male characters in the tales of the mythic West were hyper-masculine figures,[33]

Daisey would be made to fit the western-woman-as-hell-raiser stereotype. By the time of her obituary that narrative had been well established, and the author described her as "an eccentric character."[34]

Today, the monument depicting Daisey's fictional leap from the cowcatcher shows that such mythologized and stereotyped constructions of gendered identities may still have broad public purchase.

## Participatory historical geography

My writing on the Nannita Daisey monument, following established models of feminist scholarship, had endeavored to tell her story – as much as was possible given surviving historical records – in her own words. It endeavored also to fulfill my commitments to Pat and the project Pat enlisted me in, my writing also strove to address the "local history problem" Pat had alerted me to, and to set straight the distorted history long put forward about Daisey's leap from the cowcatcher. As an historical geographer interested in social memory, I hoped my scholarly skills and background might also offer a broader context for the way events of the 1880s were being commemorated in the twenty-first century. In uniting these aspirations, I was fulfilling exactly what Pat, once a stranger, had asked me, as a skilled and credentialed scholar, to do back in 2004: my published article drew to the fore the (to the best of our knowledge actual) historical details of Daisey's land claiming, and also set a new narrative context for Daisey's widely mythologized story.

But it has been challenging. Partly just because participatory research takes a great deal of time: as others have noted, commitment to a project or group over the sustained time period required of participatory-action-oriented projects, where long-term relationships must be built and sustained, can involve more than most academics can work into their schedules;[35] partly also because it involves fulfilling the needs of at least two groups – needs that don't always go together. It is also challenging because of the glacial speed of academic publishing, which meant, in this case, that despite our efforts the article didn't appear until *after* the monument debuted. At first, I was disappointed that my published writing had not even had a chance to change the monument – it seemed that my collective efforts with Pat had failed to achieve the social change Pat had enlisted me to help bring about. And so I began to conceptualize this as a "failed research project."[36]

However, it's also true that my involvement came too late for real changes in the monument – at my first trip to Edmond in 2005 its design had already been approved, funding had been secured, and parts were on their way to the foundry. All I could accomplish would be a change in the way people would come to *understand* Nannita Daisey's story and the monument, *after* the monument had been built. So perhaps my initial sense of failure came through my own engagement with a particular notion of participatory research where the goal is immediate, visible, proactive, and community-motivated liberatory change.[37]

Instead, I've now come to understand what I call *participatory historical geography* as an effort to work with communities to help them understand their past(s), and

have come to see that the impacts of such new understandings may not take immediate effect. For perhaps our efforts (mine and Pat's) to shape Edmond's social
memory based upon a more accurate telling of Daisey's story will be reflected in
books and other publications yet to come. And perhaps my efforts to set our work
in the narrative context of the mythologizing of the US West and the works of
feminist scholars will offer context also for those in the future who seek to understand the Daisey monument. Perhaps, I've come to think, it's not for me to rule our
participatory historical research either a success or a failure, and it's not for me to
decide what the impact of our work is or will be. For perhaps, unlike in much
contemporary action research, in the case of participatory historical research like
ours, the effects of our research and interpretations of Daisey's story and the Daisey
monument will be left for others to assess. Perhaps, too, the impacts of our efforts
have not yet been fully achieved, for though history is understood as events of the
past, its understanding in social memory is the result of always-ongoing efforts, and
always-emergent narratives.[38] And if our work revealed the narrative efforts in
constructing Daisey's mythologized story, it also deliberately fashioned a new narrative, though one no less culturally constructed and contextualized. Who knows,
then, what still-newer narratives about Daisey may emerge in the future.

## Notes

1 Pat did not represent a formal community group, but the project Pat spearheaded had
 support from historically oriented community members and eventually also local journalists, and Pat was well networked in local historical circles. The monument also had
 broad community support. Thus, both sides of the issue had community advocates.
2 W.G. Bowen and H.T. Shapiro, "Foreword," in W.G. Bowen and H.T. Shapiro (eds),
 *What's Happened to the Humanities?*, Princeton, NJ: Princeton University Press, 1997,
 pp. vii–viii.
3 D. Browmwich, "Scholarship as Social Action," in Bowen and Shapiro (eds), *What's
 Happened to the Humanities?*, pp. 220–43; quoted p. 220.
4 M. van der Riet, "Participatory Research and the Philosophy of Social Science: Beyond
 the Moral Imperative," *Qualitative Inquiry* 14.4, 2008, 546–65; quoted p. 551.
5 Ibid., quoted p. 555.
6 D. DeLyser, "'Thus I Salute the Kentucky Daisey's Claim': Gender, Social Memory, and the
 Mythic West at a Proposed Oklahoma Monument," *Cultural Geographies* 15.1, 2008, 63–94.
7 See P. Reason and H. Bradbury (eds), *Handbook of Action Research*, London: Sage, 2001;
 the aims of action research are outlined in their "Preface," pp. xxiii–xxxi.
8 See, Burke, "Private Griefs, Public Spaces" and Burke, "In Sight and out of View" for
 examples of those who have. A number of others have insightfully used participant
 observation and interviewing to more broadly study places of memory as they are being
 created, including, L. Lees, "Towards a Critical Geography of Architecture: The Case of
 an Ersatz Colosseum," *Ecumene* 8, 2001, 51–86; D.H. Alderman, "A Street Fit for a King:
 Naming Places and Commemoration in the American South," *The Professional Geographer*
 52, 2000, 672–84; A. Charlesworth, "A Corner of a Foreign Field that is Forever
 Spielberg's," Till, *The New Berlin*.
9 D. Harvey, "Monument and Myth," *Annals of the Association of American Geographers*
 69, 1979, 362–81; N. Johnson, "Sculpting Heroic Histories: Celebrating the Centenary
 of the 1798 Rebellion in Ireland," *Transactions of the Institute of British Geographers* NS 19,
 1994, 78–93; and N. Johnson, "Cast in Stone: Monuments, Geography, and Nationalism,"

*Environment and Planning D: Society and Space* 13, 1995, 51–65. James Duncan's book, *The City as Text: The Politics of Landscape Interpretation in the Kandyan Kingdom*, Cambridge: Cambridge University Press, 1990, also explores monuments though not as its primary focus.

10  See, for example, R. Peet, "A Sign Taken for History: Daniel Shays' Memorial in Petersham, Massachusetts," *Annals of the Association of American Geographers* 86, 1996, 21–43; Å. Boholm, "Reinvented Histories: Medieval Rome as Memorial Landscape," *Ecumene* 4, 1997, 47–272; D. Atkinson and D. Cosgrove, "Urban Rhetoric and Embodied Identities: City, Nation, and Empire at the Vittorio Emanuele II Monument in Rome, 1870–1945," *Annals of the Association of American Geographers* 88, 1998, 28–49; B.S. Osborne, "Constructing Landscapes of Power: The George Etienne Cartier Monument, Montreal," *Journal of Historical Geography* 24, 1998, 431–58; K.E. Till, "Staging the Past: Landscape Designs, Cultural Identity and *Erinnerungspolitik* at Berlin's *Neue Wache*," *Ecumene* 6, 1999, 251–83; B. Forest and J.E. Johnson, "Unraveling the Threads of History: Soviet-Era Monuments and Post-Soviet National Identity in Moscow," *Annals of the Association of American Geographers* 92, 2002, 524–7; J. Leib, "Separate Times, Shared Spaces: Arthur Ashe, Monument Avenue and the Politics of Richmond, Virginia's Symbolic Landscape," *Cultural Geographies* 9, 2002, 286–312; A.L. Burke, "Private Griefs, Public Places," *Political Geography* 22, 2003, 317–33; D.C. Harvey, "'National' Identities and the Politics of Ancient Heritage: Continuity and Change at Ancient Monuments in Britain and Ireland, c. 1675–1850," *Transactions of the Institute of British Geographers* NS 28, 2003, 473–87; P. Gough, "Sites in the Imagination: The Beaumont Hamel Newfoundland Memorial on the Somme," *Cultural Geographies* 11, 2004, 235–58; K.E. Till, *The New Berlin: Memory, Politics, Place*, Minneapolis: University of Minnesota Press, 2005.

11  Her biography is told in more detail in DeLyser, "Thus I Salute the Kentucky Daisey's Claim."

12  "Patent record," Oklahoma County, Territory of Oklahoma, book 14, page 246, 21 May 1900; "Administrator's or executor's deed record" (of Daisey-Svegeborg quarter section), book 19, page 400, Oklahoma County Court House, Oklahoma City, Territory of Oklahoma, 16 January 1905.

13  "Happy in Adventure: Strange Preferences of a Woman," *Chicago Herald*, 2 May 1892, p. 10; reprinted in abridged form in "She's a Remarkable Girl," *Utica Saturday Globe*, 7 May 1892, p. 5; "A Fair Boomer," *Dallas Morning News*, 21 April 1899, p. 2; "The Lady Boomer's 'Scoop'," *Oklahoma State Capitol*, Guthrie, Oklahoma Territory, 27 April 1889; "An Infant Metropolis," *Dallas Morning News*, 6 May 1889, with byline date 1 May 1889. Only one credited byline is known of, but she likely authored other articles as well; see Nanitta [sic] R.H. Daisy [sic], "The City of Guthrie," *Dallas Morning News* 30 April 1889, p. 4.

14  "Daisy and her Gang," *Kansas City Times*, 15 April 1892, in Franklin A. Root Collection, vol. III, Oklahoma Historical Society, Oklahoma City, Oklahoma; "Annetta Daisy's Amazons," *New York Times* 16 April 1892, p. 1.

15  Her married name was first reported in the *Edmond Sun* on 7 August 1890; no official record of her marriage survives. In 1894 when she filed the final paperwork on her homestead she was described as "intermarried with Andreas EJ Ueland Svegeborg," *Edmond Sun-Democrat*, 14 December 1894, p. 2.

16  "Patent record," Oklahoma County, Territory of Oklahoma, book 14, page 246, 21 May 1900; "Administrator's or executor's deed record" (of Daisey-Svegeborg quarter section), book 19, page 400, Oklahoma County Court House, Oklahoma City, Territory of Oklahoma, 16 January 1905. The house no longer stands, but photographs of it exist, see DeLyser, "'Thus I Salute the Kentucky Daisey's Claim.'"

17  Probate records, Territory of Oklahoma, Oklahoma County, book 19, page 88, recorded 31 December 1904. Her obituary, "Nannette [sic] Daisy [sic] is Dead," appeared in the Oklahoma City *Daily Oklahoman* on 18 October 1903. She and her husband had likely divorced (see DeLyser, "Thus I Salute the Kentucky Daisey's Claim").

18  "Half Has Not Been Told," *Dallas Morning News*, 25 April 1889.
19  "Happy in Adventure," *Chicago Herald*, 2 May 1892, p. 10; reprinted in abridged form in "She's a Remarkable Girl," *Utica Saturday Globe*, 7 May 1892, p. 5.
20  "Nanette [*sic*] Daisy [*sic*] is Dead," Oklahoma City *Daily Oklahoman* 18 October 1903; *Chicago Herald*, "Happy in Adventure" and *Utica Saturday Globe*, "She's a Remarkable Girl" (both from 1892) first mention the cowcatcher.
21  In addition to the above, see also "Women to the Fore in Oklahoma," *Arkansas Gazette*, Little Rock, Arkansas, 20 March 1910; "Woman and the Land," *Mountain Democrat*, Placerville, California, 25 June 1910.
22  See Thoburn, *A Standard History of Oklahoma*, Hoig, *The Oklahoma Land Rush of 1889* (cites only the 1889 *Dallas Morning News* eyewitness account, which does not mention either the cowcatcher or the petticoat; the others cite no sources); Hoig, *Edmond: The Early Years*; Hoig, *Edmond: The First Century*; G. Carlile, *Buckskin, Calico, and Lace: Oklahoma's Territorial Women*, Oklahoma City: Southern Hills Publishing, 1990; Linda W. Reese, *Women of Oklahoma, 1890–1920*, Norman: University of Oklahoma Press, 1997, and J.L. Crowder, Jr. *Historic Edmond, an Illustrated History*, San Antonio, Texas: Historical Publishing Network for the Edmond Historical Society, 2000.
23  The monument's design was the result of a Parks Foundation design competition with Daisey as its theme. See, R. Hibbard, "Kentucky Daisey Stakes Claim in Edmond Again," *Edmond Life and Leisure*, 25 March 2004, p. 2; L. Shearer, "Commission's Wish List," *Edmond Sun*, 14 December 2005, p. 1.
24  See, for example, S.L. Myers, *Westering Women and the Frontier Experience 1800–1915*, Albuquerque: University of New Mexico Press, 1982; S.H. Armitage and E. Jameson (eds), *The Women's West*, Norman, OK: University of Oklahoma Press, 1987; P.N. Limerick, *The Legacy of Conquest: The Unbroken Past of the American West*, New York: W W Norton, 1987; L. Schlissel, V.L. Ruiz, and J. Monk, *Western Women, their Lands, their Lives*, Albuquerque: University of New Mexico Press, 1988; J. Kay, "Western Women's History," *Journal of Historical Geography* 15, 1989, pp. 302–05; S. Deutsch, *No Separate Refuge: Culture, Class, and Gender on an Anglo-Hispanic Frontier in the American Southwest, 1880–1940*, Oxford: Oxford University Press, 1989; J. Kay, "Landscapes of Women and Men: Rethinking the Regional Historical Geography of the United States and Canada," *Journal of Historical Geography* 17, 1991, pp. 435–52; P.N. Limerick, C.A. Milner, and C.E. Rankin, *Trails: Toward a New Western History*, Manhattan, KS: University Press of Kansas, 1991; S. Daniels, "Frances Palmer and the Incorporation of the Continent," in S. Daniels, *Fields of Vision: Landscape Imagery and National Identity in England and the United States*, Princeton, NJ: Princeton University Press, 1993, pp. 174–99; V. Norwood and J. Monk, *The Desert is No Lady: Southwestern Landscapes in Women's Writing and Art*, Tucson: University of Arizona Press, 1997; K.M. Morin and J.K. Guelke, "Strategies of Representation, Relationship, and Resistance: British Women Travelers and Mormon Plural Wives, ca. 1870–1890," *Annals of the Association of American Geographers* 88, 1998, pp. 436–62; V. Ruiz and E. DuBois (eds), *Unequal Sisters: A Multicultural Reader in U.S. Women's History*, New York: Routledge, 1999, 3rd edn; E. Jameson, "The History of Women and the West," in W. Deverell, *A Companion to the American West*, London: Blackwell, 2004, pp. 179–99; K. Morin, *Frontiers of Femininity: A New Historical Geography of the Nineteenth-Century American West*, Syracuse, NY: Syracuse University Press, 2008.
25  See, for example, D.A. Brown, *The Gentle Tamers: Women of the Old Wild West*, Lincoln, NE: University of Nebraska Press, 1981; C.W. Breihan, *Wild Women of the West*, New York: New American Library, 1982; G. Carlile, *Buckskin, Calico, and Lace*; A. Seagraves, *Soiled Doves: Prostitution in the Early West*, Hayden, ID: Wesanne Publications, 1994; E.C. Flood and W. Manns, *Cowgirls: Women of the Wild West*, Santa Fe, NM: ZON International Publishing, 2000; and K.E. Krohn, *Wild West Women*, Minneapolis, MN: Lerner Publications, 2005.
26  See M. Heffernan and C. Medlicott, "A Feminine Atlas? Sacagawea, the Suffragettes, and the Commemorative Landscape in the American West, 1904–1910," *Gender, Place and*

*Culture* 9, 2002, pp. 109–31. Perhaps the most famous is the "Pioneer Woman" statue erected at Ponca City, Oklahoma in 1930, see http://www.pioneerwomanmuseum.com/thestatue.htm Accessed 18 February 2010.

27  See Thoburne, *A Standard History of Oklahoma*; A. Debo, *Oklahoma: Foot-loose and Fancy Free*, Norman: University of Oklahoma Press, 1949; A.M. Gibson, *The History of Oklahoma*, Norman: University of Oklahoma Press, 1984; D.W. Meinig, *The Shaping of America: A Geographical Perspective on 500 Years of History, vol. II: Continental America, 1800–1867*, New Haven, CT: Yale University Press, 1993; J.C. Walker, "The Difficulty of Celebrating an Invasion," in D.D. Joyce (ed.), *"An Oklahoma I had Never Seen Before": Alternative Views of Oklahoma History*, Norman: University of Oklahoma Press, 1994, pp. 15–26; D.W. Meinig, *The Shaping of America: A Geographical Perspective on 500 Years of History, vol. III: Transcontinental America, 1850–1915*, New Haven, CT: Yale University Press, 2000; and DeLyser, "Thus I Salute the Kentucky Daisey's Claim."

28  See, for example, "Oklahoma's Colonists," *Dallas Morning News*, 20 April 1889; "Driving out Intruders," *New York Times,* 20 April 1889; *Dallas Morning News,* "Oklahoma Colonies," 22 April 1889; "Into Oklahoma at Last," *New York Times,* 23 April 1889, p. 1; "In the Promised Land," *Dallas Morning News* 25 April 1889; "Half Has Not Been Told," *Dallas Morning News,* 25 April 1889; "In the Promised Land," 26 April 1889.

29  Jameson, "The History of Women and the West"; Schlissel, Ruiz, and Monk, *Western Women*.

30  J.M. Jensen and D.A. Miller, "The Gentle Tamers Revisited: New Approaches to the History of Women in the American West," in Irwin and Brooks (eds), *Women and Gender in the American West. Jensen-Miller Prize Essays from the Coalition for Western Women's History*, Albuquerque: University of New Mexico Press, 2004, pp. 9–36; quoted pp. 12–14.

31  R. Slotkin, *It's Your Misfortune and None of My Own: A New History of the American West*, Norman: University of Oklahoma Press, 1991, p. 613. See also, R.G. Athearn, *The Mythic West in Twentieth-Century America*, Lawrence, Kansas: University of Kansas Press, 1986; R. Slotkin, *Gunfighter Nation: The Myth of the Frontier in Twentieth Century America*, New York: Athenaeum, 1992; H.N. Smith, *Virgin Land. The American West as Symbol and Myth*, Cambridge, Mass.: Harvard University Press, 1950; A. Barra, *Inventing Wyatt Earp: His Life and Many Legends*, New York: Carroll and Graf Publishers, Inc., 1998.

32  "Robbed Miss Daisy," Guthrie Oklahoma *Daily Leader*, 21 December 1893, p. 3.

33  See Irwin and Brooks, *Women and Gender in the American West*.

34  "Nannette [*sic*] Daisy [*sic*] is Dead," Oklahoma City *Daily Oklahoman* 18 October 1903.

35  See, for example, A. Impey, "Culture, Conservation, and Community Reconstruction: Explorations in Advocacy Ethnomusicology and Participatory Action Research in Northern Kwazulu Natal," *Yearbook for Traditional Music* 34, 2002, pp. 9–24; and Y. Lincoln, "Engaging Sympathies: Relationships between Action Research and Social Constructivism," in P. Reason and H. Bradbury (eds), *Handbook of Action Research*, London: Sage, 2001, 124–32.

36  Examples of other possibly failed research projects include Kim England, "Getting Personal: Reflexivity, Positionality, and Feminist Research," *The Professional Geographer* 46.1, 1994, 80–90; J. Reger, "Emotions, Objectivity and Voice: An Analysis of a 'Failed' Participant Observation," *Womens Studies International Forum* 24.5, 2001, 605–16.

37  See, Peter Reason and Hilary Bradbury (eds), *Handbook of Action Research*; and Ernest T. Stringer, *Action Research*, 3rd edn, London: Sage, 2007.

38  James Fentress and Chris Wickham, *Social Memory*, London: Blackwell, 1992.

# 18

## STILL-LIFE, AFTER-LIFE, *NATURE MORTE*

### W.G. Sebald and the demands of landscape

*Jessica Dubow*

What then is time? If no-one asks me, I know; if I want to explain it to a questioner, I do not know.

*St Augustine*[1]

I am rebegot, of absence, darkness, death – things which are not.

*John Donne*[2]

As Karl Rossmann stood on the liner slowly entering the harbor of New York, a sudden burst of sunshine seemed to illuminate the Statue of Liberty; so that he saw it in a new light, although he had sighted it long before. The arm with the sword (*sic*) rose up as if newly stretched aloft.[3]

As if to foreclose on the false promise of mimesis, Kafka's novel *Amerika* displaces, disfigures, estranges. The proper names of places fail to signify and become mute apostrophes – a "bridge leads from Manhattan to Boston (rather than to Brooklyn) and hangs over the Hudson (rather than the East) River,"[4] a hotel only five storeys high has forty-seven elevators, a site never functions for itself or for introspection but emerges with the violence of an inexplicable accident. From accident to accident, orientation and destination elude the traveler, yet still drag him forward. Here, perpetual movement not only falls short of any kind of momentum but becomes an index of melancholia, or is the form that melancholia takes.

Kafka, as we know, never went to New York; nor did he need to. "Those are pictures, only pictures," he once said. "One takes a photograph of things in order to forget them. My stories are ways of closing my eyes."[5] Nor was *Amerika* – the unfinished narrative of a turn-of-century emigrant to the New World – Kafka's intended title. It was instead *Der Verschollene*: in German, "the lost ones," "the missing," the un-nameable fate of the un-named adrift, washed up[6]; but also the ineluctable progress toward silence, oblivion, fatality (from the verb *verschallen*, "to die out").

En route to the Italian resort town of Riva in October 1987, journeying circuitously from his adopted home in England to the small Bavarian village of Wertach im Allgäu where he was born, the narrator of W.G. Sebald's first prose work, *Vertigo*, thinks he sees the ghost of a young Franz Kafka who in the late summer of 1913 undertook a working holiday traveling from Vienna to Northern Italy.

Yet if *Vertigo*, Sebald's fictional reconstruction of Kafka's journey, suggests the affective genealogies of literary filiation or even the longing for a more immediate communion with a voice no longer present, it is also about a time that is ultimately indefinable and a space whose location is always in question. Indeed, if, as Kafka says, one takes an image of something in order to forget it – or, relatedly, if memory belongs to no place; has no referential *taking-place* – the same anti-mimetic gesture names the condition of *Vertigo*. Thus, Sebald, or the fictional narrator with which he is partly identified, sees the specter of Kafka not to resurrect him or bring him to light but to suggest that all places are somehow posthumous; that it is not through the common measure of empathy or identification that the spectral shades of the dead may appear but only a perception born of interruption, detachment and self-estrangement. The effect of the misdirection is as much a temporal predicament – the running confusions of past and present – as it implicitly calls forth a quite different order of space and sight. Sebald's characters are not travelers who see sites but those who can no longer make sense of the seen and try make visible this loss. Whether moving in the direction of remembering or its grey fading away, it is the erosion of a scene and not its semblance that erupts into presence. Here, time and space, deprived of their usual imaginary consistency, open onto a certain void or invisibility. No longer capable of securing the foundations for reflection and recollection, they become passages to a kind of depthless forgetting, to a ground in which experience cannot take root.

With time and space so displaced, we turn toward a reading whose alienation from discourses of a representative presence raises questions about the categories of history and landscape. For if to "history" is conventionally tied the task of being faithful to the past, and to "landscape" is attached the topology of a visible site, then Sebald asks us to think an alternative: to annul the "bright region of memory"[7] which would redeem a past referent, to inhabit the blind underside of the claim that what has occurred has *taken place*. At this point, which is also the estrangement from a point of view or, better yet, is the point of view of the estranged, Sebald's characters are no longer the inhabitants of a world but its lost travelers.[8] With time thought outside the redemptive temporality of lost and found, recollection and recovery, journeys distend into enduring delays in which the present ceases to pass and the "ghosts of repetition [...] haunt with greater frequency."[9] Alternatively, we may say that Sebald's travelers wander through the frame of the still-life, the after-life, of *nature-morte*: spaces in which the absent in history reveals its force only to those who see it from the allegorized ruins of time, or at least outside the unfolding locales of chronology.

No doubt, this is the function that so many placeless places have in almost all of Sebald's fiction: railway stations, dead-end corridors, obstructed openings, blind windows, boarding houses, waiting rooms. Not so much animated by a past as made neutral by all that is anonymous, these are spaces whose histories we cannot know, through which voices cannot carry, in which vision cannot illuminate but precisely where "a certain unavowability haunts the present."[10] No doubt, too, the figures that populate Sebald's spaces emerge as a response to this loss of time *in* time, and with it any promise of eschatology. Outcasts, exiles, suicides or the inmates of asylums and sanatoria, they are the subjects of time immobilized in this way, or of the arrests and repetitions that characterize the melancholia of trauma and survival. *Vertigo.* Just so: events without outcome or recuperation, figures deprived of intentionality or the possibility of "working through," journeys without respite or goal, travelers who move over huge distances only to beat the same path, turn at the same fork, stumble at the same *impasse* over and over again. Even on his daily walks through the well-defined city center of modern-day Vienna, Sebald's author-narrator feels himself inert: every point of departure a doubling-back, sites of time's absence, points "beyond which [he] cannot move, except to return"[11]:

> Early every morning I would set out to walk with no aim or purpose through the streets of the inner city, through the Leopoldstadt and the Josefstadt. Later, when I looked at the map, I saw to my astonishment that none of my journeys had taken me beyond a precisely defined sickle- or crescent-shaped area, [...] If the paths I had followed had been inked in, it would have seemed as though a man had kept trying out new tracks and connections over and over again, only to be thwarted each time by the limitations of his reason, imagination or will-power, and obliged to turn back again. [...] On one occasion, in Gonzagagasse, I even thought I recognized the poet Dante, banished from his home town on pain of being burned at the stake. For some considerable time he walked a short distance ahead of me, [...] but when I reached the corner he was nowhere to be seen.[12]

Walter Benjamin, as much a Sebaldian haunt as Kafka or the exiled Dante, was of course amongst the first to understand how difficult wandering might be, and how much it is opposed to the audacious Romantic ego or to what Massimo Cacciari calls "the 'virile' [...] disavowal of a home or dwelling."[13] To be sure, in both Benjamin and Sebald, there is what Kafka famously calls that "old incapacity"[14] to belong in the world: that errant consciousness which suggests human beings do not belong here, or if they do then they are always somehow missing from presence, just as they are always missing from place. In both writers, a commonality of themes – image and impression, trace and remainder – are conjoined to meditations on the relation of progress to ruination, and to the long *durée* of European culture enjoined to the violence of capitalist modernity. (Hence, the metonymic figure of Kafka: harbor of the essential archive of modern European history, prophet of an age in which, paradoxically, the fate of the prophet was to be unheard, untenable, anachronistic). In both Benjamin and Sebald too, there is a certain shared *methodos* or path – by which I mean, literally, the inseparability of walking and of thinking – that convulses with the tiniest of details while it elides transitions and brims over with the impassable particularities of every fact: the accidental configurations of an immediate environment, the apprehension of every instance as an *instant* – each time, distinct, singular – but caught in the excess of perpetual repetition.

But if Sebald can be said to be Benjamin's doppelganger,[15] this is not just due to any network of affinities with respect to disposition or method that washes over their biographical and literary personae. More fundamentally, both are borne along with what I'd like to term a "*negative phenomenology*." By this I mean a perceptual capacity that answers not to the appearance of anything or enfolds any passage from the unknown to the known, but which demands a constant dispersion, a temporal abyss into which immediate visibilities slip the moment they are glimpsed. Thus buildings, cities, streets in Sebald's fiction, are places lost – barred, burnt, abandoned. Views of landscapes, coastlines and shorelines emerge only to be engulfed by opacity, the perspectival arabesques of the visual world fall into what he calls a "night-side vanishing point"[16] or into a darkening hole beyond all action, position and force. This is, of course, not to say that the places which Sebald describes so meticulously are emptied of history; to the contrary. Deadened by the weight of all that is *not* recoverable, held within the grasp of the inexplicable and unremembered, they are not so much places constituted by a persistent lifelessness as ones in which lifelessness is the signature of historicity *as such*. Thus, for example, the city of Manchester, in the fourth story of Sebald's *The Emigrants*, is "built of countless bricks and inhabited by millions of souls, dead and alive."[17] In the suburbs of Moss Side and Hulme,

> ...there were whole blocks where the doors and windows were boarded up, and whole districts where everything had been demolished [...] One might have supposed that the city had long since been deserted, and was left now as a necropolis or mausoleum.[18]

When the traces of the great nineteenth-century metropolis reappear in *After Nature* (1988), a triptych of prose poems held together by the recurring motif of Unheimlichkeit,[19] Sebald finds himself "in a sort of no-mans-land/Behind the railway buildings, in a terrace/Of low houses apparently due for demolition, with shops left vacant,/On whose boards the names Goldblatt, Grünspan, Gottgetreu/Speilhalt, Solomon, Waislfish/and Robinsohn could be made out."[20]

But if such inscriptions are the after-images of lives that have disappeared (the stranded objects of post-industrial Manchester and the remnants of what had once been its Jewish community), it is in Sebald's descriptions of the natural world that all which is irrecoverable in history becomes most acute.[21] Here, indeed, Sebald's landscapes have that quality of obscured exposure that marks a "negative phenomenology": each is a place that does not so much confirm loss as "absorb[s] and dissolve[s] all presences into itself,"[22] as Jean-Luc Nancy puts it; or empty spaces become all the more empty precisely because they "cannot be confused with mere nothingness."[23] Thus, in *The Rings of Saturn*, as he walks the plains of rural East Anglia on the outskirts of the town of Orford and details a landscape of concealed military installations and abandoned estates (their manorial houses torn down or transformed into asylums, old-age homes and reception camps for refugees from the Third Reich), Sebald is assailed by a sudden sandstorm. It blasts out of nowhere and assaults him completely:

> As darkness closed in from the horizon like a noose being tightened, I tried in vain to make out, through the swirling and ever denser obscurement, landmarks that a short while ago still stood out clearly, but with each passing moment the space around became more constricted. Even in my immediate vicinity I could soon not distinguish any line or shape at all. The mealy dust streamed from left to right, from right to left, to and fro on every side, rising on high and powdering down, nothing but a dancing grainy whirl [...] When the worst was over [...t]here was not a breath, not a birdsong to be heard, not a rustle, nothing. And although it now grew lighter once more, the sun, which was at its zenith, remained hidden behind the banners of pollen-fine dust that hung for a long time in the air. This, I thought, will be what is left after the earth has ground itself down.[24]

But if the anonymous violence of a natural catastrophe – like the blinding of sight and the effacement of place – registers as the emptying out of historical presence,[25] and thus underwrites the Sebaldian view that memory work has no redemptive qualities, it is also guided by the knowledge of the Holocaust as that "*absolute* event of history"[26]: the singular decimation which invalidates all identity and concept, the unspeakable limit which left nothing intact, the event which henceforth demands that we hold to the crisis it creates – crisis alive and abiding.[27] For, while Sebald deals as frequently with the deep time of the European Renaissance and Enlightenment as he does with the specific period of Hitler's Germany, it is nonetheless the Holocaust and the burden of being born-after that focuses the problem of history, place and memory. At issue here, however, is not the usual question of whether representing Auschwitz is aesthetically possible or impossible, ethically bearable or forbidden.[28] Rather, representation, I suggest, is precisely what results when one insists on the present weight of all that is irrecoverable and anonymous, that the real force of the image has less to do with what is available to be seen in it than with the way it insists on the fact of absence. Moreover, it is in the representation of landscape, I will argue, that the force of loss is exemplary.

Crucially, what I am calling "representation" here obliges us to think of the term in a way that differs significantly from its conventional usage – as from its customary critique. On the one hand, it is not any idea of representation born of the aesthetic tradition which consists in the shaping of a visual field, in the organization of perspectives, or the contemplative abstractions of the gaze – the very symbolic arrangement of sight and space which, after all, gives us "landscape" in its generic sense. On the other hand, my reading has nothing Heideggarian about it. Here the "*re-*" of representation, or the mediatory work that this "*re*" indicates, is neither the image as a fatally reduced and weakened double of what it would imitate; nor, in its belated distance from an original scene, does it assume any objective dominance in point of view. Rather, with Sebald, we might come at representation from another direction: one which enfolds within itself the very fact of loss, which avows an unavowable past in the presentation of always missing presence.[29] For Louis Marin, analyzing the semiology of seventeenth-century European painting, to extract representation from the enlivening laws of mimesis means to situate it within a particular metaphoric of death, a move which demands that we renew our attention to the force and virtue of the *"re"*:

> ...the prefix *re* – brings into this term the value of substitution. Something that *was* present and *is* no longer is *now* represented. In place of something that is present *elsewhere,* there is *here* a present...[30]

Similarly, for Jean-Luc Nancy, the iterative value of *re*-presentation "is not repetitive but intensive,"[31] it is less the mimetic copy of a given thing than the heightened exposure of what is absent from presence. If such accounts seem to simply elaborate some standard post-structuralist insights, however, they offer a stronger understanding of the image of which ontology can give little account. In this respect to understand the "*re-*," as Marin does, not as a reproductive move – the mere replacement or substitution of an original – but as an accentuation of force, allows us to see re-presentation as the resurgence of absence, or a regaining of loss. What "*was* present and *is* no longer is *now* represented": an absence *within* presence, a loss amplified, intensified, accosted. So defined, the act of representation has none of the disinterested "mastery" over the object-world, or the bringing of something back into being, around which a tradition of art-historical and geographic thought have both been fashioned. Rather in *re*-presenting – precisely in the activity of repetition, of recurrence – all which has been lost remains as loss, an enduring – even inert – failure brought now to acute attention. Relatedly, we might say that the force of *re*-presentation does not just turn on a constitutive absence. It also finds equivalence in that relationship to loss that we know as melancholia: a past instant, *here*, in front of us; insistently and incessantly present. Or, in Sebald's terms, it is about living amidst the inert, in that atemporal excess of time where a past instant cannot be conjoined with another future one, where the irreplaceable cannot be replaced: a *Vertigo*.

As long as we do not hold to any reproductive understanding of the term, then, the question of representation allows us to approach Sebald's investigations of history and landscape in a very specific way. Thus, his description of so many deserted cities (Napoleon's sacking of Moscow in 1812, the bleachfields and burial grounds of seventeenth-century London, the Allied bombing campaigns of Hamburg, Nuremberg and Berlin, the wastelands of the post-industrial) like his sketching of the traces of prehistoric and natural degradation (hurricane-felled trees, a broken mass of rubble, an old bone whitened on a shoreline) do not merely illustrate past events or work as motifs that might identify previous catastrophe. Nor, like "light diffused through a veil of ash,"[32] do so many of his dust-laden, mist-laden sites simply attest to the formal problems of unreliability and incompleteness linked to the phenomenon of memory. Instead, it is from the condition or *precondition* of loss that such things come to be *re*presented. Here where the history it refers to has lost all presence, here where the "small merc[ies] of our memories"[33] fail, the "*re-*" in representation allows lack to be thought, awakening absence as it intensifies all that is now given. In short, Sebald's landscape representations are not merely compatible with loss – they demand it. "Too many buildings have fallen down, too much rubble has heaped up, the moraines and deposits are insuperable,"[34] says the character of Michael Hamburger in *The Rings of Saturn* as he attempts to remember his Berlin childhood before his family's emigration to England in 1933. "If I now look back to Berlin [...] all I see is a darkened background with a grey smudge in it, a slate pencil drawing [...] blurred and half-wiped away with a damp rag."[35] In the context of this argument, Hamburger's words are about much more than the limits of historical recovery. They are about the force of loss that is the true *point of view* of representation – and the most stubborn proof of its failure. "Blurred and half-wiped away," it is a point of view that in Sebald rejects, if not the concept of "landscape representation," then at least the way it operates: not as reference to a past event – a history, a nature – that the work of imaging would recover, but as the registration of loss and the activity that accentuates it. Or, we might say that here *landscape re-presentation* engages the idea of "medium" in the double or compound sense of the word: an attempt to make present that is also a locus of intercession with the disremembered and the dead.[36]

Despite this, Sebald's writing is not an invitation to forgetting. For as long as we forgo representation-as-reproduction, Sebald opens onto a historical past, one which, however, in principle remains inaccessible. Indeed, if the "*re-*" of representation, as suggested, operates as the force of the forgotten, it also insists that we continue to *make present*, or at least, that we continue to try. Here, however, the past can neither be accessed in terms of consciousness nor measured in terms of chronology. Thus, while Sebald assiduously specifies proper names and dates, and while he locates events within real and identifiable spaces, the effect of such detailing is not the fulfillment of any historical obligation. It is to insist that the obligation remains unfulfilled. Indeed, much like representation in its conventional mimetic sense, it is precisely time conceived as chronology that, for Sebald, is the real modality of forgetting – a reprieve effected via the two-fold erasures of diachrony and reproduction, a past paradoxically allayed by memory and the objects of memorial, a trauma redeemed and all its "claimants' accounts with history cancelled."[37] As Derrida asks:

> Is not Time the ultimate resource for the substitution of one absolute instant by another, for the replacement of the irreplaceable, the replacement of this unique referent for another that is yet another instant, completely other and yet still the same?[38]

If Derrida's question, and the request that echoes through it, asks us to acknowledge the aspect of oblivion, of utter forgetfulness, that resides in the idea of history as temporal continuum – the unstoppable course that covers over one singularity with another, that expiates and assimilates by replacing and succeeding – then Sebald sounds a similar warning. As the camp survivor and essayist Jean Améry writes, in a passage that Sebald quotes, "[w]hat oppresses me is no neurosis, but rather precisely *reflected reality*"[39] (my emphasis). As such, what is dangerous about representation-as-reproduction, he suggests, is not that it simply does duty for something which has previously taken place, or refers to what is still too intolerable to remember. The danger lies precisely in admitting the past to present cognition – a will to possession more ominous than any pure forgetting.

Against this, Sebald remembers along with that "formless chaos of the forgotten"[40] which always accompanies us. In the upheaval to which he submits the "re-" of representation, absence and the interval erupt to prevent the successive or reproductive run of time, to make present what is singular and forgotten. If such a disordered temporality is the real condition of our historicity, as so many writings in the shadow of the Holocaust have argued, then it is also an exigency reinforced by Sebald's landscape representation. Given this, we might look again at the title of *After Nature* – a variable phrase which in at least one of its meanings implies the reproductive foundation of "landscape representation." In Sebald's case, however, it means the necessity of its loss.

## Notes

1  From *Confessions*, Book 11.
2  From *A Nocturnal Upon St Lucies Day*. I am indebted to Mitch Rose for this quote.
3  F. Kafka, *Amerika*, New York: Schocken Books, 1962.
4  M. Anderson, "Kafka and New York: Notes on a Travelling Narrative," in A. Huyssen and D. Bathrick (eds), *Modernity and the Text: Revisions of German Modernism*, New York: Columbia University Press, 1989, p.150.
5  F. Kafka, *Conversations with Kafka*, trans. G. Rees, New York: New Directions, 1971.
6  As Anderson explains, "ship passengers lost at sea are *verschollen*." Anderson, "Kafka and New York," p. 147.
7  P. Ricoeur, *History, Memory, Forgetting*, trans. K. Blamey and D. Pellauer, Chicago: University of Chicago Press, 2004, p. 21.
8  For a very different, but evocative, account of landscape defined by an uncanny estrangement, see J.-L. Nancy, *The Ground of the Image*, trans. F. Fort, New York: Fordham University Press, 2005, pp. 51–62.
9  W.G. Sebald, *The Rings of Saturn*, trans. M. Hulse, London: Vintage, 2002, p. 187.
10 J. Butler, "Afterword: After Loss, What Then?," in D.L. Eng and D. Kazanjian (eds), *Loss*, Berkeley: University of California Press, 2003, p. 468.
11 M. Blanchot, *The Space of Literature*, trans. M. Lester with C. Stivale, New York: Columbia University Press, 1990, p. 93.
12 W.G. Sebald, *Vertigo*, trans. M. Hulse, London: The Harvill Press, 1999, pp. 33–5.
13 M. Cacciari , *Posthumous People: Vienna at the Turning Point*, trans. R. Friedman, Stanford California: Stanford University Press, 1996, p. 142.
14 F. Kafka, *Diaries*, trans. J. Kresh, M. Greenberg and H. Arendt, New York: Schocken Books, 1964, p. 330.
15 On the burgeoning body of work that relates the two writers see, amongst others, E.L. Santner, *On Creaturely Life: Rilke, Benjamin, Sebald*, Chicago: Chicago University

Press, 2006; J. Dubow, "Case Interrupted: Benjamin, Sebald and the Dialectical Image," *Critical Inquiry* 33:4, Summer 2007, pp. 820–36; M. Zisselsberger, "Melancholy Longings," in L. Patt, *Searching for Sebald: Photography after W.G. Sebald*, Los Angeles: The Institute of Critical Inquiry, 2007, pp. 280–301.

16  W.G. Sebald and Jann Peter Tripp, *Unrecounted*, trans. Michael Hamburger, London: Penguin, 2005, 99.

17  W.G. Sebald, *The Emigrants*, trans. Michael Hulse, London: The Harvill Press, 1997, p. 150.

18  W.G. Sebald, *The Emigrants*, p. 151.

19  Literally "unhomeliness" but usually translated as "uncanniness" and which, particularly in its Freudian sense, resonates with the Sebaldian motif of dream-like spatial recurrence.

20  W.G. Sebald, *After Nature*, trans. M. Hamburger, London: Penguin, 2002, p. 99.

21  The Sebaldian trope of natural devastation as an index of the earth's participation in the violence of human history is not something I can go into here. For a sustained analysis of the concept and its relation to Benjamin and Adorno's analyses of capitalist modernity, see Santner, *On Creaturely Life: Rilke, Benjamin, Sebald*.

22  Nancy, *The Ground of the Image*, p. 58.

23  M. Blanchot, *The Infinite Conversation*, trans. S. Hanson, Minneapolis: University of Minnesota Press, 1993, p. 68.

24  Sebald, *The Rings of Saturn*, p. 229.

25  On the Benjaminian and Sebaldian idea of "earthly violence" in which natural destruction participates in the violence of human history, see Santner.

26  M. Blanchot, *The Writing of the Disaster*, trans. A. Smock, Lincoln: University of Nebraska Press, 1986, p. 82.

27  Among the seminal texts which introduced this line of interpretation see T. Adorno, *Negative Dialectics*, trans. E.B. Ashton, New York: Seabury, 1979; J. Lyotard, *Les Fins de l'homme*, Paris: Galilée, 1981; M. Blanchot, *The Writing of the Disaster*, trans. A. Smock, Lincoln: University of Nebraska Press, 1986; S. Kofman, *Smothered Words*, trans. M. Dobie, Evanston: Northwestern University Press, 1988; P. Levi, *The Drowned and the Saved*, trans. R. Rosenthal, New York: Random House, 1989; R. Antelme, *The Human Race,* trans. F. Haight and A. Mahler, Malboro, VT: The Marlboro Press, 2002; G. Agamben, *Remnants of Auschwitz: The Witness and the Archive*, trans. D. Heller-Roazen, New York: Zone Books, 2002.

28  For an extensive discussion of the misleading nature of this question and its relation to the intersection of monotheism and the Classical problem of the image as copy or simulation, see Nancy, *The Ground of the Image*, pp. 27–49.

29  My reading of this owes much to Derrida's meditations on the "politics" of death and mourning. See J. Derrida, *The Work of Mourning*, trans. P. Brault and M. Naas (eds), Chicago: The University of Chicago Press, 2003.

30  L. Marin, *Des pouvoirs de l'image: Gloses*, Paris: Seuil, 1993, p. 11. For a further analysis of the image as it exists within a paradigm of death, see Marin, *On Representation*, trans. C. Porter, Stanford, California: Stanford University Press, 2001, pp. 269–84. The frequency of depictions of dead flesh, skulls and skeletons in Sebald's descriptions of Medieval, Northern Renaissance and Reformation art (e.g. Matthias Grünewald's *Isenheim altarpiece* and Altdorfer's *Battle of Alexander at Issus* in *After Nature* and Rembrandt's *The Anatomy Lesson of Dr. Tulp* in *The Rings of Saturn*) suggest striking parallels between Sebald's fiction and Marin's aesthetic theory. For a related, but very different treatment of representation as "the return of the dead" see also R. Barthes, *Camera Lucida: Reflections on Photography*, trans. R. Howard, New York: Hill and Wang, 1981.

31  Nancy, *The Ground of the Image*, p. 35.

32  Sebald, *Vertigo*, p. 51.

33  G. Agamben, *Profanations*, trans. J. Fort, New York: Zone Books, 2007, p. 35

34  Sebald, *The Rings of Saturn*, pp. 177–8.

35  Ibid.

36 For a sustained exploration of Sebald's "spectral materialism" or the persistent suffering inscribed within living space, see Santner, *On Creaturely Life: Rilke, Benjamin, Sebald*. For a Derridean account of the ghostly in Sebald's understanding of place, see J. Wylie, "The Spectral Geographies of W.G. Sebald," *Cultural Geographies* 14, 2007, pp. 171–88.

37 R. Tiedemann, "Dialectics at a Standstill: Approaches to the *Passengen-Werk*," in G. Smith (ed.), *On Walter Benjamin: Critical Essays and Recollections*, Cambridge, Mass.: MIT Press, 1989, p. 285.

38 Derrida, *The Work of Mourning*, p. 60.

39 J. Améry, *At the Mind's Limits*, trans. S. Rosenfeld and S.P. Rosenfeld, Bloomington: Indiana University Press, 1980, p. 96.

40 Agamben, *Profanations*, p. 35.

# 19

# THE TEXTURE OF SPACE

## Desire and displacement in Hiroshi Teshigahara's *Woman of the dunes* [*Suna no onna*]

*Matthew Gandy*

Writing about film is a little like writing about music: it is difficult to translate a sensory experience into something other than itself. In a sense, however, all writing is a process of translation though the degree to which we might recognize these limitations varies from one context to another. To convey the experience of landscape directly, for example, is very different from engaging with depictions of landscape in cinema, literature or music. Cinema presents geographers with an especially rich means to explore the concept of landscape in terms of its lineage to other forms of aesthetic representation and also through the cinematic experience itself as a critical element in modern culture. The creative possibilities for the exploration of landscape in film emerged with the earliest experimental footage from rooftops and moving trains and gradually encompassed ever more ambitious attempts to extend the scope of human sensory experience. The cinematic landscape poses questions at the heart of cultural geography: the tension between phenomenological and materialist readings of space; the changing relationship between technology, aesthetics and modern consciousness; and the role of cinema as a repository of collective memory. Cinema should not be read as a mere cipher for pre-given theoretical assumptions or seen as purely illustrative of a set of predetermined cultural positions. An open engagement with film promises no less than the revitalization of cultural geography and its intersecting conceptual terrains.

One of the most striking cinematic explorations of landscape is provided by Hiroshi Teshigahara's *Woman of the dunes* [*Suna no onna*] (1964) which focuses on two people trapped at the bottom of a deep sand pit located in a desolate stretch of dunes. These shifting dunes – which are ominously described in Kôbô Abe's original novel as "bathed in a murky reddishness" – dominate the metaphorical and iconographic structure of the film.[1] *Woman of the dunes* is an intricate collaboration between the director Hiroshi Teshigahara (1927–2001), the avant-garde novelist and playwright Kôbô Abe (1924–93), who wrote the screenplay adapted from his

own novel, and the experimental composer Tōru Takemitsu (1930–96). This "cinematic collective," which made a series of films in the 1960s, is very different from
the auteurist legacy associated with other Japanese film makers such as Yasujirō
Ozu, Kenji Mizoguchi or Akira Kurosawa.[2] Their first major collaborative project
was the film *Pitfall* [*Otoshiana*] (1961), which is set in and around an abandoned
mine, and includes extensive footage of former industrial landscapes. With the
completion of *Woman of the dunes* their work became part of an internationally
recognized "new wave" in Japanese cinema that marked a break from the classic era
of Japanese film making in the 1950s and opened up new connections with
modernist culture in Europe and North America.

## I. The dunes

The opening sequence of the film creates an atmosphere of dissonance and disorientation: complex visual patterns are set to a collage of street sounds mixed with
experimental music. The images range from fingerprint swirls and twisting map
contours to clusters of official rubber stamps. The enlarged edges of calligraphy
brush strokes reveal yet more intricate features created by the fine spray of ink or
stray brush hairs so that the transformative effects of scale are repeatedly emphasized. After the elaborate credit sequence we switch to an image of a highly magnified individual grain of sand that resembles a partially hatched egg; then through a
series of steps we zoom out, at each stage revealing a different topographic effect,
until we see the man (played by Eiji Okada) for the first time entering the frame
from the bottom of the screen (Figure 19.1). As the man walks across the frame he
leaves a trail of footprints in the sand and the camera lingers on these transitory
traces so that the scale is rendered uncertain; the line of shallow indentations might
be very large or very small, something created by an insect or perhaps a vast imprint
on a distant planet.

   This man is a teacher who is also an amateur entomologist hoping to discover a
new species of beetle during his brief vacation in order to leave some kind of trace
of his own existence for scientific posterity. He stumbles along in the mid-day heat
like one of the beetles he is studying; his rucksack packed with entomological
equipment resembling a spiky carapace attached to his back. The strangeness of
nature is emphasized with close-ups of insects such as a caterpillar crawling across
the hot sand or a dragonfly head with its iridescent compound eyes. As the light
begins to change, the man comes across a village lying in the dunes. "What a
terrible place to live," he declares, as he peers down at the cluster of small houses.

   Some local villagers approach the man, and after setting aside their suspicions
that he might be a government inspector, invite him to stay as a guest in their
village, adding a sense of anthropological intrigue to his entomological reverie. He
condescendingly remarks how he would like to stay in "such a village" and reflects
dreamily on the limitations of his life back in Tokyo where "certificates are limit-
less" – not unlike grains of sand. He is led to the top of a ridge and descends
unsteadily down a rope ladder to a ramshackle dwelling that is surrounded by steep

**FIGURE 19.1** *Woman of the dunes [Suna no onna]*, 1964. Courtesy of the British Film Institute, London.

cliffs of sand. "It's really quite an adventure," he chuckles to himself, in one of the darkly humorous uses of dramatic irony that punctuate the early part of the film. The owner is a young woman (played by Kyoko Kishida), who explains that she has lived alone since her family were lost in a sand storm. Every surface in the cramped dwelling is covered with a fine layer of sand and there is an absence of even simple screens or partitions. Over dinner she describes unusual properties of the sand which can rot any material but her words are met by his rationalist incredulity. They cheerfully discuss the characteristics of insects that live in the shack and there is a sudden close-up a boiled fish head with bulging eyes. After dinner he busies himself pinning the day's insects into neatly arranged rows as if to emphasize his meticulous mastery of these small co-inhabitants of the dunes. Light, shadow and scale are used continuously to play on our perception so that at times we cannot be sure what is being shown. In a dream-like sequence the cinematography superimposes different patterns such as the ribs of an umbrella, shifting sand formations and ultimately the woman's body in profile.

## II. The erotics of confinement

The next morning the man awakes to find that the woman is still sleeping: naked except for a cloth across her face. The covering of her face radically accentuates her nakedness and a light dusting of sand has transformed her body into a landscape. He quietly leaves the hut and emerges into the glaring sunlight to find that the rope ladder has gone: the camera pans across the steep sides of the pit and there is an ominous shuddering and falling of sand. After the woman has dressed he confronts her about their predicament and she explains that unless the sand is constantly removed then the entire village will be lost: she is one of a number of captive workers kept in pits who must shovel sand every day to protect the village from the advancing dunes. The landscape that had at first seemed merely a focus for his curiosity is now suddenly menacing and overwhelming in its grasp. The man realizes that he has been lured into a trap to be used as slave labour on behalf of the village and begins a series of desperate attempts to escape. His demands to be released are met by laughter from the villagers who periodically peer into the pit or drop food, water and other supplies for their captive workers. He tries tying up the woman to prevent her from working in order to endanger the stability of the dunes and force some kind of negotiation with their captors but all is to no avail. After he unties her, their existence becomes more intense, with close-ups of her fingers while eating and his unshaven throat gulping water. "Are you living to clear sand, or clearing sand to live," he asks. "If there were no sand, no one would bother about me," she replies. Their relationship becomes increasingly corporeal so that the cleaning of sand off each other's bodies leads to their first anguished sexual pleasure; the striations of soap across the man's back resemble the rippled surface of dunes as she scrapes her hands across his body (Figure 19.2).

He makes another, this time successful, attempt at escape by hurling a makeshift pickaxe attached to a rope only to be chased by the villagers into treacherous quicksands from which he is rescued and despatched back to the pit. The woman reveals that the sand taken from the dunes is sold illegally for construction purposes but the likelihood that such buildings would collapse is of no interest to her: her confinement precludes any consideration of distant others just as society has no interest in her own plight. The abandonment of the village to poverty and the forces of nature has fostered an indifference toward the outside world that is shared by both the woman and her captors. In the novel, the man notes that although he now finds the landscape "nauseous," "there was no reason to think of life in the holes and the beauty of the landscape as being opposed to each other."[3] Both the film and the novel explore the absence of any apparent contradiction between the touristic gaze – exemplified by the man's entomological fascination with the dunes – and the inescapable poverty facing its human inhabitants so that a counter-pastoral drama unfolds through the demystification of both the landscape and its people.

Their situation then takes on a more sinister dimension. The man asks the villagers if he might be allowed an occasional glimpse of the sea but they tell him after consulting among themselves that he must first have sex with the woman in

**FIGURE 19.2**  *Woman of the dunes [Suna no onna]*, 1964. Courtesy of the British Film Institute, London.

public. A large crowd of villagers assemble with drums and menacing *nō* masks that suggest a pre-modern Japan. Frenetic *onigoroshi daiko* [demon-killing drums] are used to instil a sense of dread through the creation of a violently repetitive sound-scape.[4] The humiliation of this scene connects with a sense of brooding political unease as if the outcast community has begun to turn on itself through acts of public retribution. He drags the woman from the house but eventually gives up on his attempted rape and the spectators, who include women, drift away from the pitiful scene. After this incident the passage of time becomes more attenuated and uncertain: the woman becomes ill and is taken away for medical treatment but the man chooses to stay in their pit and continue a series of "scientific investigations" to explore the unusual water-retaining properties of the sand. Most strikingly, the rope ladder is simply left in place after the hurried rescue of the woman, and the man wanders around the dunes before returning to the pit that has now become his world. Both the novel and the film produce a temporal displacement from minutes to years: at the end of the film it is revealed that the man has been absent for seven years and is now officially classified as a "missing person."

The film brings together two different kinds of captivity: the teacher from Tokyo who is stifled within his bureaucratic environs and the woman trapped within an even more terrifying form of physical, economic, emotional and sexual confinement.

Modern Japan is presented as a spiral of entrapment: the man is caught within a face-less society that regards its workers as little more than automatons whilst the woman represents a lonely outcast from Japan's post-war miracle. With her dark complexion and marginal existence the woman might also symbolize one of Japan's reviled *bura-kumin* or "untouchable" caste who were historically relegated to the dirtiest and most dangerous tasks but now eke out their living at the edge of modern society.[5]

## III. Collage and repetition

The idea of nature presented in *Woman of the dunes* can be read as existential in the sense that nature is portrayed as a presence that is devoid of any pre-given meaning beyond its own expansion and replication.[6] Implicit within this framework is the porosity of any clearly defined boundary between human and non-human life. At first, the man cannot understand what sustains the woman's struggle for existence and he feels a mix of pity and contempt for her. Yet his Sisyphean struggle to leave the pit becomes a measure of the indifference of nature to human endeavor just as the growing eroticism of his confinement presages his reconnection to the innate corporeality of human existence.

Much of the critical discourse surrounding the film has oscillated between two poles: on the one hand, an emphasis on the putative universalism of its existential themes; and on the other hand, a search for essential elements of a Japanese aesthetic sensibility.[7] This dichotomy masks, however, the historical specificity of existential ideas within European thought and also the innate hybridity of Japanese cultural forms preceding the growing influence of international modernism in the 1950s.[8] To say that the film is influenced by existential themes in literature and philosophy is not to say that the cinematic insights are universalist in their scope but rather internationalist in both their inspiration and orientation. *Woman of the dunes* brings together two somewhat disparate strands of existential thought: one the one hand, a set of ideas developed by Heidegger, Husserl and Merleau-Ponty, for example, on the corporeal dimensions to human subjectivity; and on the other hand, an avow-edly atheistic position, associated with Sartre and especially Camus, concerned with the innate absurdity of existence.[9]

The emphasis on scale, process and transformation connects between the rippling surface of the dunes and the claustrophobic eroticism of the human rela-tionship in the pit. The relentless movement and destructive power of the sand also signals the closeness of death in the midst of life to produce an erotics of sexual defiance (Figure 19.3).[10] "The beauty of sand, in other words, belonged to death," writes Abe. "It was the beauty of death that ran through the magnificence of its ruins and its great power of destruction," just as the fine lines or blemishes on the skin of the human captives presages the inevitability of death.[11]

The film develops Kafkaesque themes of entrapment, metamorphosis and repeti-tion that are also extensively developed in Abe's other plays and novels.[12] The focus on insects raises questions of scale, movement and perception: how do these insects experience the landscape and what difference is there between the man and his

**FIGURE 19.3** *Woman of the dunes [Suna no onna]*, 1964. Courtesy of the British Film Institute, London.

entomological prey? What, after all, really separates the human and non-human denizens of the dunes? The man's transformation into little more than a human insect and the incongruity of his presence is emphasized by the growing oddity of his surroundings. The unusual landscapes portrayed in *Woman of the dunes* are different from the more usual depictions of nature in Japanese art that have tended toward a flattened or two-dimensional perspective. Traditional forms of landscape art such as *emaki* [scroll paintings] or *shōji* [painted panels] have been dominated by mountains, water and other features with strong symbolic associations.[13] By contrast, the emphasis on dunes, which are marked by a paradoxical combination of emptiness, complexity and apparent formlessness, is a significant break with traditional modes of Japanese landscape art, and underlies the interconnection between *Woman of the dunes* and the modernist avant-garde. The emerging significance of the "formless" within abstract expressionism would have been familiar to Teshigahara both from his documentary film making in New York in 1959 and also from exhibitions in Japan.[14]

The music also provides connections with modernism through Takemitsu's abstract soundscapes which draw influences from Cage, Ligeti, Messiaen and a myriad of other strands of twentieth-century music. As if to underlie the cinematic possibilities of his work, Takemitsu referred to his music as a "picture scroll unrolled."[15] Takemitsu's connections to the modernist avant-garde also stem from his role in the *Jikken Kōbō* [experimental workshop] which drew together a circle of artists, writers and others in the 1950s.[16]

*Woman of the dunes* presents a modernist collage that is hybrid in its origin and also in its object of critique: the film is suffused with aesthetic and political dissonance involving not only a rejection of the authoritarian legacy of Japanese imperialism and social conservatism but also the intolerance of the Left toward dissent: Abe, for example, had been expelled from the Japanese Communist Party in 1956 for "Trotskyist deviance" and both he and Teshigahara were members of organizations for progressive artists and film makers such as "Cinema 58" and the "Century Club."[17]

The experimental impulse behind the film parallels other contemporary directors such as Yasuzo Masimura and Nagisa Oshima who played a leading role in bringing Japanese cinema to international audiences. Although the film won a Special Jury Prize at Cannes (and even received two Oscar nominations) it was not initially well received in Japan: the casting of Kyoko Kishida, for example, riled some Japanese audiences through her untypical facial features and the sex scenes were regarded by some critics as pornographic.[18] Outside Europe, the film's overt eroticism, minimalist structure and oblique aesthetics divided opinion: critics who perceived significant parallels with experimental European cinema, particularly the French new wave, were enthusiastic whilst other reactions were marked by varying degrees of anti-intellectual or anti-modern sentiments.[19] Revealingly, the opening credits, which appear as spaces between "ripped" paper, are provided in Japanese, English and French, making explicit the film's intended international audience.

## IV. Naked landscapes

The use of dunes as a setting for the modern nude is a motif that connects the late expressionist art of painters such as Otto Mueller and Max Pechstein through to the photography of Edward Weston and Gerhard Vetter. Unlike the familiar ensembles of nature and nudity expressed in classical and religious allegory, dunes offered a more secular *tabula rasa* within which to explore the aesthetic characteristics of the human body in a natural setting.[20] The shift of emphasis from lakes or pools to the sea itself and its surrounding landscapes also marked a different kind of engagement with nature as a source of leisure rather than contrived eroticism. Though artists such as Mueller and Pechstein include both male and female figures in their landscapes, the overwhelming emphasis has been on the juxtaposition of nature with the female body. Female artists in the twentieth century such as Georgia O'Keefe and Ana Mendieta have sought to subvert or sequester representational associations between space, landscape and gender but the film *Woman of the dunes*, despite its fleeting male nudity, remains firmly rooted in the classical lineage of compositional associations between the iconographies of nature and idealized representations of the female body. Yet to reduce *Woman of the dunes* to no more than an intriguing footnote within the dominant aesthetic traditions that permeate the modernist avant-garde risks limiting a critical evaluation of the film.[21]

The depiction of the human body in *Woman of the dunes* tends toward a more ambiguous exploration of gender and sexuality than is apparent in most other films of the early 1960s. In the novel, Abe describes how the man feels as if someone had "borrowed his body" during sex, so that "Sex, of its nature, was not defined by a single, individual body but by the species."[22] Sex is treated not as a romantic path to "completeness" but rather as a form of generic repetition that is an inescapable impulse little different from eating or sleeping. The "anti-romantic" films of Hiroshi Teshigahara, Ōshima Nagisa, Imamura Shōhei and others mark a significant break with the dominance of highly restrictive codes of sexual representation in Japanese cinema. Yet the very success of such films opened up ambiguous commonalities between art-house erotica – the so-called *nikutai eiga* [film of the flesh] – and the more mainstream exploitative *pinku eiga* [pink cinema] emerging in the 1970s.[23]

Whilst the pit within which the man is trapped might in crude psychoanalytical terms be read as a form of symbolic emasculation there is also a sense in which the man's submission to the woman's sexual needs is ultimately liberating. Certainly, the novel suggests that the man has become impotent in Tokyo and only rediscovers his sexual desire in captivity. The chaotic intensity of their forced union partially displaces some of the historically produced gender roles that have stifled freedom of expression in Japanese society. Yet the apparent "primitivism" of the woman's sexuality and the "exoticism" of the semi-arid landscapes evoke a tension between an imaginary source of sexual authenticity and the possibility for female sexuality to be regarded as historically contested. The association of nature, landscape and eroticism also connects with significant counter-currents within the history of modern sexual politics that situate the body–nature nexus as a site of resistance toward social mores or the commercial exploitation of sexuality. Yet even if the subject matter can be interpreted as socially liberating in the context of the early 1960s the presence of a gendered cinematic apparatus creates new and unresolved complexities in the relationship of the body, space and power.

## Final traces

"There will be as many interpretations as there are spectators," remarked an American reviewer on seeing the film in 1964.[24] The "new wave" of Japanese cinema that emerged in the early 1960s is marked by a dense meshwork of cultural elements which link between the specific context of a post-war society in a state of flux and engagements with international aspects of modernist thought at the precise moment of its radical de-centring. In this respect the history of modernism can be read as a series of pulses of creativity emanating outwards from cities such as Paris and New York but becoming progressively fainter by the 1960s so that the teleological impulse of high modernism becomes gradually mixed with counter-currents emerging from Buenos Aires, Tokyo and elsewhere: the modernist movement during this period becomes an increasingly disparate and polycentric synthesis before its radical implosion and retrenchment in subsequent decades.[25]

*Woman of the dunes* is derived from a creative synergy that blurs individual authorship. In this respect the production of the film – despite its international orientation – is rooted in more collaborative modes of cultural production than has been associated with the archetypal figure of the director in Europe or North America. At the same time, however, the film's abstract handling of cinematic space finds parallels with cinematic modernism in Michelangelo Antonioni's *Il deserto rosso* (1964) and Alain Resnais's *Année dernière à Marienbad* (1961).

There is a precision in which both the novel and the film evoke the fatalistic entanglement between the main protagonists and the comical parallels drawn with the insect fauna in the dunes. This emphasis on detail – not unlike the novels of Nabokov or Proust – tends to deflect the possibility for any straightforward interpretation. In Nabokov's essays on literature he offers a fierce resistance to interpretation as a form of allegorical generalization and makes a case for the essential separateness of art – not in terms of its putative autonomy à la Adorno but rather through an obligation to recognize the incommensurability of artistic production with everyday discourse.[26] There is a private domain of the imagination that resists compromise or categorization and a public realm where different systems of syntactical logic prevail.[27] But films have an intended audience who are themselves involved in the production of meaning along with the shifting contours of critical interpretation over time. In Nabokov's reading of Kafka, for example, he resists the possibility of allegorical or psychoanalytical interpretation, stating that "the abstract symbolic value of an artistic achievement should never prevail over its beautiful burning life."[28] Yet *Woman of the dunes* is both idiosyncratic and symbolic; it is framed by its own simple narrative and yet strongly allegorical, whether we choose to read the work simply in terms of its own context or sketch connections with a panoply of wider themes. Contra Nabokov, can we not retain the subtlety of detail – the texture of space – and also engage with recurring motifs of human experience?

## Notes

1  K. Abe, *The Woman in the Dunes*, trans. E.D. Saunders, London: Penguin, 2006 [1964], p. 174.
2  See M. Wada-Marciano, "Ethnicizing the Body and Film: Teshigahara Hiroshi's *Woman in the Dunes* (1964)," in A. Phillips and J. Stringer (eds), *Japanese Cinema: Texts and Contexts*, London: Routledge, 2007, pp. 180–92.
3  Abe, *The Woman in the Dunes*, p. 182.
4  P. Grilli, "Teshigahara and Takemitsu: Collaborations in Sight and Sound," essay accompanying the British Film Institute's release of *Woman of the Dunes* in 2007.
5  D. Mitchell, "Introduction," in K. Abe, *The Woman in the Dunes*, pp. v–xiii.
6  On the development of existential thought see, for example, P. Roubiczek, *Existentialism: For and Against*, Cambridge: Cambridge University Press, 1964.
7  See M. Wada-Marciano, "Ethnicizing the Body and Film."
8  See S. Tadao, "Japanese Cinema and the Traditional Arts: Imagery, Technique, and Cultural Context," in L. Ehrlich and D. Desser (eds), *Cinematic Landscapes: Observations on the Visual*

*Arts and Cinema of China and Japan*, Austin: University of Texas Press, pp. 165–86. The cultural exchanges have also been significant in other fields such as architecture. See A. Isozaki, *Japan-ness in Architecture*, Cambridge: The MIT Press, 2006.

9  For contrasting examples of existential thought see, for example, M. Merleau-Ponty, *Nature: Course Notes from the Collège de France*, trans. R. Vallier, Evanston: Northwestern University, 2003 [1995]; and A. Camus, *Le mythe de Sisyphe*, Gallimard, 1942.

10  On the relationship between eros and death see, for example, J. Laplanche, *Life and Death in Psychoanalysis*, trans. J. Mehlman, Baltimore: Johns Hopkins University Press, 1976 [1970].

11  Abe, *The Woman in the Dunes*, p. 183.

12  See, for example, M. Mariotti, "La donna di sabbia (Suna no onna)," *Cineforum* 474, 2008, 90.

13  T. Rimer, "Film and the Visual Arts in Japan: An Introduction," in L. Ehrlich and D. Desser (eds), *Cinematic Landscapes: Observations on the Visual Arts and Cinema of China and Japan*, Austin: University of Texas Press, pp. 149–54.

14  On the significance of the "formless" in modernism see Y.A. Bois and R.E. Krauss, *Formless: A User's Guide*, New York: Zone Books, 1997.

15  Cited in A. Ross, *The Rest is Noise: Listening to the Twentieth Century*, London: Fourth Estate, p. 517.

16  Grilli, "Teshigahara and Takemitsu."

17  See R. Koehler, "Three Films by Hiroshi Teshigahara," *Cineaste* 33, Winter 2007, 77–9; and D. Toop, *Haunted Weather: Music, Silence and Memory*, London: Serpent's Tail, 2004.

18  R. Bergan, "Kyoko Kishida: Obituary," *The Guardian*, 8 February 2007. Kishida had garnered a reputation for controversial roles through films such as Yasuzo Masumura's *Manji* (1964) where she plays a bored housewife who has a female lover. The screenplay for *Manji* was provided by another newly emerging director Kaneto Shindo whose widely acclaimed *Onibaba* (1964) uses a desolate late-mediaeval landscape of reeds as a setting for his exploration of sexual jealousy.

19  Some reviews, particularly in Britain, were quite dismissive of Teshigahara's "esoteric" cinematic vision. See, for example, I. Quigly, "Overcome by Sand," *Spectator*, 7 May 1965.

20  M. Faass, "Utopie: Lichtgestalten in der Landschaft," in W. Hornbostel and N. Jockel (eds), *Nackt: Die Ästhetik der Blöße*, Munich: Prestel, 2002, pp. 127–36; C. Remm, "Otto Muellers Akte in der Natur," in M.M. Moeller (ed.), *Auf der Suche nach dem Ursprünglichen: Mensch und Natur im Werk von Otto Mueller und den Künstlern der Brücke*, Munich: Hirmer, 2004, pp. 9–26.

21  A similar line of argument in relation to Renoir is developed by L. Nochlin, *Bathers, Bodies, Beauty: The Visceral Eye*, Cambridge, MA: Harvard University Press, 2006.

22  Abe, *The Woman in the Dunes*, p. 142.

23  I. Standish, *A New History of Japanese Cinema: A Century of Narrative*, New York: Continuum, 2005.

24  R. Gertner, "Woman in the Dunes," *Motion Picture Herald* 232, 11 November 1964, 10.

25  On the fading of modernism see P. Anderson, *The Origins of Postmodernity*, London: Verso, 1998.

26  V. Nabokov, *Lectures on Literature*, San Diego, Harcourt, 1980.

27  See R. Rorty, *Contingency, Irony, and Solidarity*, Cambridge: Cambridge University Press, 1989.

28  Nabokov, *Lectures on Literature*, p. 283.

# 20

# RESTORATION

## Synoptic reflections

*David Lowenthal*

"O! call back yesterday, bid time return," cries Salisbury in Shakespeare's *Richard II*, bearing the king dire news. Yearning for restoration is age-old. So is faith in its advent. "Every city and village and field will be restored, just as it was," foretold a fourth-century ecclesiastic.[1] From divine fulfillment, restoration devolved into human agency. "Not a thing in the past has not left its memories," mused H.G. Wells. "Some day we may learn to gather in that forgotten gossamer, ... weave its strands together again, until the whole past is restored to us."[2]

Restoration implies going back to an earlier condition, often the pristine original. The previous is held better – healthier, safer, purer, truer, more enduring, beautiful, or authentic – than what now exists. Whether with lost or stolen property, damaged paintings, deteriorated health, reputations damaged by accusation or slander, security from danger, or undermined trust, the aim is "retrieval of an original favored condition."[3] The purpose is essentially therapeutic: to recoup physical health, to redress a grievance, to mend social wounds.[4] To restore is to make whole again, in plain defiance of "All the king's horses, and all the king's men [who] couldn't put Humpty together again." "One can no more restore an area of natural beauty – or a painting ... – to its original state than one can turn women into the little girls they once were."[5]

## Restoration ubiquitious and innate

Restorations are legion: forests, rivers, gardens, governments, buildings, furniture, sculpture, paintings, music, medicine, bygone ideologies and dynasties – Confucian precepts, Ten Commandments, Augustan, Stuart, Bourbon, Meiji regimes. Image restoration ranges from digital repair of degraded photographs to public relations repair of tarnished reputations of countries or corporations, priests or presidents.[6] Yet restoration realms are seldom viewed in concert. While "literature from ... history, anthropology, and philosophy tackles ... restoration goals," one finds "little cross-referencing among these disciplines or with the ecological literature."[7] Few would guess the journal *Contemporary Esthetics and Restorative Practice* concerns teeth, or that dentistry debates – aggressive versus conservative restoration, stability versus aesthetics – mirror other restoration realms.[8] *Restoring Nature* is subtitled *Perspectives from the Social Sciences and Humanities*, its authors mainly philosophers and social scientists; but its thrust is entirely environmental.[9]

Paul Eggert's *Securing the Past* compares painting, sculpture, and building conservation with textual editing. This "first concerted effort to examine together the linked philosophies" of building and art restoration with literary works notes that these realms share traditions of misguided confidence in non-intrusive yet definitive restoration.[10] Absent from Eggert's purview, however, is ecology. Heeding pleas for converse between restorationists of nature and of culture,[11] I here discuss their intertwined histories.

Restoration is instinctive. Young children credit restorative powers that rejoin things broken, rejuvenate the old, bring the dead back to life. Nothing irretrievably wears out, no act is irreversible.[12] To restore someone (or something) to a state prior to harm is not only achievable but obligatory. The child feels responsible for the injury and must make amends – reparation.[13] Gradually we relinquish our own and others' restorative prowess. Yet the urge to recover remains compelling: residues of restorative faith suffuse thought, speech, and behavior. Like the Victorian poet, we call back yesterday: "Backward, turn backward, O Time, in your flight,/Make me a child again just for to-night!"[14]

To regain the "sweet and curious apprehensions" of childhood innocence was Thomas Traherne's seventeenth-century dream:

> All things were spotless and pure and glorious ... I knew not that there were any sins, or complaints, or laws.... I knew nothing of sickness or death.... All time was eternity, and a perpetual Sabbath.... Boys and girls tumbling in the street, and playing, were moving jewels. I knew not that they were born or should die. But all things abided eternally ... The city seemed to stand in Eden.[15]

Traherne's dream mirrored his clerical contemporary Thomas Burnet's accolade to "Providence; which loves to recover what was lost or decayed, ... and what was originally good and happy, to make it so again." Like Burnet's restored primordial Earth, "smooth, regular, and uniform, [with] not a wrinkle, scar or fracture in all its body,"[16] Traherne's childhood Eden is a prelapsarian paradise.

Traherne and Burnet followed Renaissance and Reformation zeal to restore ancient pasts. Renaissance humanists sought to reverse the medieval retrogression that had obliterated the Classical legacy, religious reformers to cleanse the church of Satanic corruption that defiled Christianity's original innocence. Both evils must be expunged to retrieve pure inspiration, Christ or the classics. Restoring the Golden Age – the primitive church, the classical vision – were humanists' and Protestants' parallel and often conjoined aims.[17] Their restorations were archeological and medical.[18] Like antiquarians piecing together imperial Rome from vestigial temples and statuary, scholars collated remnants of classical "unearthed fragments." reuniting the Roman rhetorician Quintilian's "mangled and mutilated" *Institutes of Oratory* returned him to "dignity and ... sound health."[19] Humanists restored lacerated heroes – ancient exiled texts – to honor and safety.[20] Pursuing his life-work's restoration of theology, Erasmus, unwell, hoped the physician Paracelsus might "restore me also."[21]

## Time's cycles and arrows in terrestrial and human history

We sense time as circle and arrow. Time's circles are nature's recurrent constancies: the waxing and waning, ebbing and flowing of diurnal, lunar, and seasonal rhythms and planetary orbits, our breathing and heartbeats, sleeping and waking. Time's cycles are restorative.

Time's arrow denies restoration. The arrow flies only once from irrecoverable past toward foreign future. It targets the contingent events and sporadic vagaries of history, beyond natural law's clockwork time. The interplay of circle and arrow continually shapes our lives. Habitual customs – lawlike, regular, predictable – interact with the uncertainties of history's directional, singular events. Society overlays rhythmic stability with novel disturbances.[22]

Past and future long seemed much alike, however. In traditional societies, lived time was more circle than arrow, human annals overwhelmingly repetitive. There was no new thing under the sun (Ecclesiastes 1:9). The fixity of divine and natural law and the enduring sameness of human nature ensured what had happened before would happen again. Resurrection and re-enactment suffused religion. Secular chronicles were repetitive: like Earth reborn every New Year's Day, each ruler's inaugural year restarted the calendar at year one.[23] Politics too were recurrently repetitive. Monarchy led to tyranny, aristocracy, oligarchy, democracy, and anarchy, then restoration to monarchy.

Time's cycle dominated geology. James Hutton limned Earth's saga "as a stately series of strictly repeating events, the making and remaking of continents as regular as the revolution of planets." Continental uplift recurrently restored matter washed away from the land. Hutton's terrestrial uplift, erosion, deposition, back to uplift mirrored the monarchy–aristocracy–oligarchy–democracy–anarchy–monarchy cycle.[24] Charles Lyell depicted Earth cycling in eternal steady state, every destruction restored. Singular events – floods, earthquakes, comets – were trifling local aberrations amid overarching uniformity. Lyell posited wholesale restoration. When global warming resumed, "then might those genera of animals return, of which the memorials are preserved in the ancient rocks." Out-aging today's Pleistocene 're-wilders', Lyell fancied "the huge iguanodon might reappear in the woods, and the ichthyosaur in the sea, while the pterodactyl might flit again through the umbrageous groves of tree-ferns."[25] Only after 1860 did fossil and artifactual finds,

Darwinian evolution, and Marshian ecological history persuade Lyell to admit progress in animate life, include mankind in natural history, and relinquish time's restorative cycle for time's directional arrow.[26]

## Restoration denied: gospels, enlightenment, evolution, entropy

Time's arrow took early flight in Judeo-Christian annals. The Old Testament chronicled a contingent history following one-time Creation and Adam's Fall; the New Testament inserted into secular annals unique sacred events, the birth, life, death, and resurrection of Christ, and His eventual Second Coming. Events stemming from divine or human will happen only once. But such temporal consciousness was absent from the habitual regimen of ordinary life, rare even in scholarly thought until the eighteenth century.

Enlightenment *savants* began to see human affairs less as repetitive or cyclical than impelled by ongoing improvement. Innovation, once a threat to settled order, became a welcome harbinger of progress. Far from impious, secular advance fulfilled scriptural commands: God left the world unfinished for man, made in His image, to perfect. Progress was cumulative, each generation building on prior advances. The shift from restorative circle to directional arrow transformed Western culture. Stasis gave way to linear progress; preordained *life cycles* became self-fashioned *life courses*.[27] New World Manifest Destiny corroborated foretold perfectibility.

Natural history too was transformed. Cyclic regularities were interrupted, accelerated or retarded, warped into new trajectories by episodic catastrophes – asteroid impacts, tsunamis, heat-occluding volcanic dust. One-off dislocations reshaped continents with novel landforms, plants, and animals – incontrovertible proof of time's arrow. Natural history did not conflict with biblical faith, indeed, the Christian saga's unique and unrepeatable history stimulated awareness of nature's contingencies. Earth, like human history, shed perennial regularities for sporadic singularities.[28]

Evolutionary biology reinforced circle-to-arrow shifts. Selection generated ever-novel conditions, extinguishing old and engendering new life-forms. Nothing struck (indeed, dismayed) Darwin more forcibly than the absence, amid teeming extant species, of virtually all previous ones. He dolefully concluded that "not one living species will transmit its unaltered likeness to a distant futurity."[29] Darwin's "intolerable thought that [man] and all other sentient beings are doomed to complete annihilation"[30] was memorably echoed, the very year *On the Origin of Species* appeared, in FitzGerald's *Rubaiyat*:

> One thing is certain – *This* life flies;
> One thing is certain and the rest is Lies;
> The Flower that once has blown forever dies.[31]

Even more sobering than biological carnage was the wasting of the entire universe, notably the sun's impending heat death. "Within a finite period," Kelvin warned of entropy's grim implacability, "the earth [would become] unfit for the habitation of man." Indeed, no creatures could "enjoy the light and heat essential to their life, for many million years longer."[32] Such forecasts undermined faith in nature's ever-lasting stability. The withdrawal of the sun became, with Freud, a classic symptom of paranoia. Humanity's age-old terror lest daylight not be restored remains part of today's mind-set.[33]

## Restoration redux: nostalgic reaction to history's terrors

The anxieties unleashed by time's arrow went beyond mordant awareness of evolutionary extinctions and the second law of thermodynamics. Especially distressing was the erosion of clockwork regularities by volatile uncertainties in human affairs. "I shot an arrow into the air/It fell to earth, I knew not where."[34] Time's straying arrow, mainly exposing mankind's crimes and errors, left outcomes uncertain, unforeseeable, bewildering. Accelerated historical change fuelled fears of society spinning out of control. "The series of events comes swifter and swifter," wrote Carlyle, "velocity increasing as the square of time."[35] From Tennyson and Hardy and Ruskin to Spengler and Wells, *savants* likened the decline of the West to the death of the universe. *Fin-de-siècle* forecasts of universal winter, reprieved by Rutherford's 1904 discovery of solar radioactivity, resurfaced as fiery fate by nuclear fission. The Bomb and its dread progeny made doom-laden prognoses common coin.[36]

Such fears intensified nostalgia for restoration to earlier times when change was slow, cyclic, or imperceptible. Haunted by Mircea Eliade's "terror of history,"[37] men averted gaze from temporal chaos, hoping to restore intelligible, dependable certitudes. Against revolution made fearsome in regicidal England and France, Stuart and Bourbon monarchical restoration promised time-honored security.[38]

## Restoring what feels natural, restoring nature

Regime restorers eagerly deployed analogies with nature. Edmund Burke termed nature's stability the antidote to French revolutionary turmoil. Ingrained custom was "natural": statesmen should emulate nature's cyclical constancy. "By preserving the method of nature in the conduct of the state, in what we improve we are never wholly new, in what we retain we are never wholly obsolete." In commonwealths that respected tradition no radical break disturbed life's regular rhythms, and subjects rested content with ancient authority. Seventeenth-century English restorers had "regenerated the deficient part of the old constitution through the parts which were not impaired. They kept these old parts exactly as they were, that the part recovered might be suited to them."[39] Burke's aim to minimize change in human affairs foreshadowed the tectonic gradualism of Lyell, for whom slow and insensible mental progress mirrored nature's own way.[40]

Against convulsive innovation, traditional custom was extolled as "natural." The goal was not, however, restoration to Hobbes's state of nature, devoid of civic improvement. Domesticating raw nature – building shelters against climatic extremes, cooking food to make it clean and safe, storing grain, oil and livestock

against dearth or famine – was "natural." Restoring the fabric of nature itself gained
favor in Victorian reaction to industrial blight and urban squalor. But like biblical
Eden, English "nature" thus redeemed was agrarian or emparked or gardenesque,
thoroughly humanized, intensely managed. Bucolic nature was perfected by culti-
vation and control, idealized in the canvases of Claude and Poussin, the verses of
Goldsmith and Wordsworth.

First to exalt untouched nature as Edenic restoration were nineteenth-century
Americans aghast at their land's lost purity and vanishing wilderness.[41] Their icon
was Longfellow's forest primeval, not a Wordsworthian garden. "One is nearer God's
heart in a garden/Than anywhere else on earth," penned Dorothy Frances Gurney
in England; but across the Atlantic her "garden" literally morphed into "forest."[42]

The wilderness cult transformed symbols of triumphant conquest into emblems
of horrendous despoliation. Once betokening civilized progress, the logger's axe
and the hewn stump now bespoke the rape of nature.[43] In New York's Central Park
Frederick Law Olmsted "planted trees to look like "natural scenery" with such
success that those who accepted the scenery as 'natural', objected to cutting the
trees he had planned to cull."[44] Olmsted's naturalistic deceptions heralded land-
scapers who aimed "to create just the right look of human-free nature in each
national park."[45]

Re-wilding featured a 1908 best-seller set in Tennessee's Cumberland Gap. Fouled
by soulless loggers, a once crystal-clear stream was "black as soot." Tree-felling, "the
cruel deadly work of civilization, [meant] a buzzing monster ... biting a savage way
through a log, that screamed with pain as the brutal thing tore through its vitals." Our
hero, a mining engineer turned nature-lover, vows to restore Lonesome Cove:

> "I'll tear down those mining shacks, ... stock the river with bass again. And
> I'll plant young poplars to cover the sight of every bit of uptorn earth ... bury
> every bottle and tin can ... take away every sign of civilization..."
>
> "And leave old Mother Nature to cover up the scars," says his fiancée,
> June.
>
> "So that Lonesome Cove will be just as it was."
>
> "Just as it was in the beginning," echoed June.
>
> "And shall be to the end."[46]

Restoration redeems all: corporate greed vanquished, machine-age poisons excised,
nature healing, Edenic plenitude in everlasting tranquility.[47]

Restoration ecology's redemptive bent reflects biblical tradition. Repairing "to
the same state again" accords with Burnet's "methods of Providence." The second
Golden Age of his *Sacred Theory of the Earth* (1691) would restore all lands "to the
same posture they had at the beginning ... before any disorder came into the natural
or moral world."[48] Two centuries on, another English cleric rejoiced that St. John's
Revelation placed "the restoration of man and the restoration of nature ... side by
side."[49] Mortified by Lynn White's 1967 indictment of Judeo-Christian environmental abuse – "the most anthropocentric religion the world has seen" – restora-

tion theologians now entreat humanity to bring the created world close to that perfect restoration destined by God, and thereby gain His forgiveness.[50] 'As we are restored to relationship with God', preach ecojustice theologians, 'so must we restore this watershed.'[51]

## Decay and restoration in the arts

Restoration in architecture, sculpture, and painting reflects changing notions of originality, authenticity, purity, and sustainability. Around 1800 the visual arts began to exchange classical completeness for romantic fragmentation – letting, even helping, things decay. Before, antique sculptures were seldom exhibited or marketed unless restored. After, breakage and mutilation proclaimed a work's age and fidelity to origins. Marble limbs were hacked off, coins and canvases artificially aged, buildings artfully disarrayed, furniture distressed. Picturesque ruination patinated buildings and gardens with broken stones, moss, and lichen. The cult of ruins extended to literature and to life itself, memoirs typically termed "Fragments," suicide suggesting genius truncated.[52]

But eternal verity, not mortal decay, inspired Victorian church restorers. Dilapidated medieval structures were restored to how they *should* have been. Thousands of French and English cathedrals and churches were antiquated, with nineteenth-century materials and technical skills, back to "pure" Gothic. Aesthetic piety replaced ("Scraped") ancient builders' errors and imperfections with idealized archaisms. "To restore a building [was] to reinstate it in a condition of completeness that could never have existed at any given time."[53]

"Scrape" incited "Anti-Scrape," forbidding all renovation. "Restoration is impossible," proclaimed a historian–archivist. "You may repeat the outward form … but you cannot the material, the mortar, … and above all the tooling … There is an anachronism in every stone…. The sensation of sham is invincible."[54] William Morris held medieval buildings "monuments of a bygone art, created by bygone manners, that modern art cannot meddle with without destroying."[55] John Ruskin termed restoration not only implausible but impious. Old edifices, like living beings, deserved daily care, not artificial rejuvenation. Venerable buildings, like Burke's ancient institutions, should be interfered with as little as possible.[56]

Revealingly, ecclesiastical restorers who reiterated these precepts were shocked when shown to have contravened them. Notorious for replacing surviving Norman with neo-Gothic, George Gilbert Scott claimed he sought "the least possible displacement of old stone, [because] an original detail [however] decayed and mutilated [was] infinitely more valuable than the most skilful attempt at its restoration." Scott confessed to slippage between belief and behavior. Having "restored" a dilapidated fifteenth-century chapel, he was "filled with wonder how I ever was induced to consent to it at all, as it was contrary my own principles."[57] The architect G.E. Street likewise flouted his own precept that "in dealing with old buildings … we cannot be wrong in letting well alone." He reproached restorers of Burgos Cathedral and St. Mark's, Venice, for the same intrusive meddling that led Street himself to replace a fourteenth-century choir arm of Dublin's Christ Church Cathedral with a pastiche of the original.[58] Yet neither Scott nor Street were hypocrites. Manufacturing simulacra, they believed they were restoring the true past.

Anti-scrape tenets long ruled European art and architecture. Restoration was deemed a last resort. When forced to intervene lest a building collapse or a painting perish, conservators stressed fidelity to origins, expunging previous ill-conceived restorations, accentuating what was oldest. Unavoidable replacements contrasted in texture and color. But blatant new–old disjunctions destroyed aesthetic unity and dimmed antiquity's aura.

Hence some later restorers stressed emphasized *venerableness*, antiquating paintings with patinas of varnish, buildings with lichen, while others stripped off marks of age and wear to reveal "original" *aesthetic* qualities; retrieving supposed initial intent, many Old Masters emerged from restoration with varnish removed and colors gleaming as if new-made. Either way, aesthetic taste overruled anti-scrape purism. Replacements now matching, not clashing with, original elements were detectable only on close inspection of tooling on stonework, dates on stained glass, slightly differing tints.

The restoration history of Giotto's fourteenth-century *Life and Miracles of St. Francis*, in Santa Croce, Florence, typifies changing criteria. Partly destroyed and whitewashed over in the eighteenth century, the frescoes were restored in the mid nineteenth century by Gaetano Bianchi, who repainted Giotto's scenes, adding pseudo-Giottoesque figures in lacunae. Mid-twentieth-century restorers condemned Bianchi's "forgeries," expunged by Leonetto Tintori to highlight the remnant original. Critics "praised the recovery of the 'true', albeit fragmentary state" showing Giotto's "original intent." But the 1970s repudiated Tintori's purist restoration as "optically disruptive." Stressing "visible dialogue between past and present," restorers then recaptured the frescoes' original significance as "Franciscan stories with deep Christian meaning."[59]

Restoration increasingly embraces artifacts' total history. Besides original materials, forms, and intentions, worth inheres also in the attritions of time and the interventions of collectors, curators, conservators, even forgers. Valued objects continually accrue new meanings and values, altering or shedding older ones. Prizing the palimpsest means respecting previous restorations, historically significant for, often visually essential to, the surviving original. James Wyatt's reviled (and later replaced) restoration at Hereford Cathedral would now be seen "integral to the building's history, no more to be demolished than the medieval masonry behind" it.[60] "What de-restoration achieves" may be less informative than what can be learned from past restoration.[61]

Restoration conflicts ceaselessly embroil conservators, curators, and the public. Promoters lauded the 1990s' renovation of the Sistine Chapel and Michelangelo's *Last Judgment*, freeing the frescoes of five centuries' accumulated grime, crude repainting, and earlier restorers' darkened glue, as a "Glorious Restoration" revealing Michelangelo's coloristic genius. However, critics termed it Chernobyl-like destruction voiding the frescoes of divine inspiration.[62] For restorers, removing time's disfiguring veil, recovering "full chromatic effects," dispelled nineteenth-century misconceptions of Michelangelo as "a black and melancholy artist." But critics contended the expunged veil contained Michelangelo's *a secco* shadow-and-chiaroscuro finishing, the darkening intended, his art tonal not chromatic.[63]

Difficulties of ascertaining intention – the restoration perhaps revealed Michelangelo's first creative burst at the expense of his second thoughts – afflict most restorations. Creators change aims midstream, often again after finishing. "Any artist's intention is a complex and shifting compound of conscious and unconscious aspirations, adjustments, re-definitions, acts of chance and evasions,"[64] further affected by patrons, clients, collaborators, publishers, editors, creditors with whom creators may or may not agree, but must nonetheless contend, lest their work remain unfinished, unsung, or unsold.[65]

But injunctions about original structures, original intent, original anything are fast eroding. That no assemblage, no structure, no image can be returned to any previous state is ever more evident. Merely approximating or suggesting what once was, every restoration is filtered through and tinctured by irremediably modern minds.[66] Indeed, "the whole point of restoration is to change an object," holds a prominent conservator, whether by increasing its life-span or enhancing its appearance or history.[67]

## Divergent theory and practice vis-à-vis nature and culture

Restorers' images commingle. Aldo Leopold's maxim for conserving biota, never to "discard seemingly useless parts, [for] to keep every wheel and cog is the first precaution of intelligent tinkering,"[68] echoes Burke's constitutional restorers who "kept these old parts exactly as they were" and Scott's "least possible displacement of old stone." Ecologists' homeostatic metaphors from self-regulating machines in engineering and cybernetics gave way to aesthetic terminology, thence, consonant with man as nature's steward rather than master, to medical and health-care argot.[69]

As guardians of culture shifted from remaking things whole, to revering original fragments, to recreating palimpsests, analogous impulses led environmental stewards from regenerating gardens, to restoring degraded landscapes, to rewilding, to processual concerns. Insights from art and architecture were cautionary: "Just as faked art is less valuable than authentic art, faked nature is less valuable than original nature."[70] Bureau of Reclamation plans to dam the Grand Canyon for tourist access were likened to "flood[ing] the Sistine Chapel so tourists can get nearer to the ceiling."[71] Whether damaged landscapes can or should be restored to their "original" state became the "Sistine Chapel debate."[72] But these parallels miscarry. Appraised differently, legacies of nature and culture arouse unlike restoration aims. Culture's restorers revert either to specific moments – of inception or creation, of peak prowess or beauty or fame, of iconic persons or events – or to particular eras or life-spans. Neither suits nature: lacking creators and moments of creation, natural assemblages continually evolve. Paintings, plays, creeds, legal codes get restored to whenever most new and fresh, effective, intelligible, admired, or sacred. No such criteria motivate ecological restorers, whose desiderata often antedate human existence. Landscapes are usually returned to when least anthropogenically disturbed or most ecologically copious or sustainable.

Cultural legacy is prized mainly as individual items noteworthy for specific persons, events, or qualities – Lascaux, Stonehenge, Parthenon, Chartres, Monticello, Gettysburg, Mona Lisa, a Shakespeare First Folio – dating from finite times. Contrariwise, nature's legacies are mainly aggregates engendered over eons; valued less as single plants or animals than clusters, swarms or herds, species, genera or ecosystems.[73] Indeed, some dismiss endangered species as doomed relics deflecting concern from restoring natural selection, currently imperiled by homogenizing human forces.[74] Unlike artifacts and works of art, restoring particular species or discrete reserves moreover impinges beyond boundaries. River restorers consider downstream impacts; reintroducing wolves to national parks takes note of distant ranchers' fears; transgenic restorers of American elm and chestnut address species alteration, ecosystem imbalance, and food chain worries.[75]

Nature restorers integrating ecosystem management with individual and community healing deploy medical analogies. As a prosthesis aims "to rehabilitate the function of leg rather than to recompose original flesh and bones," so a restoration ecology stresses function over composition – "not just certain species, communities, or habitats, but all natural and anthropogenic flow processes."[76] People seemingly "relate to the analogy of restoring the human body and are intrigued by the similarities." Both feature multiply interacting parts; "environment is not something to be passively fixed like a car but … actively healed like the human body." Ecosystem intervention becomes major surgery, clearing stream and shore debris resembles aspirins for headaches, band-aids for bruises.[77] Environmental psychologists credit psychic health restoration to recuperation in peaceful natural surroundings, those restored in turn promoting nature restoration.[78]

Like physicians, ecologists give nature a helping hand. Ecological restoration is likened to setting a broken bone: after the healer resets the trajectory, nature does most of the work.[79] But fundamental difference remains. "Curing sick watersheds," claimed a botanist, "is not unlike the responsibilities demanded of doctors."[80] Nor is it quite like it. To ecologists "nature itself is the best restorer of all," as in the speedy unaided (after removing channelization and pollutants) restoration of Kissimee River sinuosity, and in tidal Thames fish recovery.[81] But physicians seldom exalt non-intervention. Dying landscapes are held to require terminal care like dying people.[82] But human death is terminal, whereas landscapes endure, however altered and transformed.

Artifacts and institutions are designed for human purpose; nature is undesigned. Restoring art alters existing artifacts; restoring nature turns it into an artifact. Hence restoration's "big lie": "Artifactual restored nature is … fundamentally different from natural objects and systems."[83] To be sure, most restorations aim to repair degraded artifactual nature; no place on Earth remains untouched by human agency. But we readily conceive pre-disturbance scenes and glorify untrammeled nature; artifacts cannot be so conceived. Nothing in art resembles the guilt-ridden restitution that drives ecological restoration;[84] geriatric medicine offers no parallel to re-wilding's spiritual epiphany.[85]

Immaculate purity, once divine and saintly, now idealizes nature more than culture. Nature is widely held to merit unshackled veneration. But that nature should repair itself – an idle fancy for humans dependent on agriculture, antibiotics, reservoirs, sewage systems – is scientifically untenable. Ecology seventy years ago abandoned equilibrium models that equated non-interference with environmental health and stable climaxes, yet many ecologists still elevate nature over culture, deploring humanity's imprint as retrogression from the untouched fundament. In the celebratory volume launching UNESCO's cultural landscapes program, essay after essay terms culture a menace to nature and ranks anthropogenic below pristine landscapes – even when acknowledging that none *is* pristine. A prime criterion for cultural World Heritage designation remains "harmony with nature."[86] Public faith in the beneficent stability of untamed nature persists in denial of untold natural disasters; hence the contrariness of gardeners who denounce exotics while happily cultivating them, of re-wilders who restore without sullying trace of human agency.

Stewards of nature and culture both know human meddling is ubiquitous and unavoidable, but react contrarily. Culture's custodians consider intervention normal and necessary, nature's feel it reprehensible, masking or conforming it to "natural" outcomes. Culture is "ours" to tamper with; nature, increasingly, is not. To be sure, many feel buildings and works of art should be left to age and perish at some "natural" tempo. But like physicians, most cultural conservators eschew the hands-off stance as untrue to history and unacceptable to their clients. In contrast, even remedial disturbance to "nature" seems distressing, notably among Americans.[87]

Consider public reactions to the Tower of Pisa, to Avebury, and to Yellowstone's Old Faithful geyser, each newsworthy for restoration to supposed stability. Few demurred against lifting the Pisa campanile, whose collapse was imminent, back to its historic incline; though nature caused it to lean, the tower is a human artifact. Underpinning and uprighting precariously tilted sarsens in Avebury's Neolithic stone circle likewise gained widespread approval;[88] not so Yellowstone. Public outrage met a Procter & Gamble television ad claiming that a dose of their laxative made Old Faithful's erratic eruptions "regular" again. Nature was *ipso facto* "faithful," butting in sacrilegious.[89]

Yet in ecology, as in art, what restorers do is often at odds with what they think or say. River restoration practitioners and stakeholders worldwide were asked what restoration meant.[90] Four out of five chose John Cairns's strict canonical definition, "complete structural and functional return to a predisturbance state."[91] But their work habitually embraced not just *return* and *recovery* but *improving*, even *creating*. Mirroring a British ecologist's "opportunity to create something new and more valuable than what was there originally,"[92] "most activities the restoration community undertakes are actually … rehabilitation or enhancement." Declaring fidelity to "the most … widely accepted definition in the restoration literature," they routinely transgressed it.[93] They willingly manipulated, but – like Victorian architectural restorers – were reluctant to say so. "We bury our efforts beneath an ecological cover, and pretty quickly a landscape that depends on or originates in extensive human contrivance becomes naturalized."[94] Hence complaints that landscape restoration is "gardening dressed up with jargon to simulate ecology," "quite literally, agriculture in reverse," "a fiction" better retitled landscape architecture.[95] "Restoration is fencing, planting, fertilizing, tilling, and weeding the wildland garden: ... afforestation, fire control, prescribed burning, crowd control, biological control, ... and much more."[96]

The phrase "Much more" means restoring not just ecosystems but "the human communities that sustain and are sustained by" them. Recognizing that "ecosystem values will shift over time as they have been doing throughout history," restorers must work with "kindred, intellectual adventurers" in the social sciences, arts, and humanities.[97]

## Embryonic restoration

Christian theologians envision restoration as preparation for future redemption. Anticipatory restoration appeals to visionaries concerned about potential apocalypse. What future generations may want is unknowable, but we can forecast what they may need to recover from global calamity. To regenerate a viable ecology or workable society, access to records, ideas, and techniques of civilized history could be crucial. An encyclopedic time capsule was constructed in China between the seventh and twelfth centuries. Fearing imminent cataclysm, Buddhists at Cloud Dwelling Monastery near Beijing inscribed the tenets and history of their faith on thousands of stone tablets sealed in caves, for survivors' potential rehabilitation. "After the days of doom, the [tablets] would emerge from the earth" to edify future people.[98]

Feeding distant posterity spawned the Svalbard seed bank in arctic Norway, storing millions of seeds and animal DNA samples in sub-freezing safety, so that survivors of global catastrophe might restart agriculture.[99] Others envisage the erosion-free moon, far enough from Earth to escape contamination from nuclear fallout, as a DNA storage locker.[100] Most durable is deep space. The science-fiction-inspired Alliance to Rescue Civilization would launch a rocket packed with data about Earth and its inhabitants for far-future retrieval in some remote galaxy.[101] Restoration as potential and incipient, as seed not fruit, as data not deed, is cumulative, inviting ongoing enrichment by future generations. Envisaged restoration deploys humanity's continuing creative along with its conserving instincts.

Creation and restoration were once divinely conjoined vocations. God created; repentant humans restored, not only repairing losses caused by sinful corruption, but also *improving* pristine nature. "Restoration heals and reveals, illuminates and demonstrates," proclaimed the twelfth-century theologian Hugh of St. Victor. Indeed, "the works of restoration are much worthier than the works of creation; because those were made for service, ... these for salvation."[102] Restoring things ruined and laid waste perfected them – "the water of creation reformed into the wine of restoration." This reading of restoration became the "Fortunate Fall" of

Milton's *Paradise Lost*, Adam's sin engendering a greater good "more wonderful than that by which creation first brought forth light out of darkness."[103] Lauding Pope Leo X for revitalizing Rome, Erasmus likewise asserted that "to restore great things is sometimes not only a harder but a nobler task than to have introduced them."[104]

This humanist sacred vision succumbed to the cult of originality. Although he lauded nature's cyclical restorations, Lyell rejected them in human affairs. He ridiculed ancient mythic reiterations wherein "the same individual men were doomed to be re-born, ... the same arts ... invented, and the same cities built and destroyed." In lieu of restoration he celebrated rediscovery. Delving into Earth's past, Lyell borrowed a celebrated aphorism from Barthold Niebuhr's *History of Rome*: "He who calls what has vanished back again into being, enjoys a bliss like that of creating."[105] Like Hamlet's father, the past is reborn in the mind's eye.

## Notes

1 Bishop Nemesius of Emesa, "On the Nature of Man," quoted in G.J. Whitrow, *The Nature of Time*, London: Penguin, 1975, p. 17.

2 H.G. Wells, *The Dream*, London: Collins, 1929, p. 236.

3 R.A. Duff, "Restorative Punishment and Punitive Restoration," in L. Walgrave (ed.), *Restorative Justice and the Law*, Cullompton, England: Willan, 2002, p. 84.

4 B.A. Weiner, *Sins of the Parents: The Politics of National Apologies in the United States*, Philadelphia, PA: Temple University Press, 2005, p. 116; C.S. Maier, "Overcoming the Past? Narrative and Negotiation, Remembering and Reparation: Issues at the Interface of History and the Law," in J. Torpey, (ed.), *Politics and the Past: On Repairing Historical Injustices*, Lanham, MD: Rowman and Littlefield, 2003, pp. 295–304.

5 M. Dekkers, *The Way of All Flesh: A Celebration of Decay*, London: Harvill, 2000, p. 94.

6 W.L. Benoit, *Accounts, Excuses, and Apologies: A Theory of Image Restoration Strategies*, Albany: SUNY Press, 1995; B.A. Miller, *Divine Apology: The Discourse of Religious Image Restoration*, Westport: Praeger 2002; T.L. Pedigo, *Restoration Manual: A Workbook for Restoring Fallen Ministers and Religious Leaders*, fifth edn, Colorado Springs: Winning Edge, 2007; J. Kauffman, "When Sorry is Not Enough: Archbishop Cardinal Bernard Law's Image Restoration Strategies in the Statement on Sexual Abuse of Minors by Clergy," *Public Relations Review* 34:3, Sept. 2008, 58–62; J.R. Blaney and W.L. Benoit, *The Clinton Scandals and the Politics of Image Restoration*, New York: Praeger, 2001.

7 R.J. Hobbs, "Setting Effective and Realistic Restoration Goals: Key Directions for Research," *Restoration Ecology* 15, 2007, 354–7 at 356.

8 G.J. Christensen, "What Has Happened to Conservative Tooth Restorations?" and "Longevity versus Aesthetics: the Great Restorative Debate," *Journal of the American Dental Association* 136, 2005, 1436–7 and 138, 2007, 1013–15.

9 P.H. Gobster and R.B. Hull (eds), *Restoring Nature: Perspectives from the Social Sciences and Humanities*, Washington, DC: Island Press, 2000.

10 P. Eggert, *Securing the Past: Conservation in Art, Architecture and Literature*, Cambridge: Cambridge University Press, 2009, i, 9.

11 M. Hall, *Earth Repair: A Transatlantic History of Environmental Restoration*, Charlottesville, VA: University of Virginia Press, 2005, p. 245.

12 J. Piaget, *The Child's Conception of the World* [1929], Paterson, NJ: Littlefield & Adams, 1960, pp. 361–7; V. Slaughter, R. Jaakola, and S. Carey, "Constructing a Coherent Theory: Children's Biological Understanding of Life and Death," in M. Siegal and C. Peterson (eds), *Children's Understanding of Biology and Health*, Cambridge: Cambridge University Press, 1999, pp. 71–96.

13 B. Hamber, "Narrowing the Micro and the Macro: A Psychological Perspective on Reparations in Societies in Transition," in P. de Greiff (ed.), *The Handbook of Reparations*, Oxford: Oxford University Press, 2005, pp. 562–3.

14 E. Akers Allen, "Rock Me to Sleep, Mother" [1859], Boston, 1883.

15 T. Traherne, "The Third Century" [*c*.1660], in his *Centuries, Poems, and Thanksgivings*, 2 vols, Oxford: Clarendon Press, 1958, pp. 110–12.

16 T. Burnet, *The Sacred Theory of the Earth* [1691], Carbondale, IL: University of Illinois Press, 1965, pp. 53, 64.

17 A. Kemp, *The Estrangement of the Past*, London: Oxford University Press, 1991.

18 T.M. Greene, *The Light in Troy: Imitation and Discovery in Renaissance Poetry*, New Haven, CT: Yale University Press, 1982, p. 92.

19 Poggio Bracciolini to Guarino of Verona, 1446, in *Petrarch's Letters to Classical Authors*, Chicago, IL: University of Chicago Press, 1910, p. 93.

20 A.B. Giamatti, "Hippolytus among the Exiles: The Romance of Early Humanism," in his *Exile and Change in Renaissance Literature*, New Haven, CT: Yale University Press, 1984, pp. 24, 26.

21 Desiderius Erasmus to Theophrastus Paracelsus, March 1527, in J. Huizinga, *Erasmus and the Age of Reformation* [1924], New York, NY: Harper & Row, 1957, pp. 242–3.

22 S.J. Gould, *Time's Arrow, Time's Cycle: Myth and Metaphor in the Discovery of Geological Time*, Cambridge, MA: Harvard University Press, 1987, pp. 196–8.

23 Not until the eighteenth century was regnal dating replaced by the global chronology that subjects us all to time's arrow.

24 J. Hutton, *Theory of the Earth with Proofs and Illustrations*, 1795; Gould, *Time's Arrow*, pp. 77–9, 129.

25 C. Lyell, *Principles of Geology, Being an Attempt to Explain the Former Changes of the Earth's Surface by Reference to Causes Now in Operation*, London: 1830, 1: pp. 75–6, 141–2, 165–6, 473; Gould, *Time's Arrow*, pp. 105–45. Whereas Lyell relied on nature to resurrect extinct megafauna, current re-wilding requires human agency to transplant extant proxy species from other continents.

26 C. Lyell, *The Geological Evidences of the Antiquity of Man*, 1863; Gould, *Time's Arrow*, p. 168. G.P. Marsh's *Man and Nature*, New York, NY: 1864 forced Lyell to abandon his view that human impact on nature was negligible: Lyell to Marsh, Sept. 22, 1865, cited in D. Lowenthal, *George Perkins Marsh, Prophet of Conservation*, Seattle, WA: University of Washington Press, 2000, p. 302.

27 J. Demos, *Circle and Lines: The Shape of Life in Early America*, Cambridge, MA: Harvard University Press, 2004, pp. 61–77.

28 M.J.S. Rudwick, *Bursting the Limits of Time: The Reconstruction of Geohistory in an Age of Revolution*, Chicago, IL: University of Chicago Press, 2005, pp. 188–93, 642–51.

29 Charles Darwin, *On the Origin of Species by Means of Natural Selection* [1859], Oxford: Oxford University Press, 1998, p. 395.

30 Francis Darwin (ed.), *The Life and Letters of Charles Darwin, Including an Autobiographical Chapter* [1876/1887], Chestnut Hill, MA: Adamant, 2001, p. 282.

31 E. FitzGerald, *Rubáiyàt of Omar Khayyám*, London: 1859, p. lxiii.

32 W. Thomson (later Lord Kelvin), 'On a universal tendency in nature to the dissipation of mechanical energy,' *Proceedings of the Royal Society of Edinburgh*, 18 Apr. 1852; Thomson, 'On the age of the sun's heat,' *Macmillan's Magazine* 5 (5 Mar. 1862), 288–93; H. von Helmholtz, "Observations on the Sun's Store of Force," 1854, cited in R.C. Smith, *Observational Astrophysics*, Cambridge: Cambridge University Press, 1995, p. 240.

33 G. Beer, "'The Death of the Sun': Victorian solar physics and solar myth," in J.B. Bullen (ed.), *The Sun Is God: Painting, Literature, and Mythology in the Nineteenth Century*, Oxford: Clarendon Press, 1989; P.T. Davies, *The Last Three Minutes: Conjectures about the Ultimate Fate of the Universe*, New York, NY: Basic Books, 1994, pp. 9–13; "Rotation of Earth Plunges Entire North American Continent into Darkness," *The Onion*, 27 Feb. 2006, 1.

34 H.W. Longfellow, "The Arrow and the Song," 1845.

35  T. Carlyle, "Shooting Niagara: And After?" [1867], in his *Critical and Miscellaneous Essays*, London: 1887–88, vol. 3, p. 590.

36  J.H. Buckley, *The Triumph of Time: A Study of the Victorian Concepts of Time, History, Progress, and Decadence*, Cambridge, MA: Harvard University Press, 1967, pp. 55–70; Beer, "'The Death of the Sun," pp. 171–3.

37  M. Eliade, *The Myth of the Eternal Return*, New York, NY: Bollingen/Pantheon, 1954.

38  Geoffrey Cubitt, "The Political Uses of 17th-Century English History in Bourbon Restoration France," *Historical Journal* 50:1, 2007, 73–95.

39  Edmund Burke, *Reflections on the Revolution in France* [1790], Stanford, CA: Stanford University Press, 2001, pp. 184–5, 181, 170.

40  Lyell, *Principles of Geology*, vol. 1, p. 72.

41  C. Merchant, *Reinventing Eden: The Fate of Nature in Western Culture*, New York, NY: Routledge, 2003.

42  Sign on tree in Mianus River Gorge Preserve, Bedford, NY, quoted in J. Duncan and N. Duncan, "Aestheticization of the Politics of Landscape Preservation," *Annals of the Association of American Geographers* 91, 2001, 405.

43  Thomas R. Cox et al., *This Well-Wooded Land: Americans and Their Forests from Colonial Times to the Present*, Lincoln, NE: University of Nebraska Press, 1985, pp. 144–7; N. Cikovsky, "'The Ravages of the Axe': The Meaning of the Tree Stump in 19th-century American Art," *Art Bulletin* 61, 1971, 611–16 at 613.

44  A.W. Spirn, "Constructing Nature: The Legacy of Frederick Law Olmsted," in W. Cronon (ed.), *Uncommon Ground: Toward Reinventing Nature*, New York, NY: W.W. Norton. 1995, pp. 111–12.

45  Hall, *Earth Repair*, pp. 141–3.

46  J. Fox, Jr., *The Trail of the Lonesome Pine*, New York, NY: Grosset & Dunlap, 1908, pp. 201–2.

47  Federal restoration reified Fox's fiction: in 1940 Daniel Boone's famed Wilderness Road was reborn as Cumberland Gap National Historical Park.

48  Burnet, *The Sacred Theory of the Earth*, pp. 376, 257.

49  Regius professor of divinity B.F. Westcott, *The Gospel of Life: Thoughts Introductory to the Study of Christian Doctrine*, London: 1892, p. 243.

50  L. White, Jr., "The Historical Roots of Our Ecologic Crisis," *Science* 155:3767, 1967; D. J. Moo, "Nature in the New Creation: New Testament Eschatology and the Environment," *Journal of the Evangelical Theological Society* 49, 2006, 449–88.

51  W. Jenkins, *Ecologies of Grace: Environmental Ethics and Christian Theology*, New York: Oxford University Press, 2008, pp. 232–3.

52  I detail this transition in D. Lowenthal, *The Past Is a Foreign Country*, Cambridge: Cambridge University Press, 1985, pp. 145–82, and in "The Value of Age and Decay," in W.E. Krumbein et al. (eds), *Durability and Change: The Science, Responsibility, and Cost of Sustaining Cultural Heritage*, London: John Wiley, 1994, pp. 39–49.

53  E.-E. Viollet-le-Duc, "On Restoration," quoted in P. Eggert, *Securing the Past*, p. 54.

54  Francis Palgrave to Dawson Turner, 19 July 1847; F. Palgrave, *History of Normandy and England*, 1851; both in Lowenthal, *George Perkins Marsh*, p. 278.

55  W. Morris, "Repair not Restoration," 1877, in S. Tschudi-Madsen, *Restoration and Anti-Restoration: A Study in English Restoration Philosophy*, 2nd edn, Oslo: Universitetsforlaget, 1976, Annex VI.

56  J. Ruskin, *Modern Painters*, 1846 and *Seven Lamps of Architecture*, 1849, cited in Lowenthal, *George Perkins Marsh*, pp. 164–8, 278–80. To be sure, Burke *commanded* restoration, whereas Ruskin and Morris *condemned* it; but they treasured alike a past perfected by venerating its integrity.

57  G. Gilbert Scott, *Plea for the Faithful Restoration of Our Ancient Churches*, 1850, and *Recollections*, 1879, quoted in Lowenthal, *George Perkins Marsh*, p. 326.

58  G.E. Street, "Destructive Restoration on the Continent," 1857, "Report to the S.P.A.B.," 1880–1886, and *Some Account of Gothic Architecture in Spain*, 1865, quoted in Lowenthal, *George Perkins Marsh*, pp. 151, 278, 327.

59 C.S. Hoeniger, "Aesthetic Unity or Conservation Honesty? Four Generations of Wall-Painting Restorers in Italy and the Changing Approaches to Loss, 1850–1970," in A. Oddy and S. Smith (eds), *Past Practice – Future Prospects*, British Museum Occasional Paper No. 145, London: British Museum Press, 2001, quoting U. Baldini, 1978, B. Cole, 1976, and C. Brandi, 1963.

60 P. Wilkinson, "Restoration: A Dynamic Process," in *Restoration: The Story Continues*, London: English Heritage, 2004, p. 22.

61 J. Podany, "Restoring What Wasn't There: Reconsideration of the 18th-Century Restorations to the Lansdowne *Herakles* in the Collection of the J. Paul Getty Museum," in A. Oddy (ed.), *Restoration: Is It Acceptable?* British Museum Occasional Paper 99, London: British Museum Press, 1994, p. 15.

62 C. Pietrangeli et al., *The Sistine Chapel: A Glorious Restoration*, New York, NY: Abrams, 1999; P. L. Arguimbau, 'Michelangelo's Cleaned Off Sistine Chapel', blog, 5 Oct. 2006; R. Serrin, "Michelangelo and the Destruction of the Sistine Chapel," lecture, Classical Design Foundation, Jan. 5, 2006, incorporating his "Lies and Misdemeanors: Gianluigi Colalucci's Sistine Chapel revisited." Pietrangeli's "Glorious Restoration" patently recalls the biblically prophesized return of the Israelites and the ensuing millennial kingdom and England's Stuart Restoration; see J. Morrill, "The Later Stuarts: A Glorious Restoration?" *History Today* 38:7, July 1988, 8–16. "Glorious" also designates the 1990s restoration of Mission San Xavier del Bac in Tucson, Arizona, 2000s renovation of Mobile, Alabama's 1850s Cathedral-Basilica of the Immaculate Conception, the renovated seventeenth-century Nether Auchendrane House in Ayrshire, Scotland, and several architectural heritage projects in Ulster.

63 James Beck, with Michael Daley, *Art Restoration: The Culture, the Business and the Scandal*, New York: Norton, 1996, pp. 88–100; Eggert, *Securing the Past*, pp. 90–3.

64 M. Kemp, "Looking at Leonardo's *Last Supper*," in P. Booth et al. (eds), *Appearance, Opinion, Change: Evaluating the Look of Paintings*, London: UK Institute for Conservation, 1990, p. 18.

65 Hence the difficulty of deciding the *Urtext* for D.H. Lawrence's *Sons and Lovers* and for Theodore Dreiser's *Sister Carrie*, both original publications much altered by author-sanctioned editorial intercession, Eggert, *Securing the Past*, pp. 192–4.

66 S. Palazzi, "Restoration: Dealing with a Ghost," in A. Oddy and S. Carroll (eds), *Reversibility – Does It Exist?* British Museum Occasional Paper 135, London: British Museum Press, 1999, p. 176.

67 Salvador Munoz Vinas. 'Minimal intervention revisited', in A. Bracker and A. Richmond (eds), *Conservation: Principles, Dilemmas, and Uncomfortable Truths*, Oxford: Elsevier, 2009, pp. 47–59, at 52–4.

68 A. Leopold, "The Round River," in his *A Sand County Almanac, with Other Essays on Conservation from Round River* [1949/1953], New York, NY: Oxford University Press, 1966, p. 177.

69 J. Keulartz, "Using Metaphors in Restoring Nature," *Nature and Culture* 2, 2007, 27–48.

70 R. Elliot, *Faking Nature: The Ethics of Environmental Restoration* [1982], London: Routledge, 1997, p. vii.

71 Sierra Club advertisements, June 1966, Hall, *Earth Repair*, pp. 1–2, 13–15.

72 P. Losin, "Faking nature – A Review," *Restoration & Management Notes* 4, 1986; "The Sistine Chapel Debate: Peter Losin Replies," *Restoration & Management Notes* 6, 1988, p. 6.

73 There are exceptions: Niagara Falls and Yosemite are valued for their singular scenery, not as examples of generic waterfalls and canyons. But restoring culture usually features unique particulars, restoring nature composite amalgams. See D. Lowenthal, "Natural and Cultural Heritage," in K.R. Olwig and D. Lowenthal (eds), *The Nature of Cultural Heritage and the Culture of Natural Heritage: Northern Perspectives on a Contested Patrimony*, London: Routledge, 2006, pp. 79–90.

74 S.M. Meyer, *The End of the Wild*, Cambridge, MA: MIT Press, 2006.

75 M.C. Buckley and E.E. Crone, "Negative Off-Site Impacts of Ecological Restoration: Understanding and Addressing the Conflict," *Conservation Biology* 22:5, 2008, 1118–24; S.A. Merkle et al., "Restoration of Threatened Species: A Noble Cause for Transgenic Trees," *Tree Genetics & Genomes* 3:2, April 2007, 111–18. Restoration ecology is in this sense at odds with the emerging field of reintroduction biology's focus on single species, usually charismatic vertebrates: P.J. Seddon, D.P. Armstrong, and R.F. Maloney, "Developing the Science of Reintroduction Biology," *Conservation Biology* 21:2, 2007, 303–12.

76 Y. Choi, "Restoration Ecology to the Future: A Call for New Paradigm," *Restoration Ecology* 15, 2007, 352; Z. Naveh, *Transdisciplinary Challenges in Landscape Ecology and Restoration Ecology – An Anthology*, Dordrecht: Springer, 2007.

77 V. Schaefer, "Science, Stewardship, and Spirituality: the Human Body as a Model for Ecological Restoration," *Restoration Ecology* 14:1, 2006, pp. 1–3. Ecosystem health even figures in medical schooling: D.J. Rapport et al., "Strange Bedfellows: Ecosystem Health in the Medical Curriculum," *Ecosystem Health* 7:3, Sept 2001, 155–62.

78 S. Kaplan, "The Restorative Benefits of Nature: Toward an Integrative Framework," *Journal of Environmental Psychology* 15, 1995, 169–82; T. Hartig and H. Staats, "The Need for Psychological Restoration as a Determinant of Environmental Preferences," *Journal of Environmental Psychology* 26, 2006, 215–26; A.E. van den Berg, T. Hartig, and H. Staats, "Preference for Nature in Urbanized Societies: Stress, Restoration, and the Pursuit of Sustainability," *Journal of Social Issues* 63:1, 2007, 79–86; T. Hartig, F.G. Kaiser, and P.A. Bowler, "Psychological Restoration in Nature as a Positive Motivation for Ecological Behavior," *Environment & Behavior* 33:4, July 2001, 590–607. However, participants were urban North Americans and northern Europeans predisposed to idealize nature to begin with.

79 W. Throop and R. Purdom, "Wilderness Restoration: the Paradox of Public Participation," *Restoration Ecology* 14, 2006, 497, citing H. Rolston III, *Conserving Natural Value*, New York, NY: Columbia University Press, 1994.

80 W.P. Cottam, [1958], quoted in Hall, *Earth Repair*, p. 125.

81 J. Cairns, Jr., "Restoring Damaged Aquatic Ecosystems," *Journal of Social, Political & Economic Studies* 31:1, 2006, 53–74.

82 L.J. Krisjanson and R.J. Hobbs, "Degrading Landscapes: Lessons from Palliative Care," *Ecosystem Health* 7:4, Dec. 2001, 203–13.

83 E. Katz, "The Big Lie: Human Restoration of Nature," *Research in Philosophy and Technology* 12, 1992, 231–41.

84 Elliot, *Faking Nature*, pp.111–12. Ecological restoration is a "ritual of atonement for living in a culture that is responsible for causing morally unacceptable environmental degradation, ... imbu[ing] the practitioner with optimism and a sense of expiation." The hope of redemption makes "ecological restoration ... especially attractive to those whose cultural roots stem from the Protestant Reformation," A.F. Clewell and J. Aronson, "Motivations for the Restoration of Ecosystems," *Conservation Biology* 20, 2006, 420–8.

85 "Mainstream doctors are turned off by geriatrics. The Old Crock ... has high blood presure. He has diabetes. He has arthritis. There's nothing glamorous about taking care of any of those things," F. Silverstone, quoted in A. Gawande, "Annals of Medicine: The Way We Age Now," *The New Yorker*, 30 Apr. 2007, 53.

86 B. von Droste, H. Plachter, and M. Rössler (eds), *Cultural Landscapes of Universal Value – Components of a Global Strategy*, Jena: Gustav Fischer Verlag/UNESCO, 1995.

87 J.E. Dizard, *Going Wild: Hunting, Animals Rights, and the Contested Meaning of Nature*, rev. edn, Amherst: University of Massachusetts Press, 1999; D. Lowenthal, "Environment as Heritage," in K. Flint and H. Morphy (eds), *Culture, Landscape, and the Environment: The Linacre Lectures 1997*, Oxford: Oxford University Press, 2000, pp. 198–217; M. Hall, *Earth Repair*, pp. 11–13, 138–49, 195–99.

88 S. de Bruxelles, "Ancient Stones to Regain True Standing," *The Times* (London), 8 Apr. 2003; J. Burland, 'Solving the 800-year Mystery of Pisa's Leaning Tower,' *Daily Telegraph* [London], 28 July 2010; R. Scott, "The Accidental Rainforest, the Leaning Tower of Pisa, and Making the Most of Opportunity," in I.D. Rotherham (ed.), *Loving the Aliens*.

*Ecology, History and Management of Exotic Plants and Animals: Issues for Nature Conservation*, Special Ser. no. 4, *Journal of Practical Ecology and Conservation*, June 2005, 83–4.

89 E. Fotherington, "Stick to Prunes," *Audubon* 6, 2003; A. Opel, "Corporate Culture Keeps Nature Regular: the 'Super Citizen,' the Media, and the 'Metamucil and Old Faithful' Ad," *Capitalism, Nature, Socialism* 17(3), Sept. 2006, 100–13.

90 J.M. Wheaton, S.E. Darby, D.A. Sear, and J.A. Milne, "Does Scientific Conjecture Accurately Describe Restoration Practice? Insight from an International River Restoration Survey," *Area* 38, 2006, 128–42. The 2003–04 website survey enlisted 300 respondents from 36 countries.

91 J. Cairns, Jr., "The Status of the Theoretical and Applied Science of Restoration Ecology," *The Environmental Professional* 13, 1991, 186–94.

92 A.D. Bradshaw, "Alternative Endpoints for Reclamation," in J. Cairns, Jr. (ed.), *Rehabilitating Damaged Ecosystems*, 2 vols., London: CRC Press, 1988, 69–85.

93 Wheaton et al., "Does Scientific Conjecture Accurately Describe Restoration Practice?"

94 E.S. Higgs, "Restoration Goes Wild: A Reply to Throop and Purdom," *Restoration Ecology* 14, 2006, 500–3.

95 P. Del Tredici, "Neocreationism and the Illusion of Ecological Restoration," *Harvard Design Magazine*, no. 20 (Spring/Summer 2004), 1–3; W.R. Jordan III, "Restoration, Community, and Wilderness," in Gobster and Hull, *Restoring Nature*; M.A. Davis, "'Restoration' – A Misnomer?" *Science* 287, 2000, 1203.

96 D. Janzen, "Gardenification of Wildland Nature and the Human Footprint," *Science* 279, 1998, 1312–13.

97 E. Higgs, "The Two-Culture Problem: Ecological Restoration and the Integration of Knowledge," *Restoration Ecology* 13, 2005, 159–64.

98 L. Ledderose, "Carving Sutras into Stone Before the Catastrophe: The Inscription of 1118 at Cloud Dwelling Monastery Near Beijing," *Proceedings of the British Academy* 125, 2004, 381–454. Excavated in 1957, the stones were again reburied in 1999.

99 J. Seabrook, "Annals of Agriculture: Sowing for Apocalypse," *The New Yorker*, 27 Aug. 2007, 60–71; *Svalbard Global Seed Vault*, Norwegian Ministry of Agriculture and Food, www.regjeringen.no/en/dep/lmd/campain/svalbard-global-seed-vault/news/summary-of-the-svalbard-conference.html?id=504663.

100 R. Morgan, "Proposed Use for Moon: Storage Locker for DNA," *International Herald Tribune*, 2 Aug. 2006; W.E. Burrows, *The Survival Imperative: Using Space to Protect Earth*, New York: Forge Books, 2006, pp. 208–35.

101 G. Benford, *Deep Time: How Humanity Communicates Across Millennia*, New York: Avon, 1999, pp. 93–127.

102 De Sacramentis [c.1134] and Sententiae divinitatis, quoted in B.T. Coolman, *The Theology of Hugh of St. Victor: An Interpretation*, Cambridge: Cambridge University Press, 2010, pp. 12–15, 99, 132, 144, 172, 205, 218.

103 J. Milton, *Paradise Lost* [1667], XII:471–4. London: Penguin, 2000, p. 283.

104 Erasmus to Pope Leo X, 1 Feb. 1516, letter 384, *Collected Works* of Erasmus, vol. 3, *Letters 298 to 445*, Toronto: University of Toronto Press, 1976, pp. 221–2.

105 Lyell, *Principles*, p. 74; Gould, *Time's Arrow*, p. 155; B.G. Niebuhr, *The History of Rome* [1811–12], rev. edn, Cambridge, England, 1831, vol. 1, p. 5.

# 21

# OVERLAPPING AMBIGUITIES, DISCIPLINARY PERSPECTIVES, AND METAPHORS OF LOOKING

## Reflections on a landscape photograph

*Joan M. Schwartz*

> Anything more beautiful than the photographs of the Valley of Chamouni, now in your printsellers' windows, cannot be conceived. For geographical and geological purposes, they are worth anything; for art purposes, worth – a good deal less than zero.[1]

Underlying Ruskin's 1872 observation is a fundamental assumption about photographic meaning at the intersection of geography and the humanities: the value of the photograph as an object of scholarly concern is not universal: it is contingent upon disciplinary priorities. For Ruskin, the "worth" of the photographs of the Valley of Chamonix varied; the meanings they generated were relative; the criteria against which they were assessed were discipline-dependent. The landscape photographs which were worthless to the art critic possessed great value for the geographer and the geologist. For Ruskin, the value of these photographs for geographical purposes lay in their topographical facts about the physical landscape of the valley; for art purposes, it lay in the aesthetic qualities of the image as a pictorial composition. This suggests that, in Ruskin's time, art critics and geographers (physical geographers, presumably, in 1872) were not only using different criteria, and asking different questions, but they were also looking at different things. Clearly, critical engagement with the photograph first requires an acknowledgment of what it is we focus on when we study photographs.

Now, more than a century and a quarter later, we are inclined to ask how photographs of the Valley of Chamonix expressed and mediated the relationship of people to place – the subjective experience of space, attitudes to Nature, the development of tourism, or the perception of mountains; how the facts they communicated were part of efforts to conceptualize geographical knowledge, visualize geographical space, and construct imaginative geographies. Of course, we can never know all the meanings invested in and generated by the photographs of the Valley of Chamonix.

Ruskin's assessment, which sets aesthetic against documentary as a measure of value, is unduly limiting. These mountain landscape images entered and influenced the geographical imagination of artists, naturalists, travellers, printsellers, Thomas Cook agents, alpinists, and other nineteenth-century observers. There, they assumed many roles and performed many functions: as inducements to travel, a stimulus to observe the world first-hand; as instruments of travel, a means of engaging with place; as souvenirs of travel, a way of bringing places and experiences home; as surrogates for travel, a medium for seeing across space and time; as tools of the geographical imagination, a vehicle for expressing, constructing, and visualizing imaginative geographies.

It has been two decades since the "crisis of representation in the human sciences"[2] began to undermine the certainties by which modernism defined knowledge. With the destabilization of the basic notions of fact, truth, and reality, concern for the mediated nature of representation spread from ethnographic representation to other forms of representation – museum, landscape, visual – and academic inquiry into both photography and landscape turned attention to the construction of meaning. Calling this conceptual shift "'tectonic' in its implications," and using an overtly geographical metaphor, James Clifford argued, "We ground things now on a moving earth. There is no longer any place of overview (mountaintop) from which to map human ways of life, no Archimedian point from which to represent the world."[3] At this very time in cultural geography, with this denial of "brute" or "bedrock" reality, interest in landscape shifted to investigations of the "meaning" of landscape between representation and reality. At the forefront of this shift was Denis Cosgrove and Stephen Daniels' seminal edited collection *The Iconography of Landscape: Essays on the Symbolic Representation, Design and Use of Past Environments*, which acknowledged multiple interpretations of landscape-as-text and set a new agenda for cultural geography.[4]

Landscape photography is, thus, a doubly contested terrain, a place where overlapping discourses further complicate the dynamics of meaning-making. Taking up the challenges of the late 1980s from a different disciplinary perspective, John Taylor approached the ambiguities of landscape photographs thus:

> It is illusory to develop beliefs and responses to landscape photography: it does not exist. This assertion does not contest that people walk about the land and sometimes take photographs or look at them. What it does contest is that there is an underlying reality to either landscape or photography and that we can know them by studying them directly or in isolation. Photographs do not offer a simple window on to the world 'out there': nor is landscape anything and everything seen from the top of a hill. To imagine their relationship in landscape photography we have to study practices.[5]

Taylor's denial of the existence of "an underlying reality to either landscape or photography" is key to any understanding of landscape photography within geographical inquiry. Geographers preoccupied with the meaning of landscape would certainly concur with Taylor's assertion that landscape is not "anything and everything seen from the top of a hill." They should also be receptive to Taylor's contention that, "to imagine their relationship in landscape photography we have to study practices."

## Metaphors of looking

The idea that the photograph is, at once, both an objective re-presentation of some pre-photographic referent, and a creative representation of some portion of material reality is given visual expression in René Magritte's well-known canvas, *La condition humaine*.[6] It shows a landscape painting placed on an easel in front of a window in such a way that the painted and actual vegetation, topography, and clouds align, and the painting blends seamlessly into the landscape it depicts.[7] Magritte argued that *La condition humaine* demonstrated "how we see the world."[8] The superimposition of painted canvas – revealed by the unfinished edge where it is tacked to the stretcher – and the landscape beyond suggests the transparency and truthfulness of pictorial representation. Admittedly a painting and not a photograph (although one could imagine such a photograph of a photograph of a landscape), *La condition humaine* symbolizes how we approach visual images as "a simple window on to the world out there." However, Magritte's painting itself demonstrates three ways of looking central to the analysis of the photograph in geographical inquiry: looking *at*, looking *through*, and looking *with*. The content of the painting on the easel set before the window suggests the conflation of surface and subject, representation and reality, but the edges of the canvas, subtly betraying its presence in the scene, reveal that the painting – however mimetic – is, in truth, a mediated representation extracted from a larger material reality. And one step removed from the tensions within the canvas, *La condition humaine*, itself, is a painted commentary on the act of looking.

It has been seventy-five years since René Magritte produced this painting, and more than twenty years since the "crisis of representation" laid bare the mediated nature of visual, indeed all, representation, but if, as John Taylor and others have urged, we are to study practices, then *La condition humaine* suggests that the practices we need to study are not only those involved in the act of representation but also those which frame the act of looking. In approaching *La condition humaine*, the question *What do we see?* is clearly predicated upon the answer to another question, *How are we looking?* Are we looking *at* or *through* the canvas on the easel? These two ways of looking return us to Ruskin and disciplinary perspectives. Are we wearing the hat of a Ruskinian art critic looking *at* the surface of the canvas, primarily interested in this painted landscape as a visual image, a form of aesthetic expression, a product of creative genius, with an emphasis on pictorial qualities? Or shall we don the cap of Ruskin's geographer to look *through* the painting on the easel to the scene beyond, using it as a record of landscape elements, a surrogate for first-hand observation to study the nature of field and forest? And finally, can we look *with* Magritte's *La condition humaine* to learn about the very act of looking, a topic with significance beyond the painting's surface and subject and frame?

The commentary on "how we see the world" embedded within Magritte's canvas is reiterated in distinctly photographic terms by Roland Barthes' observation that,

> the Photograph belongs to that class of laminated objects whose two leaves cannot be separated without destroying them both: the windowpane and the landscape....[9]

Barthes' notion of the photograph as a laminated object is a useful conceptual device for several reasons. First, it offers a way to grasp the nature of the photograph as a simultaneously mimetic and mediated visual image, acknowledging both the visual realism of the photograph as an agent of sight and the rhetorical power of the photograph as a site of agency. Second, it emphasizes that the meaning of a photograph must not be conflated with its content, that there is more to the photograph than meets the eye. Third, Barthes' notion of the photograph as a laminated object highlights the simple, important, and analytically powerful distinction between what the photograph is *of* and what the photograph is *about*. It establishes two, parallel and compatible, ways to interrogate the photograph in geographical inquiry: as the "landscape" for the visual facts contained, and as "the windowpane" for the human choices embedded in the photograph as a representation. Fourth and most important here, in shifting attention away from the traditional focus on the "landscape," and drawing attention to the "windowpane," it demands interrogation of the photograph as an act of communication.

Barthes' use of "windowpane" can be confusing. The photograph has frequently been conceived as a "window" – on the world or on the past, implying that the photograph furnishes a transparent view of the landscapes of that "foreign country" we call the past.[10] This "look-out into a vanished world" has been accepted as precise and unmediated, an effective surrogate for "being there." It treats the image as a window in a literal sense, suggesting that, through it, we can see the scene portrayed as if first-hand. The image carrier disappears and the photograph becomes conflated with the material reality (landscape) recorded in it: as Barthes puts it, "the referent adheres." But windows are made up of windowpanes, pieces of glass which may exhibit flow marks, or may come in colors or shapes or sizes, which themselves catch the eye and distort the landscape beyond. Barthes' windowpane is the surface of the photograph; his landscape is the subject recorded in it.

The metaphors do not end there. In his landmark 1978 exhibition of American art photography, Museum of Modern Art photography curator John Szarkowski adopted a different metaphorical polarization, proposing that photographic images could be viewed in two ways – as mirrors and as windows – corresponding to "expressionist" as opposed to "realist" modes of artistic representation. James Guimond, in his book *American Photography and the American Dream*, suggests yet another metaphor: doors – "that enable us, if only briefly, to step out of our old ideas about ordinary realities and experience them in a fresher, more vivid way."[11] And even these metaphors are differently employed; for example, Simon Schama uses *La condition humaine* as an interpretive device in his study of myth, memory, and the meaning of landscape, to explore "what lies beyond the windowpane of our apprehension."[12] However, what Magritte and Barthes, Szarkowski, Guimond, and Schama are doing is employing metaphorical "openings" in order to draw attention to practices of looking. Let me then reflect upon a single landscape photograph by focusing on the practices of looking suggested by these metaphors, practices which help us to distinguish between surface and subject, and, ultimately, resolve Ruskin's differential assessment of the mutually exclusive worth of landscape photographs for art and for geography.

## Reflections on a landscape photograph

Consider the landscape photograph entitled, *The Wigwam, a Canadian Scene at Penllergare* (Figure 21.1).[13] This photograph, now in the Metropolitan Museum of Art in New York, was taken by John Dillwyn Llewelyn (1810–82), on his estate in Glamorgan, South Wales, likely sometime in 1855. Married to Emma Thomasina Talbot (1806–81), a cousin of William Henry Fox Talbot, the inventor of the negative–positive photographic process, Llewelyn was a pioneer of British photography whose photographic

**FIGURE 21.1**   *The Wigwam, a Canadian Scene at Penllergare,* by John Dillwyn Llewelyn (1810–1882) ca.1855; albumen print, Metropolitan Museum of Art, New York.

circle included his sister, daughter, and son-in-law Nevil Story Maskelyne, as well as some of the most prominent names in the first generation of photographic practitioners.[14]

*The Wigwam, a Canadian Scene at Penllergare* presents a tightly cropped view of a wigwam-like structure made of shakes rather than the traditional skins or bark. A half-dozen steps cut into the bank lead past fishing nets set out to dry, down to the water's edge where a birchbark canoe sits at the ready. Smoke rising in a funnel of light from a campfire and the reflection of the nets in the pond define a space of human presence between water and wilderness. A variant print, *The Birch Bark Canoe*, shows Llewelyn's young son Willy (William Mansel Dillwyn Llewelyn, 1838–66) sitting bankside, arranging the fishing nets; a chair sits conspicuously at the entrance to the wigwam, outlined against the dark interior. A companion print, *North American Wigwam*,[15] offers a wider-angle view from across the estate's Upper Lake. All three present fantasy impressions which are more photographic fiction than visual fact. In light of Ruskin's assessment of the photographs of the Valley of Chamonix, how can attention to practices of looking help us to understand Llewelyn's "New World" images for both art and geographical purposes?

Looking *at* this landscape photograph, the image on its surface, as a picture, we privilege particular aesthetic qualities and art-historical concerns. Both the photograph and Llewelyn, himself, occupy a place of some significance in art histories of photography because of Llewelyn's domestic and intellectual connections to pillars of early British photography. *The Wigwam, a Canadian Scene at Penllergare* merits attention for its aesthetic qualities as well as its place in Llewelyn's *oeuvre*. *The Wigwam* (#511) was shown at the Exhibition of Photographs and Daguerreotypes organized by the Photographic Society [of London] at the Gallery of the Society of Water Colour Painters, 5 Pall Mall East, from the first week of January to the end of March 1856.[16] It would also have likely circulated to Llewelyn's immediate circle of family and friends, which included luminaries of British science whose intellectual reach encompassed the Royal Society, the Linnean Society, the Swansea Literary and Philosophical Society, the Photographic Society, the Photographic Exchange Club, and the Amateur Photographic Association. A hundred and fifty years later, it was raised to the pinnacle of the fine art photography canon in 2005 when it was acquired by the Metropolitan Museum of Art in New York as part of the Gilman Paper Company collection of photographs. Amassed by the late Howard Gilman, its more than 8,500 photographs, dating primarily from the first century of the medium, was "widely regarded as the world's finest collection of photographs in private hands" and has been credited by the MET as playing "a central role in establishing photography's historical canon" and with setting "the standard for connoisseurship in the field."[17]

Looking *through* Llewelyn's photograph, we focus on the forest scene. *The Wigwam, a Canadian Scene at Penllergare* documents a corner of one of the great garden estates of nineteenth-century Wales. For "geographical and geological purposes," it presents a portion of the physical landscape and the way in which it was shaped by Llewelyn. The photograph shows one of the two lakes created as part of the landscaping of the country house on the outskirts of Swansea. A popular feature of the estate, the Upper and Lower Lakes furnished a favorite photographic subject for Llewelyn, his friends[18] and family, and in their photographs, the topography, flora, and fauna of Penllergare figure prominently. Described as "a fine example of the seriousness with which the Victorians took their amusements,"[19] this wilderness prospect, assembled in all likelihood for the enjoyment of family and friends, also reveals something of the lifestyle and interests of the educated, leisured class in South Wales at the height of its prosperity in the mid nineteenth century. Llewelyn has been credited with the creation of an outstanding example of a picturesque, romantic landscape, and *North American Wigwam*, in particular, furnishes a glimpse of the transformation of the South Wales landscape and Llewelyn's contribution to garden design.

But, what do *The Wigwam, a Canadian Scene at Penllergare* and *North American Wigwam* have to offer to contemporary cultural-historical geographers interested in the ways in which notions of place and identity are constructed and sustained? Beyond surface and subject, how can geographers engage these photographs, of a landscape folly in Wales, created and exhibited as art, to gain a greater understanding of nineteenth-century perceptions of North American wilderness or prevailing ideas about Canada as a outlying colony of Empire?

The inspiration for Llewelyn's carefully constructed tableau is not known. One can speculate that it was the Canada court at the Great Exhibition of 1851 where a birchbark canoe was suspended over displays in which fur pelts, stuffed animals, and forest products figured prominently. Perhaps it was travel literature, a globe-trotting house guest, or Henry Wadsworth Longfellow's recently published *Songs of Hiawatha*. Certainly, *The Wigwam, a Canadian Scene at Penllergare* bears more than a passing resemblance to the frontispiece illustration by William Henry Bartlett in N.P. Willis' *Canadian Scenery*, a copy of which has been located in Llewelyn's library.[20] Published in 1842, the engraving entitled, *Wigwam in the Forest*,[21] includes all the elements in Llewelyn's scene: shore, trees, canoe, and wigwam against a dense forest backdrop. The same set of visual cues reappears six years later in the lithographic view, *Indian Wigwam in Lower Canada*[22] after a painting by Cornelius Krieghoff, published under the patronage of Lord Elgin, Governor General of British North America, as part of the series entitled *Scenes in Canada*.

Given such visual, material, and literary precedents, *The Wigwam, a Canadian Scene at Penllergare* is clearly an essential piece of Canadian iconography. Returned to the contexts in which it was originally created, circulated, and viewed, it emerges as an imaginative geography of land and life in mid-nineteenth-century Canada. Of course, Llewelyn's *Canadian Scene at Penllergare* is not a record of facts about the Canadian forest or the aboriginal peoples who dwelt there. Rather, it is a scene where geographical knowledge and photographic technology conspire to portray a place known only second-hand to its creator. In this combination of art, artifact, and artifice, a perception of place is physically constructed and photographically preserved. This visual articulation of ideas about Canada, reinforced by the title, suggested and sustained a message about the essential nature of Canada. By looking *with*, rather than simply *at* or *through*, Llewelyn's photograph, *The Wigwam, a Canadian Scene at Penllergare* functions as a conduit to another time and place. Understood as *both* aesthetic expression and geographical description, the photograph serves to bridge disciplinary perspectives and shed light on the ways in which the ambiguous meanings of landscape and photograph overlap in the landscape photograph.

The photographs of the Valley of Chamonix in the windows of booksellers in 1872 were likely the work of William England or perhaps the Bisson Frères, nineteenth-century landscape photographers whose work is now held in major fine art photography collections around the world. Within the images themselves, their exposure in the printsellers' windows, their rejection as art, and their role in geography we can discern larger issues. The definition of art has changed since Ruskin's day. So have our understandings of the nature, worth, and meaning of photographs and of landscape. Ambiguities embedded in the landscape and the photograph, respectively, overlap in the landscape photograph. Conflicting differential assessments of worth, grounded in disciplinary perspectives, can be reconciled when they are freed from the fetters of disciplinary perspectives and understood as the result of practices of looking. In geography, the meaning of space, place, and landscape, as we have come to recognize, is contingent on context, but context itself is not a totalizing or stable concept. Equally in the humanities, the meaning of photographs, indeed all visual images, is dynamic: historically grounded, socially constituted, culturally constructed, gendered, raced, and classed, and often technologically determined. Over the last 150 years, the meaning and value of the photographs of the Valley of Chamonix, to which John Ruskin drew attention, and *The Wigwam, a Canadian Scene at Penllergare*, created and exhibited by John Dillwyn Llewelyn, have changed with the contexts in which they have been circulated, viewed, and preserved. Disciplinary perspectives collide in Llewelyn's landscape photographs, but, when situated at the intersection of geography and the humanities, their "artistic" worth and "geographical" value, clearly considered mutually exclusive by John Ruskin, are reconciled through attention to our changing practices of looking. Looking *at*, looking *through*, and looking *with* Llewelyn's work, we can recognize it, simultaneously, as an act of creative expression, factual documentation, and visual imagination.

Much of what we now take for granted about our photographically mediated relationship with the world around us was, in the mid-nineteenth century, new and tentative and even revolutionary. It was in this period that photography as a technology and geography as a discipline came of age together. Open to interpretation squarely at the intersection of geography and the humanities, landscape photographs now assume their most powerful rhetorical role in the geographical imagination as tools in the practice of looking *with*. In his essay, "Historical Geography," H.C. Darby suggested that, in the mid nineteenth century, "men were asking new questions not only about their own age but also about the past."[23] Ruskin's observation might be used to suggest that photographs were, in part, responsible for both the new questions and the new answers.

# Notes

1 Ruskin, by this time, had become disillusioned with photography as an aid to art. J. Ruskin ... excerpted from *The eagle's nest; ten lectures on the relation of natural science to art, given before the University of Oxford, in Lent term, 1872*. Lecture VII. "The Relation to Art of the Sciences of Inorganic Form," 29 February 1872; in *Selections from the Writings of John Ruskin*. Second Series, 1860–1888. Orpington & London: George Allen, 1893, p. 101.

2 G.E. Marcus and M.M.J. Fischer, "A Crisis of Representation in the Human Sciences," chapter 1 in *Anthropology as Cultural Critique: An Experimental Moment in the Human Sciences*. Chicago: The University of Chicago Press, 1986.

3 J. Clifford, "Introduction: Partial Truths," in J. Clifford and G.E. Marcus (eds), *Writing Culture: The Poetics and Politics of Ethnography*, Berkeley and Los Angeles, CA: University of California Press, 1986, p.22.

4 D. Cosgrove and S. Daniels (eds), *The Iconography of Landscape: Essays on the Symbolic Representation, Design and Use of Past Environments*, Cambridge: Cambridge University Press, 1988.

5 J. Taylor, "The Alphabetic Universe: Photography and the Picturesque Landscape," in S. Pugh (ed.), *Reading Landscape: City – Country – Capital*, Manchester: Manchester University Press, 1990, p.177.

6 *La condition humaine*, by René Magritte (1898–1967), 1933: oil on canvas, National Gallery of Art, Washington, DC.

7 Other paintings by Magritte which show what Sarah Whitfield calls "one of Magritte's simplest yet most disconcerting paradoxes," that is "the idea of showing a canvas on which the landscape behind is reproduced," include *La belle captive* (1931), *L'appel des cîmes* (1943), and *Les promenades d'Euclide* (1955). See S. Whitfield, *Magritte*, London: The South Bank Centre, 1992; also, D. Sylvester, *Magritte*, London: Thames and Hudson, 1992.

8 Lecture, R. Magritte, "La Ligne de vie," 1938, quoted in Whitfield, *Magritte*, text accompanying Plate 62, *La condition humaine*.

9 R. Barthes, *Camera Lucida: Reflections on Photography*, New York, NY: Hill and Wang, 1981, originally published as *La chambre claire*, Paris: Editions du Seuil, 1980, p.6.

10 D. Lowenthal, *The Past is a Foreign Country*, Cambridge: Cambridge University Press, 1985.

11 J. Szarkowski, *Mirrors and Windows, American Photography since 1960*, New York, NY: Museum of Modern Art, 1978; J. Guimond, *American Photography and the American Dream*, Chapel Hill, NC: University of North Carolina Press, 1991, p.17.

12 S. Schama, *Landscape and Memory*, Toronto: Random House, 1995, p.12.

13 This is a revised, much expanded version of a discussion of Llewelyn's work in J.M. Schwartz, "Photographic Reflections: Nature, Landscape, and Environment," *Environmental History* 12, 4, October 2007.

14 For example, Calvert Richard Jones, Philip Henry Delamotte, Robert Hunt, and Hugh Welch Diamond.

15 *North American Wigwam*, by John Dillwyn Llewelyn (1810–1882), ca. 1855: albumen print, Library and Archives Canada, Ottawa (PA-164777).

16 *The Birch Bark Canoe* (#518) was another of the twenty-two photographs exhibited by Llewelyn. See R. Taylor, *Photographic Exhibitions in Britain 1839–1865: Records from Victorian Exhibition Catalogues*, Ottawa: National Gallery of Canada, 2002; available online at: http://peib.dmu.ac.uk/

17 Master Photographs from the Gilman Collection: A Landmark Acquisition, Metropolitan Museum of Art, New York, 28 June – 6 September 2005. For more on the Gilman Paper Company acquisition, see: http://www.metmuseum.org/special/Gilman/gilman_more.htm (accessed 23 November 2008).

18 James Knight exhibited views taken at Penllergare at the 1856, 1857, and 1859 exhibitions of the Photographic Society, London, as well as in the Photographic Gallery of the Exhibition of the Art Treasures of the United Kingdom held in the Central Library in Manchester from May to October 1857. Philip Henry Delamotte exhibited views taken at Penllergare in 1855 in London and 1857 in Manchester.

19 H.P. Kraus, Jr., *Sun Pictures: Llewelyn – Maskelyne – Talbot, A Family Circle*, Catalogue Two, New York, NY: Hans P. Kraus, Jr., n.d., p.65.

20 I am grateful to Sue and Richard Morris for this information.

21 *Wigwam in the Forest*, title page engraving from a watercolour by William Henry Bartlett, in Nathaniel Parker Willis, *Canadian Scenery, Illustrated in a series of Views by W[illiam] H[enry] Bartlett*, London: George Virtue, 1842. National Library of Canada, Ottawa (NL-19384).

22 *Indian Wigwam in Lower Canada*, lithograph executed by Andreas Borum, Munich; printed by Th. Kammerer, and published R. & C. Chalmers, Montreal, from an original painting by Cornelius Krieghoff (1815–72), 1848. National Gallery of Canada, Ottawa (no. 30820).

23 H.C. Darby, "Historical Geography," in H.P.R. Finberg (ed.), *Approaches to History: A Symposium*, Toronto: University of Toronto Press, 1962, p.152.

**PART IV**

# Performing

# 22

# INVERTING PERSPECTIVE

## Icons' performative geographies

*Veronica della Dora*

This is an icon of the Mother of God (Plate VIII). I stand before her and stare at her. She stares back at me. No matter how many steps back I take, or whether I move to her right or to her left, her gaze continues to follow me – intensely, insistently. A silent dialogue is started. Sacred space is configured between us – me and the icon. I become her vanishing point.

The image does not aim at realism. The hands of the Mother of God are dispro-portionately large, as if to remind me of the incommensurability of what she is holding: God Himself. Her Son portrayed in a gesture of blessing looks less like an infant than a king in miniature. The two figures are set still against a golden back-drop. Their faces are not illumined by a beam of light, as in Giotto's or Caravaggio's paintings. Light comes from within; it quietly embraces me; it intimately draws me into contemplation.

My gaze does not wander through an illusionist three-dimensional space, as in Western Renaissance sacred images ruled by linear perspective; it is simply captured by the gaze of the Mother of God. It is not allowed to go beyond the surface of the icon. But her gaze is allowed to do so – in the opposite direction. My body becomes a surface penetrated by the intensity of her gaze. I become part of the composition.

Geographers and art historians have traditionally identified the Western Renaissance invention of linear perspective with the beginning of modernity. Leon Battista Alberti's own emblem, a free-floating "winged eye", spoke of this new "way of seeing" as the ultimate disembodied act, whereby "one of the five senses, sight, becomes autonomous and is no longer obliged to work together with the other four senses".[1] *Perspicere* ("to see clearly") implies distancing, and distancing in turn enables conceptual control over the world.[2] While Marxist and feminist geographers have not failed at pointing out the power implications of the mercantilist, bourgeois, masculine "distanced gaze" over landscape,[3] and more recently "more-than-representational" experimenters at its limits,[4] linear perspective has nevertheless been largely naturalized as *the* way through which we "look at" landscape.

In this chapter, I would like to take one step back, and explore a different way to look at and represent the world. Unlike Western linear perspective, the so-called "inverse" perspective of Byzantine and post-Byzantine icons and frescoed (or mosaic) cycles collapses the distance between seer and seen; it consciously resists naturalism and literally "wraps" the beholder, making her its vanishing point. An integrating part of liturgical performance and private worship, these images cannot be fully appreciated, or understood separately from their numinous materialities, nor from the complex spatialities they help articulate *outside* of their surface. As such, they call for an approach that is at once iconographic and phenomenological – and thus inherently geographical.

## Western and Eastern traditions

The warm glow of golden luminescence that forms the central element of every icon could not be more different from the optically corrected, mercantile light of Renaissance religious painting – a difference as great as that between the insistence of St. Gergory Palamas upon the appearance of the uncreated light in nature (its appearance ... to the three disciples at the Transfiguration on Mount Tabor and to Palamas' monastic contemporaries on Mount Athos chanting prayers as the golden sun sinks into the surrounding Aegean Sea) and the equally emphatic insistence of his English contemporary William of Ockham, set forth under the wan northern light of Oxford, that when it comes to nature, what your eyeballs see is precisely what you get.[5]

Linear perspective was theorized and practically developed in early fifteenth-century Tuscany, even though it had been studied in the West as an optical science since the twelfth century, and by the late thirteenth century painters like Cimabue and Giotto were already experimenting with new ways to achieve a greater realism than their predecessors.[6] The development and application of linear perspective were part of a broader spatial revolution and were paralleled by the rediscovery of Ptolemaic cartographic science. Just as geographers and historians of cartography have traditionally defined the translation of the spherical globe on a bi-dimensional gridded surface as the point of passage from a place-based to a geometrical model of the Earth, so have art historians envisaged linear perspective as the main turning point in the representation of place.[7] As Franco Farinelli notes, however, the former was simply the copy of the latter: "the only difference was that Ptolemy's projection worked vertically, whereas modern linear perspective horizontally."[8] Both rested on Euclidean geometry and both, the Italian geographer argues, made for the most terrible and pervasive spatial model, "one which shall wrap the whole globe in the modern era."[9]

Linear perspective produces the realist illusion of three-dimensional space on a two-dimensional surface. Objects are drawn smaller as their distance from the observer increases, and visual axes converge in a vanishing point in (or beyond) the canvas. Linear perspective creates depth and deceives haptic experience. It implies detachment.[10] It has been deemed to be the visual culmination of a longer Western tradition of progressive deification and disenchantment with nature initiated in the thirteenth century with new technological inventions.[11] This attitude has been in turn ascribed to a distinctively Western religious sensibility rooted in Scholasticism: a "rational," "humanistic," almost "scientific" sensibility, which found visual expression in Giotto's "imitation of nature" and its apotheosis in the Renaissance. By then, the Western painter of religious art had become a master of mathematical techniques who depicted themes "which happened to be religious" in the same manner as secular ones.[12]

In the Christian East, by contrast, icons continued to be produced in prayer and to find their fulfillment in prayer. Nature continued to be conceived "as a symbolic system through which God spoke to men."[13] Icons too worked through symbols. In the patristic tradition the term "symbol," however, is employed in a strong sense, different from its usage in the modern West. For the Fathers of the Church, "a symbol actually participates in some sense with the spiritual reality it symbolizes, whereas for most people today the term 'symbol' tends to imply something essentially different from the thing it symbolizes."[14] The patristic view attributes much greater significance to the symbol, for it does not see it as completely separate, "other" from its referent. Through symbols, icons seek "to incarnate the visible of the invisible, God beyond everything and through everything."[15]

In the fourth century, Saint Anthony the Great told the philosophers that "we [Christians] do not reach the acceptance of the mystery of faith through logics, but as guided by Divine Grace."[16] Divine Grace is not experienced through speculation. It does not speak *of* things, but *from* things. It does not call for "rational distancing," but rather for communion, for a "more-than-rational" immersion; it calls for penetration – for inner transformation. In the Eastern Christian tradition, materiality and multi-sensorial experience play a central part in this process of transformation. Orthodox worship exalts all the senses. It uses and appreciates the material world: the wood and the painting, the materials for writing, the bread and the wine, the blessed basil and the grapes, the incense and the wax... .[17] Unlike words, icons, chants, and fragrances are means of direct communication: they reach the heart without necessarily being mediated through the mind.[18] As Saint John the Damascene wrote in the seventh century,

> We are double, made of soul and body. Our soul is not naked; it is wrapped up in a mantel; it is impossible to reach the spiritual without the bodily. Listening to sensible words with our bodily ears, we hear spiritual things; in a similar fashion, through bodily contemplation, we reach spiritual contemplation.[19]

## Icons' materialities

Icons are paradoxical objects. They problematize, or rather collapse the binary distinctions through which Western thought has been traditionally articulated. They blur the boundaries between referent and symbol, same and otherness, temporal and eternal, immanent and transcendent. Icons bring the Lord, the Mother of God, the angels, the saints, and the prophets "here now." "As you stand before the icon of the Saviour," the famous nineteenth-century Russian spiritual writer and saint Ignatiy Bryanchaninov wrote, "stand as if before the Lord Jesus Christ Himself."[20] Icons, however, are not holy in themselves, but holy by participation: they are statements and embodiments of the truth, vehicles of Grace, making the intangible tangible, the invisible visible, divinity accessible to man, like the incarnated Christ. As such, they are not concerned with imitation of the material world, but with its transfiguration. They demand a personal involvement in the act of perceiving, even to the extent of being changed by what one sees, or touches. "Because God having been made flesh in Jesus Christ, humans are able to glimpse the very face of God in matter itself."[21]

Western sacred art, especially under the spell of modern philosophy and empirical sciences, followed the inverse path: the visible of nature is the prototype. Where the Eastern Church envisages in the icon a sign of the uncreated, the Western Church firmly opposes the veneration of sacred images. Condemning the iconoclasm of Protestantism, the Council of Trent (1545–63) exalted the didactic function of visual arts, but at the same time – unlike in the Second Council of Nicea (787 AD) – the mysteric value of the image was contested, arguing that "the image is a representation, not a presentation."[22] The experience of "physicality" is consequently different in the two traditions. In Western sacred art, plastic mobile bodies are always observed from a distance – through linear perspective. In the Christian East physicality is experienced through the direct contact with the icon, through the bodily act of veneration.

Icons are hybrid creations of organic and inorganic matter. They are physical objects made of wood, egg-based tempera, golden leaf, fish-based glue, mineral pigments, and other materials. Icons perform and are performed. During the holy liturgy they are incensed, touched, and kissed; candles are lighted and prayers are chanted in front of them; they are offered flowers. In the house of the faithful, they occupy a seat of honor as illustrious guests.[23] Icons can suffer and rejoice. Miraculous icons may cry, bleed, and speak. They may dictate their own will to the faithful, or reveal in dream their secret location of burial, be it under the ground, or in the rock. Sometimes, they may even generate political action, or be cause of contention.[24]

Icons are repositories of untold stories. They are sites of gathering and stratification of human and non-human agencies. Exposed to the incense of decades or even centuries of long night services, their surface might be covered with a dark patina. Watermarks might serve as joyful reminders of past festivities in which the entire church was blessed with holy water. Other layers might add over time, inscribing different "micro-geographies" of memory on wood. Scratches, cracks, and other injuries might tell us stories of iconoclasts, vandals, or simply sacristans' hasty movements. A silver coverage might serve as testimony of some abbot's gratitude. Traces of lipstick and candle wax, as well as jewels and other *támata* (ex-votos) might speak of incessant flows of devout pilgrims asking (or thanking) for miracles, whereas woodworms' holes and mould, warps and splits, and other signs of non-human action simply remind us that, as artefacts, icons are not discrete entities, but "material form[s] bound into continual cycles of articulation and disarticulation;"[25] that, as sacred objects, they are an "indissoluble whole" with ritual (Plate IX). They remind us that icons move across space and time.[26]

## Inverse landscape

The agency of icons does not exhaust itself in their materiality, or in the penetrating gazes of their holy persons. It also works through their compositional rhetoric – namely through what art historians termed "inverse perspective." Unlike linear perspective, this technique does not draw the eye of the viewer through the picture frame into the natural world depicted therein; it rather concentrates his attention on the image itself. The size of the figures is determined by importance and not receding view, so that figures in the background are often larger than figures in the foreground.[27] Visual axes do not converge at some distant vanishing point within the image's scenic background. Usually formed by curved mountains or buildings symmetrically arranged, perspective lines converge at a point several inches in front of the icon's surface (and slightly below it), "as if aimed at the heart of the viewer standing in its presence."[28]

Landscape is thus no longer simply a "way of seeing," but a "way of being." The beholder becomes essential to the contemplation of the icon, and is in turn transformed by it. "The essence of the exercise is to establish a communion between the event or persons represented in the icon and those who stand before it."[29] Landscape features are thus doubly symbolic, in the strong, ontological sense suggested by the Fathers of the Church. Take the example of mountains and caves. On the Nativity icon, the infant Christ is born in a cave, indicating the earth itself taking the glory of divinity through Incarnation. In the Theophany, the banks of the river Jordan feature in a cave-like shape. The baptismal water of the river topped by the Holy Spirit echoes the imaging of "the Spirit moving upon the face of the waters" in the first Creation, and thus signifies the restoration of Paradise. In the Crucifixion, a small cave under the cross on Golgothà shelters the skull of Adam whose bones are still part of the dust of the earth from which they were formed, yet whose Fall is being redeemed in that moment. Finally, in the Resurrection Christ is shown in a vast cave deep within the earth, triumphantly raising up Adam and Eve from the dark pit of Hell.[30] Mountains too operate as signposts for the life of Christ, from Mount of Temptation to the Mount of Transfiguration (Tabor) and the Mount of Crucifixion (Golgothà). The rocky element thus serves as a powerful iconographic link and can be "read" as a part of a (sacred) text (Plate XI). At the same time, however, it also plays an active role in how the icon invites the viewer into the reality of his transfigured humanity:

> Slabs of stone stacked in ascending pinnacles and separated from each other by deep crevasses ... bend slowly away from the spectator in the foreground until at the top of the painting they curve back once again toward the viewer

as the rock reaches ever higher. The effect of this illusion is to concentrate the light at the icons' center out toward the viewer standing before it. The surrounding mountains function as the concave surface of a large reflecting mirror. This is what desert travellers do, by analogy, in using a curved lens to focus sunlight on dry grass to ignite fire. In such a way, the spectator himself is virtually "set aflame" by consuming the divine light brought into the focus of the icon.[31]

The effect is further magnified in the space of the church, through frescoed (or mosaic) cycles of images of the Life of Christ and other biblical events. The curved surfaces of cupolas and chapels work together with images to literally "wrap" the faithful and redirect her attention to the central figures (see Plates X and XI).

These scenes play a double role. Individually, they serve as *loci* for embodied contemplation, ultimately leading to self-transformation. As a sequence, they operate as powerful *loci memoriae* of historical happenings (stations of the Life of Christ, for example); as reminders of feasts embedded in the cyclical time of the liturgical calendar and in the eternal sacred time of divine liturgy.

Both functions might be re-conducted to the roots of the modern Greek word for landscape (τοπίο) and to the Byzantine ekphrastic tradition. Landscape (in the modern Western sense) refers to "a portion of land or territory which the eye can comprehend in a single view, including all the elements it contains."[32] Philologically, the term landscape denotes a collective relationship between a bounded patch of cultivated land and a local community. As a concept, landscape rests on a series of tensions: between space and place, the gaze and territory, reality and representation. Landscape's Germanic and Romance variants (*Landschaft, landschap; paysaje, paysage, paesaggio*) lean toward either pole to different degrees and can be associated with related traditions of scenic painting ruled by linear perspective (especially Flemish and Italian Renaissance painting).[33] *Topío*, by contrast, finds its origins in a Classical mnemonic tradition of *loci memoriae* and is philologically connected to place (*topos*), rather than to the land, or the community. Place here is intended both in the sense of memory place (a site or object used for mnemonic associations) and of place for embodied meditation.[34]

In the Hellenistic and Roman culture landscapes were perceived as systems of singularities, of *loci memoriae* which would have had the same effect in paint (or print).[35] Plants, trees and animals served as mnemonic devices as well as decorative elements. Individual *loci* were privileged over the holistic order constructed by linear perspective in Renaissance landscapes. In Byzantium, hunting parks and pleasure gardens were the *topía* par excellence. They were both places for self-edification and containers for discrete natural *loci memoriae* (trees, flowers, animals, caves, etc.). Along with works of art, these spaces constituted privileged subjects for *ekphráseis*, or literary descriptions whose goal was to bring an object (or a place) before the reader's eyes; to turn the listener into spectator, and the poet into painter. Like sacred icons, *ekphráseis* were not necessarily realistic. They made present not the actual picture, but the spiritual reality behind it. Descriptions of real and imagined *loci amoeni* served as tools for inner elevation, as useful fictions for self-edification.[36]

The landscape elements of Byzantine and post-Byzantine sacred images rest on this double principle. Taken individually, the Mount of Transfiguration, the burial cave of Lazarus, the tree climbed by Zachaios, the walls of Jerusalem (Plate XI) are all markers (and thus reminders) of events, *loci memoriae* magnified (or "diminished") by importance (rather than distance). Taken as a whole, they also enable the viewer to become part of the scene and reach spiritual truth. They offer a transfigured perception of the world and the self to which our modern eyes, vitiated by the certainties of linear perspective, have perhaps become unaccustomed.

## Notes

1 F. Farinelli, *L'Invenzione della Terra*, Palermo: Sellerio, 2007, p. 78.
2 See D. Cosgrove, "Prospect, Perspective and the Evolution of the Landscape Idea," *Transactions of the Institute of British Geographers* New Series 10, 1985, 45–62. The common use of the phrase "I see" (meaning both the physical act of vision and reasoned understanding) epitomizes the Western tendency to align sight with knowledge and reason. See D. Cosgrove, *Landscape and the European Sense of Sight-Eyeing Nature*, 2003. In K. Anderson, M. Domosh, N. Thrift and S. Pile, *Handbook of Cultural Geography*, New Delhi: Sage Publications, 2003, pp. 249–68.
3 See, for example, D. Cosgrove, *Social Formation and Symbolic Landscape*, Totowa, NJ: Barnes and Noble, 1984, D. Cosgrove, *Prospect, Perspective* ...; G. Rose, *Feminism and Geography: The Limits of Geographical Knowledge*, Cambridge: Cambridge University Press, 1993; etc.
4 J. Wylie, "An Essay on Ascending Glastonbury Tor," *Geoforum* 33, 2002, 441–54.
5 B. Foltz, "Nature Godly and Beautiful: The Iconic Earth," *Phenomenology* 1, 2001, 124.
6 Cosgrove, *Social Formation and Symbolic Landscape*, p. 47.
7 L. Nuti, "Mapping Places: Chorography and Vision in the Renaissance," in D. Cosgrove (ed.), *Mappings*, London: Reaktion Books, 1999, pp. 90–108.
8 Farinelli, *L'Invenzione della Terra*, p. 78.
9 Farinelli, *L'Invenzione della Terra*, p. 76.
10 Linear perspective has been "blamed" for dissociating sight from the body. Follow with your gaze the binaries of a train converging on the horizon. And then follow them with your feet – they will not converge.
11 L. White, *Medieval Theology and Social Change*, Oxford: Oxford University Press, 1964, p. 134.
12 Foltz, *Nature Godly and Beautiful*, pp. 121–3.
13 Heidegger, quoted in ibid., p. 120.
14 S. Brock, *Saint Efrem, Hymns on Paradise*, Crestwood, NY: St. Vladimir's Seminary Press, 1990, p. 42.
15 M. Zibawi, *Icone. Senso e storia*, Milano: Jaca Books, 1995, p. 13.
16 E. Voulgarakēs and K. Kyriakidēs, *Neró apó tēn erēmo*, Athens: Apostolikē Diakonia, 1996, p. 23.
17 P. Chiaranz, *I riferimenti tradizionali con i quali l'Ortodossia affronta il problema ecologico*, Trieste: Pontificio Ateneo Antoniano (unpublished paper), 2001, p. 42.
18 L. Martin, *Sacred Doorways: A Beginner Guide to Icons*. Brewster, MA: Paraclete Press, 2001, p. 204.
19 Quoted in Zibawi, *Icone*, p. 26.
20 Quoted in O. Tarasov, *Icon and Devotion: Sacred Spaces in Imperial Russia*, London: Reaktion Books, 2002, pp. 67–8.
21 B. Lane, *The Solace of Fierce Landscapes: Exploring Desert and Mountain Spirituality*. New York and Oxford: Oxford University Press, 1998, p. 125. See also A. Kartsonis, "The

Responding Icon," in L. Safran (ed.), *Heaven on Earth: Art and the Church in Byzantium*, University Park, PA: The University of Pennsylvania Press, 2002, p. 58; L. Ouspensky, *Theology of the Icon*, Crestwood, NY: St. Vladimir's Seminar Press, 1982, p. 35.

22 Zibawi, *Icone*, p. 12.

23 L. Danforth, *Firewalking and Religious Healing*, Princeton, NJ: Princeton University Press, 1989.

24 C. Stewart, "Oneirevomenoi: The Events in Koronos, 1930," *Archaiologia kai technēs* 80, 2001, 8–14.

25 C. DeSilvey, "Observed Decay: Telling Stories with Mutable Things," *Journal of Material Culture* 11, 2007, 355.

26 Tarasov, *Icon and Devotion*, p. 53.

27 For this reason Constantinos Cavarnos refers to inverse perspective also as "psychological perspective." See C. Cavarnos, *Guide to Byzantine Iconography*, Boston, MA: Holy Transfiguration Monastery, 2001; see also L. Ouspenky and V. Lossky, *The Meaning of Icons*, Crestwood, NY: St. Vladimir's Seminar Press, 1982.

28 Lane, *The Lure of Fierce Landscapes*, p. 126.

29 Baggley, quoted in ibid.

30 Foltz, *Nature Godly and Beautiful*, p. 138. "From a cave, You shone forth the world, into a cave You sink from the world. Then a cave was found when there was no room in the inn. And now You are lodged in another cave where You lack a tomb," so goes a thirteenth-century sermon composed by Germanos II, patriarch of Constantinople.

31 Lane, *The Lure of Fierce Landscapes*, p. 126.

32 Webster's Dictionary.

33 K. Olwig, *Landscape Nature and the Body Politics*, Madison: The University of Wisconsin Press, 2002; D. Cosgrove, "Modernity, Community and the Landscape Idea," *Journal of Material Culture* 11, 2006, 49–66.

34 Epicurean memory gardens, or the porches under which peripatetic philosophers strolled as they taught are good examples of this meditative and mnemonic tradition.

35 Roman *topiarii* were both the gardeners and those who painted the constitutive elements of a landscape and their singularities. See G. Mangani, *Cartografia Morale*, Modena: Franco Cosimo Panini, 2006, p. 59.

36 On Byzantine gardens and parks, see H. Maguire, "Gardens and Parks in Constantinople," *Dumbarton Oaks Papers* 54, 2000, 251–64 and H. Maguire, "Paradise Withdrawn," in A. Littlewood et al. (eds), *Byzantine Garden Culture*, Washington, DC, Dumbarton Oaks, 2002, pp. 23–35. On the ekphrastic tradition, see L. James and R. Webb, "To Understand Ultimate Things and Enter Secret Places: Ekphrasis and Art in Byzantium," *Art History* 14, 1991, 1–17.

# 23

# LITERARY GEOGRAPHY

## The novel as a spatial event

*Sheila Hones*

## Introduction

English-language geographical work with literary texts has traditionally concentrated on representation, focusing on setting, plot, and description in the novel. Geographers have looked at the insights fiction provides into real-world geographies, the fictional worlds that stories generate, the location of plot events, and the ways in which places, regions, and landscapes have been depicted.[1] Work in the histories of science and of the book meanwhile has concentrated on issues of publication, dissemination, and reception, with an emphasis on the geography of textual communities.[2] So far, however, not much attention has been paid to the ways in which the reading process itself might be understood as a spatial practice. The purpose of this chapter is therefore to suggest some of the ways in which geographers might think of the novel as a relationally generated event in itself, constantly emerging and re-emerging at the intersection of social practices and geographical contexts.

The idea that the novel can be understood as a geographical event is based on the assumption that even the lone reader absorbed in a novel is engaging with space as the dimension of difference and distance, of "relations-between."[3] This means that "the lone reader" is never really alone: he or she is spatially connected not only to a story, to a book, and to a text, but also to a narrator, an author, and a multitude of other readers, known and unknown, present and absent, near and far. Unique readings, as a result, emerge in the here and now of a particular engagement with text in much the same way that unique places emerge in space as "localized knots in wider webs of social practice."[4]

In approaching the idea of reading as a geographical process, I am drawing on three of Doreen Massey's propositions about space: first, that it is the product of interrelations; second, that it is the dimension of coexistence; and third, that it is always in a state of becoming.[5] Reconfigured to refer to literary space and to the reading process as a spatial event, these propositions become: first, that the geography of the novel can be understood to emerge out of highly complex spatial interrelations which connect writer, text, and reader; second, that multiple writings, re-writings, readings, and re-readings of any one novel will always coexist in space at any one time; and finally, that the novel itself should be understood in geographical terms not as a stable object of analysis but as a permanently unfolding and unfinished event. In order to explore this way of thinking about literary space I am taking as a case-study text Henry James's *The Portrait of a Lady*.[6]

## The event of the novel

By including in his influential work on literary geography a strong case for the inclusion of "a much greater focus on the text itself," Marc Brosseau enabled a productive shift in emphasis away from the study of the geographical content of literary subject matter and setting, and toward the study of geographical style, drawing attention to the ways in which "the literary text may constitute a 'geographer' in its own right as it generates norms, particular modes of readability, that produce a particular kind of geography."[7] This argument usefully draws attention to the ways in which agency has been distributed in the reading process: on the one hand, as Brosseau insists, it is important to accord agency to the literary work and "how it defines its reader, how it creates an 'eye'."[8] On the other hand, as James Kneale has pointed out in his work on science fiction, literary works become activated only in the context of a "relationship which joins authors, texts and readers."[9] As work in reader-response criticism also implies, text and reader together produce the event of the novel.[10] Agency is in this way conventionally distributed among the various participants in the event, as the work becomes regenerated and renegotiated in the process of being read and discussed.

In these terms, the spatial event of the novel is an event which (like the event of place) is a contingent achievement: while the literary work can still be understood as a major participant in the event it cannot be taken as the singular, self-contained source of meaning. *The Portrait of a Lady*, for example, is widely understood to have achieved a permanent position in the canon of US fiction on the authority of its intrinsic literary qualities. Nonetheless, as Jane Tompkins remarks of Nathaniel Hawthorne's similarly canonical novel, *The Scarlet Letter*, while it is "an object that … has come to embody successive concepts of literary excellence," it is not in itself "a stable object possessing features of enduring value."[11] The status of both texts has in practice been sustained over time by their being located and relocated within powerful social and academic networks: the texts themselves, Tompkins insists, are "not durable at all," at least "in any describable, documentable sense." The sustained conviction that these are works of great literary art is therefore "a contextual matter," which is to say that their reputation depends upon the situations (in other words, geographies) within which they are encountered. Readings of these novels arise "within a particular cultural setting (of which the author's reputation is a part) that reflects and elaborates the features of that setting simultaneously."[12]

In the case of *The Portrait*, contemporary readers can with relative ease access the novel in a wide variety of scholarly and mass-market paperback and hardback editions, illustrated editions, annotated editions, graded readers for language learners, translations, and dramatizations. They can also turn to the original serializations in *Macmillan's Magazine* and *The Atlantic Monthly*, as well as to vintage copies of the original one-volume and three-volume hardback editions. These coexisting editions and formats offer the reader two main textual variants: the 1881 original, and the text of the revised 1908 New York Edition. While these versions originally appeared in historical sequence, with the second being a revision of the first, they are nonetheless equally present in contemporary literary space; today, the two *Portraits* are both readily available and not always clearly distinguished. Two of the more scholarly editions currently available, for example, the Norton Critical Edition and the Library of America edition, print different versions. The Norton edition uses the 1908 New York Edition text; the Library of America volume reprints the text of the 1881 first book edition. Although these two texts both include detailed notes on textual variants, it is nonetheless possible in both cases to open the book and start reading without realizing that another version exists.

Not only does the actual text of the *Portrait* come in different formats and different versions, in some important ways the two versions are the products of different authors. The young "Henry James" of the original serialized versions is a very different writer (and a different person) to "the master" who took up the text again twenty-seven years later. For many critics the later James is a better writer who wrote a better book; it is nonetheless the case that the later version cannot be unequivocally distinguished from the earlier. It is a product of co-authorship: the early James and the late James working (harmoniously or not) together. In one version of a book chapter on autobiography and geography, the geographer Ian Cook makes this general point as he addresses the reader directly:

> So what do you think is going on in the relationship between my writing and your reading? Are "we"/"they" getting on OK? But, please don't think that you now know "me" really well…. Please note that I make no claim that I will be the same by the time this book is published, anyway. I wholeheartedly subscribe to the view that people are processes, not things. I will have changed. I am bound to.[13]

As this suggests, both versions of Henry James, in company with their readers, are currently active in the spatial event of the novel. And the instability of the author figure is compounded by the fact that a published work is never, really, the work of a single individual at all, but rather a product co-authored to a greater or lesser extent with publishers, editors, proof-readers, designers, and typographers, a product that has also frequently been influenced by comments from reviewers, colleagues, and readers. Cook draws our attention to this point also when he writes as "Ian Cook, et al." by emphasizing his author-identity as "a collective" even when he has no named co-authors.

When the text is viewed in this way, as a spatial event, an always emerging collaboration in space–time, there is thus no single entity that can unproblematically be identified as *the* text and no figure that can be unproblematically identified as *the* author. It is equally evident that there is no such real-life entity as *the* reader. Literary critics who refer to "the reader" tend to use the term to refer to "an interpretive (not a natural) category," one that, like "the text" operates "as a hermeneutic device in practical criticism and other areas of literary study."[14] Real readers, meanwhile, engage with the text in innumerable variations and – as with the author – even "the same person" can be a different reader under different spatio-temporal circumstances. Online discussion groups and blogs in which readers record their experiences as they read a novel chapter by chapter provide useful evidence of the ways in which reader reactions to a text may change even during the course of a first reading. While these changes may appear to be primarily a matter of time, they also no doubt result from a reader's continuous process of relocation in literary space, a relocation which might, for example, be affected by the very process of engaging with other readers or potential readers, even by the process of articulating a response. As Jane Tompkins argues, there is "never a case in which circumstances do not affect the way people read and hence *what* they read – the text itself."[15]

Posting on the website *43things*, reader Meliflower, for example, first wrote that she found *The Portrait* "absolutely charming," remarking that "Henry James is a wonderful story teller." In response, Dazee commented: "Thanks for posting this. I need something new to read," to which Meliflower then replied: "Ohhh … let me know what you think if you start getting into it … we'll have an online 'book club.'" Several months later, Meliflower updated her comments: "I am not so terribly taken with this book as I was. And I have 330 pages more to trudge through." A week after that, her interest in the novel had begun to return – "I started getting into it again" – and after another week she wrote: "I only have about 80 pages left and the last quarter of the book is an incredible scandal. I'm glad I stuck with it."[16]

Writing in a formal context, the novelist Jeffrey Eugenides has described the different experiences he had with the *Portrait* in the course of two differently situated complete readings:

> [T]he first time I read *The Portrait of a Lady* was under duress. Our English teacher … assigned it in the fall of 1976. Our class divided in a clean gender split: the girls loved it; the boys, myself included, gagged all the way through. In graduate school, I read *The Portrait* again, and fell out of my malodorous thriftstore armchair. This time through, the book bore no relation to the one I'd so masterfully misread seven years earlier. Not only did the writing now strike me as charming and elegant, but the book's action, which had seemed so quiet and repressed before, revealed itself as passionate and gripping.[17]

As James Kneale has noted, the different experiences Eugenides had with the same novel provide us with "more than just an example of the way that readers (and hence texts) change."[18] In other words, the point is not so much that his attitude to the novel changed over time, but that it changed in response to a shift in the social context in which his readings took place. The second time he engaged with the novel, differences in the reader position and in the context of reading would have actively reconfigured the author/text/reader nexus, altering the various kinds of conventions and expectations that were contributing to the shaping of the event.[19] In the high school reading, it seems likely that one of the effects of the immediate social context was to encourage a binary division ("a clean gender split") in reader response. In the graduate school reading, the socio-spatial situation and power geometries of the reading context would have been entirely different. Traces of multiple literary frameworks and their related conventions, including "school teaching, the tropes of criticism, biographical cults of the author" would almost certainly have contributed to this second reading event. To repeat the point made above: these would have been conventions that were not only embedded to some extent within the literary work itself but were also part of the literary space of the reader, generated by his encounters with "pedagogy, newspaper reviewing, audiences, etc."[20]

## Texts, authors, and readers in space and time

To complicate matters farther, in considering the text in this way as a spatial event we have to go beyond the simple acknowledgment that there is more than one *Portrait*, produced by more than one author, and known to more than one reader. We also have to take into account the fact that these textual variants are now coexistent, that the authors are entangled, and that readers influence each other (unevenly). For example, as Nina Baym points out in her essay on revision and thematic change in the *Portrait*, the manner in which the preface for the later edition "instructed the reader how to interpret, what to admire, and what to deplore in the work," has had a significant impact on subsequent critical readings of *The Portrait*. For many readers today, the 1908 version and its preface haunt the 1881 version, despite the fact that the earlier version "is a different work" and the earlier James "a masterful writer with his own interests."[21] The different versions of the *Portrait* that coexist today, literally sitting next to each other on shelves in book-shops and libraries, are in fact both internally multiple products of complex spatio-temporally extended processes of co-authorship.

Significantly different readers also coexist. At times, individual readers may be physically co-present and engaged in real-time interaction, actually sitting next to each other in book groups, seminar discussions, or conferences. At other times, they may be engaged in mutual exchange at a physical distance, participating in online book discussions or in the academic processes associated with book or journal article publication. But of course this interaction is not uniformly multilateral or mutual. To take a somewhat banal example, not everybody in a book group will be paying equal attention to every comment made, equally able to comprehend the suggestions of others, or even literally able to hear what everybody in the room is saying. It seems likely, for example, that some members of the high-school English class attended by Jeffrey Eugenides were paying more attention than others. Even within professional literary criticism it would be practically impossible today for an individual scholar to read everything ever written on James. Not only is there a vast body of critical work available, but also accessibility fluctuates according to variables such as institutional affiliation and competence in different languages, and the potential for scholarly collaboration is unevenly distributed across time, and space. For a variety of reasons it is, for example, much more likely that Japan-based James scholars will access work produced in the US than the other way around. And while Nina Baym, writing in 1976, was able to refer to F.O. Matthiessen's 1944 discussion of the revisions,[22] Matthiessen himself was writing at a time when Baym's work did not yet exist and her very different approach to the revisions was at least unthought, if not unthinkable.

Interactions between readers, between texts, and between authors thus occur in complex ways in literary space–time, and as a result, texts, authors, and readers will always effectively be both internally multiple and reciprocally influential. Each version, each edition of the *Portrait* is equally and independently "the text;" and yet, at the same time, the *Portrait* is a collective text, made up of all of its variants taken together. The early James and the late James are equally independently Henry James. The later version is clearly distinct from, but nonetheless inseparable from, the earlier and the two currently coexist in literary space. Similarly, six people participating in a seminar on the *Portrait* will be engaging with their shared topic in various ways and with various different motivations. To a varying degree, their readings of the text and their contributions to the discussion will be unique, a uniqueness which may seem either brilliant or idiotic to their colleagues. Nonetheless, at the same time, their ways of engaging with the text, of articulating ideas, and of producing collective knowledge will be to some inevitable extent be mutually co-productive.

In proposing that individual readings are thus never in any strict or pure sense absolutely individual, I am relying on the assumption that identity itself, while in each individual case unique, is nonetheless generated in social contexts and relational by nature.[23] If we take this view of identity, and work with it in defining the reader of the *Portrait* in any given situation, then we can see how even a single individual, reading and reflecting internally on the *Portrait*, is always reading alone and yet at the same time reading in a social context that includes, as Massey indicates, not only the significant (if unacknowledged) presence of other readers, but also a wide range of significant (if even less recognized) absences, gaps, and misunderstandings.

This productive tension between individuality and collectivity, unique readings and social contexts, is of course even more evident in the performance of readings that are literally articulated. In terms of understanding how readers actually engage with and develop particular reactions to particular texts, the individual reader is a black box: we have no reliable objective access to individual reading processes. All we have is what readers say about their readings, and in the process of communicating ideas and opinions about a text the reader/speaker will to a certain extent have to organize and simplify a fluid, contradictory, and inchoate reading experience into a comprehensible argument or narrative. Furthermore, if the unarticulated reading process is always itself relationally produced, then the articulation of any particular reading will tend to be even more context-sensitive.

To a certain degree, in the case of *The Portrait* the extent of this entanglement, this multiplicity within separable elements, results from the novel's particularly complex publication history, with its two distinct, coexisting, and inseparable variants. In comparison, the text (and hence the author) of most other works of fiction might appear to be much more stable. But the instability of the text (the fact that there is no such thing as *the* text) is not really a product of the particular history of the *Portrait* and its revisions. The significant point is not so much that various different versions of the text exist simultaneously and cannot be disentangled; the point is that the text, the author, and the reader – and their various space-time trajectories – cannot be unambiguously untangled either. In other words, these different texts, editions, authors, and readers are mutually co-productive in very complex ways.

It is in this sense that a work of fiction comes to life – *happens* – in the interaction of various elements, conventionally stabilized in the tripartite division "author–text-reader," but with each element within that structure characterized by internal variation and multiplicity. Through their interaction in space-time, these individually multiple elements collectively generate particular contexts within which the novel emerges as a spatialized and transitory event. In referring to contexts here I am not using the term to indicate something static, anything like a backdrop or frame. I am not suggesting that contexts come first, are static, or exist independently of interaction, and I certainly do not want to suggest that particular readings evolve or emerge naturally in and from particular situations. My understanding of contextual milieu here is instead based on Nigel Thrift's definition of context as "a performative social situation," "a parcel of socially constructed time-space."[24] Crucial to this use of the concept of context is that it does not imply a fixed time and place: it is not necessarily local, not does it necessarily imply simultaneity. In other words, while I am emphasizing the way that particular, individual readings (engagements with text) always have a unique "where and when," taking place in the flow of an always emerging "here and now," this location in space-time cannot in any sense be fixed or explained by reference to particular times and places.

## Literary geography: from setting to event

While the *Portrait* is a particularly complex case, it is only an extreme example of the general point that all fiction comes to life at the intersection of multiple participants, including author(s), editor(s), publishers, texts, teachers, critics, and readers. The text, when it happens, comes into being in the interaction of differently contextualized processes, and these processes are each in themselves generated in the context of countless interactions in space and time. Obviously, the historical

Henry James did exist; copies of the various versions of the *Portrait* exist; actual readers exist. The point is that it is only when they come together, in real-time interaction, that *The Portrait* can happen as an event. Without readers, the text cannot happen; without the text/reader interaction that makes the novel happen, the author figure does not exist; and without the author figure, there would be neither text nor reader. None of these figures exists outside space and time: even the "ideal" or "implied" reader that a real reader may identify "inside" the text is really an interpretation emerging contingently at the intersection of text, author, and critic.

In actual messy practice, the borders between these three elements cannot be maintained. Once we start to think of the author and the reader, for example, as relational effects rather than independent entities, then we can start to conceptualize the text event, too, as a contingent achievement. And one of the reasons that this is important is because it shifts the focus from assessment to engagement. It diminishes the significance of distinguishing good readings from bad readings, brilliant analyses from naïve reactions, and emphasizes, instead, the way in which different kinds of reading perform different kinds of contextual appropriateness. A brilliant academic reading might be very productive in a situation – say, a graduate seminar – in which the participants shared common terminologies, references, purposes, ambitions, and conventions of interaction. It might be far less useful (it might even be counter-productive) in the context of an online reading group or a book review.

Taking all of this into account, it becomes clear that literary geography does not have to limit itself to the practice of looking for the geography in novels or reading a literary work geographically. Rather, the novel can be understood as a geographical phenomenon in itself, an event that emerges in individual readings which are nonetheless highly relational. The novel *as it happens* can in this way be understood as something that is constituted in and through its engagements, interactions, and relations, not just with other texts but with various authors and various readers.[25] The novel happens in space, is the product of interrelations, emerges in the dimension of coexistence, and is always in a state of becoming. Like place, the novel happens at "the coming together of the previously unrelated": it is a constellation of processes rather than a thing": it is "open and internally multiple," not "intrinsically coherent."[26] In this sense, the novel is not so much a representation of geographies as it is a geographical event.

## Notes

1 Anonymous, "Geography in Literature," *Geographical Review*, 14 (4), 1924; Anonymous, "Bibliographical Sources for Geographical Fiction," *Geographical Review*, 28 (3), 1938; W.E. Mallory, and P. Simpson-Housley, *Geography and Literature: A Meeting of the Disciplines*, Syracuse University Press, 1987; D.C.D. Pocock, "Geography and Literature," *Progress in Human Geography* 12 (1), 1988; M. Brosseau, "Geography's Literature," *Progress in Human Geography* 18, 1994; J. Sharp, "Towards a Critical Analysis of Fictive Geographies," *Area* 32 (3), 2000.

2  J.A. Secord, *Victorian Sensation: The Extraordinary Publication, Reception, and Secret Authorship of "Vestiges of the Natural History of Creation,"* Chicago, IL: University of Chicago Press, 2000; D.L. Livingstone, "Science, Text and Space: Thoughts on the Geography of Reading," *Transactions of the Institute of British Geographers* 30 (4), 2005; I.M. Keighren, "Bringing Geography to the Book: Charting the Reception of *Influences of Geographic Environment,*" *Transactions of the Institute of British Geographers* 31 (4), 2006.

3  D. Massey, *For Space*, London: Sage Publications, 2005, p. 9.

4  D. Gregory, "Edward Said's Imaginative Geographies," in *Thinking Space*, M. Crang and N. Thrift (eds), London: Routledge, 2000, p. 231.

5  Massey, *For Space*.

6  H. James, "The Portrait of a Lady," *Macmillan's Magazine* 42–45 (Oct. 1880–Nov. 1881); *Atlantic Monthly* 46–8 (Nov. 1880–Dec. 1881); *The Portrait of a Lady*, Boston, MA: Houghton, Mifflin and Co.; and London: Macmillan and Co., 1881; *The Portrait of a Lady*, New York, NY: Charles Scribner & Sons, 1908.

7  Brosseau, "Geography's Literature," p. 349.

8  Brosseau, "Geography's Literature," p. 347.

9  J. Kneale, "The Virtual Realities of Technology and Fiction: Reading William Gibson's Cyberspace," in *Virtual Geographies*, M. Crang, P. Crang, and J. May (eds), London: Routledge, 1999, p. 208.

10  J.L. Machor and P. Goldstein (eds), *Reception Study: from Literary Theory to Cultural Studies*, New York, NY: Routledge, 2001; P.P. Schweickart, and E.A. Flynn (eds), *Reading Sites: Social Difference and Reader Response*, New York, NY: Modern Language Association, 2004.

11  J. Tompkins, [1984], "Masterpiece Theater: The Politics of Hawthorne's Literary Reputation," in Machor and Goldstein, *Reception Study*, p. 150.

12  Tompkins, "Masterpiece Theater," p. 149.

13  I. Cook et al., "You Want to Be Careful You Don't End up Like Ian. He's All Over the Place," in "Autobiography in/of an Expanded Field (the director's cut)," *Research Paper 34*, Brighton: University of Sussex, 1998.

14  S. Mailloux, *Interpretive Conventions: The Reader in the Study of American Fiction*, Ithaca, NY: Cornell University Press, 1982, p. 13.

15  Tompkins, "Masterpiece Theater," p. 137.

16  "Meliflower." "43 Things I'd Like to Do: Read *Portrait of a Lady* by Henry James," http://www.43things.com/things/view/1360035, 2007 (last accessed 26 September 2008).

17  J. Eugenides, "Personal Best: *The Portrait of a Lady*," 1996 http://www.salon.com/weekly/james960930.html (last accessed 26 September 2008).

18  J. Kneale, Personal email communication to the author, 30 September 2008.

19  Kneale, "The Virtual Realities," see note 9.

20  Kneale, Personal email, see note 18.

21  N. Baym, "Revision and Thematic Change in *The Portrait of a Lady*," 1976, reprinted in R. Shelston (ed.), *"Washington Square" and "The Portrait of a Lady": A Selection of Critical Essays*, London: Macmillan, 1984, p. 185.

22  F.O. Matthiessen, "The Painter's Sponge and Varnish Bottle," 1944, reprinted in R.D. Bamberg (ed.), *The Portrait of a Lady: An Authoritative Text/Henry James and the Novel/Reviews and Criticism*, New York, NY: W.W. Norton & Co, 1995, 577–97.

23  D. Massey, "Geographies of Responsibility," *Geografiska Annaler: Series B, Human Geography* 86 (1), 2004, 5.

24  N. Thrift, *Spatial Formations*, London: Sage Publications, 1996, p. 41.

25  S. Hones, "Text as it Happens: Literary Geography," *Geography Compass* 3, 2008; S. Hones, "Text as it Happens: Literary Geography – Teaching and Learning Guide," *Geography Compass* 4, 2009.

26  Massey, *For Space*, p. 141.

# 24

# MATERIALIZING VISION

## Performing a high-rise view

*Jane M. Jacobs, Stephen Cairns, and Ignaz Strebel*

Vision in the sense of active seeing is inescapable in the practice of geography.[1]

## Introduction

Geographer Denis Cosgrove had a long but tragically never fully realized interest in what he referred to as the "icon of the window."[2] His concern was primarily in how the window operated as a technology of vision. For example, in his 2005 Hettner Lectures, he argued that the restricted space of the Southern Californian suburban lot was compensated for by "picture windows," which both "frame[d] a picture view and erase[d] the boundaries of internal and external living."[3] Cosgrove's emphasis in this instance was on the en-framed view, and his wider scholarship always spoke to his pursuit of geographies of a world in which the virtual and practiced are understood to operate simultaneously in the production of vision. For Cosgrove, the window was an icon of vision because it encapsulated both seeing and being. The English term "window" expresses a version of such gathering together, deriving as it does from the Middle English (Old Norse) *vindauga*, eye of the wind ("vindr" – wind and "auga" – eye). In this chapter we wish to take further these suggestive invocations of the relationship between practices, technologies and vision. We do so not in relation to a suburban window and its view, but by considering the windows and views of a residential high-rise building. The framing work that the high-rise window performs in the making of a high-rise view is complex and contingent. It depends upon an array of other factors, both material and immaterial: existing window discourses, design specifications, regulations, residents, maintenance workers, smoke, vertigo, curtains, worry, wonder, asbestos, glass, steel, water and, of course, the wind and the eye. Through the example of the high-rise window we wish to inhabit the embodied and materialized ways in which a "high-rise view" happens. Our aim is to re-present the view (and thus vision) as a practiced event – vision-ing/view-ing

– embedded, in this case, in a specific building technology, a moment in the life-cycle of high-rise housing, and a peculiar instance of camera-based research interaction. Furthermore, and in the wake of a cultural history of the window that has privileged the "eye" (vision), we wish to reacquaint the story of the window with "wind" (atmosphere). In doing so we return the term "window" to its other, more technical, label of "fenestration." Fenestration refers in a narrow sense to the arrangement of windows in a building, and more popularly to the various materials and technologies that comprise window fittings. But the word derives from a Latin term meaning "to furnish with a hole or opening."

The windows that have garnered our interest belong to a British social housing residential high-rise estate: Red Road in Glasgow (Figure 24.1). The estate comprises two 26–28-storey slab blocks and six 32-storey point blocks that were constructed between 1964 and 1968. Red Road was a product of the post-war, state-led, programs of mass housing provision in Britain. Despite being commissioned and built to be a generic high-rise housing template, Red Road's construction was marked by the personality of its architect, the idiosyncrasies of his vision, and the contingencies of the construction process.[4] Today the estate is an emblematic example of the dramatic slide from utopian vision to dystopian reality that has marked so many post-war high-rise modernist social housing programs in Britain. It has suffered a history of dis-investment. Its residents include many long-term households, but also a range of tenants whose housing needs are urgent or who have not been willing or able to exit social housing through right-to-buy offerings. In 2005 the Glasgow Housing Association made a determination that Red Road was no longer a "sustainable" or "viable" social housing development and announced a £60m redevelopment strategy, the first stage of which is to be the demolition of the block with the address 213/183/153 Petershill Drive. It is the windows of 213/183/153 Petershill Drive that we look at and through.

**FIGURE 24.1**  The Red Road estate, Glasgow. Photograph by the authors.

Our interest in the high-rise window has not been confined to the historical and cultural production of "a view." As our opening quote from Denis Cosgrove suggests, we wished to consider the "active seeing" that produced something that might be labelled as a "high-rise view." In telling this story we are alert to the ways in which the taking in of a view through a window involves not only the human viewers who "do" the view and the human designers and builders who "make" the view, but also a range of other participants who, although not human, are nonetheless still active in the viewing event. These others participants may include immaterial things such as safety regulations or thermal comfort specifications, but they also include the materials of the building themselves. Our start point for thinking this way about the view is work inspired by sociologist of science and technology, Bruno Latour. That tradition of thinking has been dubbed, somewhat awkwardly, Actor-Network Theory, a term that Latour admits is not especially useful and has resulted in considerable confusion.[5] Latour conceives of the world as a set of local effects, produced by the contingent relationships between all kinds of "actors" (objects, subjects, human beings, machines, animals, nature, ideas, organizations, rules, laws) operating through folded and flattened scales (not micro *or* macro, but micro *and* macro as circulating entities) and through a range of proximities (what is visible in the situation under analysis as well as what is not visible but which made the situation what it is). In particular, Latour has been interested in what he calls the "*summing up* of interactions through various kinds of devices, inscriptions, forms and formulae, into a very local, very practical, very tiny locus."[6] The analytical work (methodology) of this project is not to diagnose an essential character of such a locus, but to follow the circulations, distributions, connections of which it is an effect. The label Actor-Network Theory seems a long way from the conventional concerns of the humanities, but it is worth registering that Actor-Network Theorists often also consider themselves to be engaged in a semiotics of materiality (or a *relational materiality*). As John Law puts it: "[Actor-Network Theory] takes the semiotic insight ... of the relationality of entities, and applies this ruthlessly to all materials ... and not simply those that are linguistic."[7] Finally, Actor-Network Theory is interested in *performativity*: "How ... things get performed (and perform themselves) into relations that are relatively stable and stay in place."[8]

In what follows we take these insights from Actor-Network Theory and draw upon them to understand the socio-technical achievement of a high-rise view and high-rise view-ing. In doing this work we chart the diverse and interacting practical orders through which the technological and the human co-orchestrate, in our case, the elevated view of the high-rise block.[9] Interactions such as these are always productive, although that productivity may incline toward either "success" or "failure." Like all building technologies, a window-in-action may perform what it was programmed to do or it may err toward that which was not programmed: it may become an opening to jump from or through which to throw rubbish, activities that Hand and Shove call "anti-programming."[10] It may also, through technology failure or redundancy, become de- or re-programmed (a window that does not open, or a window with broken glass that lets in too much of the outside). In this sense performativity and (building) performance are linked.

## Modernism's windows

Although the high-rise mass housing of Britain's post-war housing boom is a distant relation of avant-garde architectural modernism, Dunleavy noted that state housing and planning authorities found sustenance in the theoretical speculations of figures such as Le Corbusier.[11] In Le Corbusier's original vision of the high-rise city, the window/view had a very specific place. This may be illustrated by a sketch he did in his 1933 book *La Ville radieuse*.[12]

In this sketch a high-rise vision is stripped of its formal features and represented simply as a set of gathered-together services (water, electricity, gas, telephone). Atop this unclad high-rise block is a single apartment, occupied by a tiny figure and a disconnected, distended eye looking outwards. Colomina argues that this eye serves to communicate Le Corbusier's commitment to the window as an aperture for light and seeing.[13] This commitment to light and seeing was set against the nineteenth-century domestic interior. That earlier interior turned inward from the world outside as if a refuge.[14] For example, architect Adolf Loos saw the ideal window to be glazed with ground glass so that light was let in but there was no view out.[15] As Walter Benjamin put it, the transparency and openness of Le Corbusier's modernism "put an end to dwelling in the old sense" and to its "nihilistic cosiness."[16] The modernist window shuns curtains and other "hampering objects" and "throws the subject towards the periphery of the house."[17]

Le Corbusier famously broke from the tradition of the "vertical window" and developed the expansive "horizontal window." This transformation did not simply alter the parameters by which light or air was let into the interior. As Friedberg notes, the horizontal window also encased "a panorama that ... brought more of the outside in."[18] Modernism's window restructured the relationship between occupant and exterior. The vertical window positions the occupant/viewer as central to the framing and, by taking in foreground, horizon and sky, replicates perspectival depth.[19] In contrast, Colomina argues, modernism's horizontal window replicates the vision of the "camera, the mechanical eye" in which (especially with the movie camera) there is no central, centered viewing position.[20] Furnished with the horizontal window the apartment becomes a "viewing mechanism" in which the views taken in are choreographed by the occupant. As Friedberg notes, "the frame of the window and its glass expanse become only one element in the inhabitor's ... mobile view."[21] The relationship of window to view has been taken up by Sandy Isenstadt in his study of the emergence of the large-scale, plate glass suburban window in America.[22] His scholarship shows that the window's relationship to viewing the landscape has been both elaborated through design (bigger and more carefully placed windows) and consolidated by real estate advertising into a domestic asset called "a view."

As this preceding discussion shows, much of the cultural historical work on the window has emphasized its role in the ordering of the gaze and en-framing

views. But the work a window does as a building technology extends beyond
that of vision. The window is at core an opening between inside and outside. In the
modern window, glass is central to this achievement, its transparency keeping the
"outside out and at the same time bringing it in."[23] But glass is not alone in the
making of the modern window. It is joined by other materials (steel, wood,
aluminium, PVC, fabric), working in other ways (as frames, casings, jambs, hinges,
locks, decoration, insulation), to manage the opening of which they are a part.
Furthermore, an opening is always relational: the work a window opening
does is relative to the walls around it (window to wall ratios and window place-
ment). In unison with a human user (and in some "smart" buildings even without
that human) the window assemblage controls what effects the opening has on
the interior of the building: how much light is let in, the amount of air and
noise that is admitted or released, levels of safety and security. In short, windows
modulate atmosphere and comfort. Although there is a vast technical literature
on the relationship between windows and indoor thermal comfort (and also security), this has not featured in cultural histories of the window. An exception is
Isenstadt's already mentioned account of the American suburban window. Part of
the history of the suburban plate glass window is the contentious relationship
between its visual and atmospheric effects. Isenstadt shows that during the late
nineteenth and early twentieth centuries in both Britain and America there was
considerable discussion among architects, designers and home-owners about the
relationship between windows as a mechanism for delivering a view into a home
and the effects of expanses of glass on the look and feel of domestic spaces.
Specifically, he presents evidence of the ways in which a move to larger windows
was seen to erode a sense of warmth in the home. Interestingly, the examples
that Isenstadt gives are not about the impact of large plate glass windows on actual
temperatures and air circulation (atmosphere in narrow sense), but more on the
"sense" of warmth and enclosure (atmosphere in the wider sense). For instance,
Isenstadt cites a quaint example derived from an 1864 issue of *Harper's Weekly*.
A Mr Rogers was reported to have caught a cold simply from dining near a plate
glass window. It was not a draught or chill from the (apparently closed) window
that produced the cold-inducing atmosphere, but simply Mr Rogers' "force of
imagination."

One cannot separate off the ability of the window to attach itself evermore
emphatically and narrowly to the experience of the "view," from the elaboration of
other non-window technologies given over to the management of atmosphere
(temperature and circulation). For example, if we return to Le Corbusier's distended
eye atop his flayed high-rise apartment, we see that it is supported by a range of
other technologies, one of which is called "exact air" (a mechanical delivery system
of ventilation, heating and cooling). The emergence of the modernist window as a
viewing machine occurred in unison with the development of other technologies
intent on creating the home as a "climatic fortress."[24] Indeed, it may well be that the
rise of the window as a viewing technology was dependent upon transferring its
role as an aperture of atmosphere to other technologies of comfort.

## High-rise windows

We have already noted how Le Corbusier's vision of a clean and aerated high-rise city was the distant template for many of Britain's state-sponsored, post-war high-rise housing developments, including the likes of Red Road. Yet what was built in the name of modernism in such mass housing programs usually dispensed with key design elements. Nonetheless, when the first and tallest of the Red Road flats was formally opened on 28 October 1966, the window was enrolled into the publicity efforts that marked the materializing of this housing vision. Front-page news of the day was a picture of William Ross, the then Secretary of State for Scotland, and his wife, looking out of a top-storey window at the construction site.

At this celebrated moment an occupant/window/view assemblage was mobilized to speak for the wider project of post-war mass housing provision. This was not an uncommon iconography: an early image from an official promotional brochure for the now infamous Pruitt-Igoe housing development in St. Louis shows a recently housed family gazing out from a large window to the view beyond. Here, and in Red Road, viewer/window/view are assembled to represent the hope of the new life offered by high-rise housing.

But there are some significant differences evident in these two historical depictions of the occupant/window/view assemblage. In the Pruitt-Igoe image an unnamed African American family were depicted from behind a full length group

**FIGURE 24.2** William Ross, then Secretary of State for Scotland, and his wife, at a Red Road window on the occasion of the estate's official opening, 28 October 1966. Image courtesy of *The Herald & Evening Times* picture archive.

shot gazing through a large picture window (it appears to be a common space window as opposed to the window of an apartment). The viewer/window/view image for Red Road is staged differently. For a start, we see clearly the faces of these important people who are taking in this view of an estate under construction and in so doing sanctioning its arrival as a housing solution. More significantly, the entire occupant/window/view composition is closer in and includes the action of opening the window and looking and pointing out. This "inaugural" image of a Red Road view both confirms the view as an asset of living high, but also illustrates some of the peculiarities of how a Red Road view has to be performed. The Red Road view is made not by passively standing in front of a glass expanse (as in the Pruitt-Igoe image), nor even (as Colomina suggests) through some newly mobile viewing practice. It is made, so it would seem, by getting up close to the window, opening it and looking out.

In the final section of the chapter we wish to open up current thinking about the relationship between the modern window and "the view." We do this by taking literally Cosgrove's invocation of "active seeing" and examine in detail how residents in Red Road "do" their high-rise views. Relatedly, we reflect upon the many things, including windows, that come together to produce Red Road views. Finally, we reflect upon the relationship of the Red Road window to atmosphere and comfort.

## Re-viewing Red Road's windows

In the course of our research on Red Road's past and present fortunes, we asked residents to engage in an exercise we dubbed "Show us your home." This was an invitation that sought to solicit from the resident narratives about, as well as interactions with, their flat. One sub-set of this activity was asking residents to "show us your view."[25] In making this request (as with all our discussions with residents) we were not interested in determining resident satisfaction levels, as might a researcher working within the conventions of post-occupancy evaluation studies. Rather, we were interested in how a view was done, and what constituent elements had to come together to produce a view. Usually, but not necessarily, the request "show us your view" induced an action, that of residents moving to the window technology. This first action was usually followed by a sequence of other actions: in some cases gestures that swept across the scope of the scene outside, in other cases the pulling back of curtains, in most cases the opening up of the window and a poking of the head outside, just as the Secretary of State for Scotland and his wife had done some 40 years earlier. As one of our participants aptly put it, as she showed us her view; "No, you got to get right up into it [the window] and you will see it all." This tells us immediately that there is something different between the visioning practices depicted in cultural historical accountings of the window in modernism, and the practices occurring through and with these middling modernist high-rise windows.

In thinking this through we must address, and not merely in passing, that the high-rise view-ing events we recorded involved at least four viewers (the resident, the interviewer, the camera person and the camera). This was not a "normal" view-ing sequence: it was staged, self-conscious and crowded, not unlike the event that led to the "inaugural" image of a Red Road view (involving as it did politicians, architects, government officials and journalists). It is likely that everyday viewing events vary: from active showings, through to contemplative lookings out, and on to the glimpsed view. It is also likely that some, but not all, view-ings involve opening the window. What is clear is that generally when something called "a view" is being "shown" (as well as seen) in Red Road then one must "get up into" the window, open the window and poke one's head out.

We must also address, and again not merely in passing, the design and materials of Red Road's windows, their surrounds and the many other objects that come to attach themselves to the technology and space of the window. When built, the prefabricated, single glazed, steel-frame windows of Red Road were at the fore-front of window technology. Furthermore, being steel-framed they were stitched into local initiatives that sought to link the building of the "all steel" Red Road to efforts to revive the fortunes of the local steel industry which was, post-war, suffering a slump.[26] Although they are, relatively speaking, more horizontal than vertical in dimensions, they are also, relative to wall area, modest apertures with the window sill located at a "usual" height. They also, despite the height of the building, open and close like normal domestic windows. And, because they had such opening devices, Red Road windows also had safety devices such as locks and child safety catches. These additional technologies ensured that under normal circumstances the window opened only a small way, such that people (and especially children) and things could not fall out. Finally, the Red Road windows were not supported by a collectively delivered window cleaning service, such as one might see on commer-cial high-rise buildings. Red Road windows were domestic windows and were to be cleaned by residents themselves. To facilitate this they featured relatively novel tilt-and-turn hinge technology for ease of cleaning thereby ensuring that residents did not place themselves at risk in the interests of a clean window.

All these attributes shape the ways in which the Red Road view is done. For example, the size of the windows of Red Road means that they were a long way from any modernist ideals of a transparent, glass screen. As an aging technology that is imprecisely lived with, these windows interrupt the smooth alignment of window and view in many ways. They have curtains, hold ornaments, carry stickers, get dirty and cracked, and even, in some cases, have smoke damage. They also carried lots of talk. Often enough that talk was about the view to be seen ("my view here is lovely," "it is just too beautiful," "see doun there, look at that, in' that beautiful?" "see at night ... its really terrific," "It's not a very nice view ... except perhaps with the empty space. Here at least got that ..."). Sociologies of technology have shown that talk always goes with technology in action. This is often more so with failing technolo-gies which solicit commentaries of various kinds to mediate their deficient perform-ances (inquiries, apologies, excuses, explanations, complaints).[27] For example, one resident when asked to show us her view took us immediately to the window and started pointing, but her first directive to us was not for us to "look at the view" but to look at the window itself: "See this doun here ... see the damp on the windae, don't look hen, no. It's really mank."

During the course of showing us her view, this aged resident repeatedly apologized for the state of her window, demonstrating how heavy the windows were to open and turn in order to clean and explaining that, because she was about to leave Red Road, she had stopped window-cleaning altogether. In a sense her apologies and explanations were themselves a kind of discursive maintenance – doing the work of repairing a window whose performance as a viewing technology had been compromised. In another Red Road flat, accessing "the view" was equally difficult. Rather than show us the view, the resident initially stayed on the sofa with window curtains drawn and simply described to us the unseen view outside. Only after a while did this resident go to the window and pull back the curtain, and only after a direct request to show the view to the person holding the camera does she start to do what most Red Road residents find they need to do to "show us their view," open the window (Figure 24.3).

**Res:** Petershill Road we actually look onto. And from here you can just see the other flats. You can see Sighthill flats and the receration [sic] centre which is to the left there [goes to the window and partly pulls back curtain - **still a** -, camera moves in to the window to look out - **still b**]

**Int:** Oh yeah.

**Res:** Oh and the little playing field where I used to have my picnics, er and we've got…

**Int:** Just show Emma where you used to have your picnics [lets go of one side of curtains and moves to the other side of window and draws back the other curtain - **still c**].

**FIGURE 24.3** The talking and showing of a Red Road view. Photograph by the authors.

In this case the delays and interruptions to (the showing of) the view are the result of the resident feeling as if it were not worth showing and looking, not because what was outside was not worth seeing but because the quality of seeing had been impaired by smoke damage to the glass as a result of two serious chip-pan fires in the house. As the resident explained: "It was just ... well you can't really see this. Again [as she puts her hand up to the window to gesture to the view], with the fire, this window smoked up." In Red Road, the showing of a view happens through an active seeing, in which the window's part is far more than a technology of transparency.

## Draughts and other atmospherics

While the windows of Red Road cannot but help be viewing devices, they are also called upon to play an important part in the control of atmosphere, for there is no "exact air" system in Red Road. The management of interior temperature and ventilation happens by way of orchestrated activities involving the in-built heating system, ad hoc (resident supplied) technologies of heating and cooling, and the windows, which as we have noted can open and shut with a tilt-and-turn system. How to manage atmospheric comfort has been a concern to both housing managers and residents alike. This is nowhere more evident than in the quest to keep warm. In the history of the estate, three different domestic heating systems have been provided by the local housing authority. These systems have been subsidized by residents themselves, with technology ranging from portable electric and oil heaters, to kerosene burners, and including the construction of elaborate fake hearths.

As an opening, a window does its essential work in relation to that which is not window, the wall. In the case of Red Road this relationship has shaped much, not only about how its windows can perform, but also the entire building. This was given explicit expression by one Red Road resident whose first (unsolicited) mention of the windows of Red Road was inextricably linked to talking about another Red Road material.

> Res: We had asbestos, ay. See, that's all asbestos up there. Up there, next to the blinds. That ceiling. They've not took it away yet. They say because ... you see we always wanted [new] windaes, see.

The cladded steel-frame structure of Red Road was, for its time, considered novel by UK standards, given that most tower block construction was steel-reinforced concrete. It was this novel building technology that resulted in the introduction of asbestos, which was used as the fire insulating material. As a result of the presence of asbestos, repair and replacement work of various kinds has produced specific safety issues for professionals and residents alike. It has above all meant that the windows in Red Road cannot economically or safely be replaced and so remain the original single glazed steel-frame windows. As the extracts from resident interviews above show, residents are forced to live with an aging and worn window technology. This does not simply interfere with the view. Indeed, for many residents there was a genuine appreciation of and attachment to their high-rise views despite any interference that might be generated by the windows themselves. The "problem" of the Red Road window is more its impact on the atmosphere of the flats. As one resident (who also thinks her view is wonderful) says:

> Res: The draught that comes through thay windaes. Now, they're up ... how long did I say I was in here?
> Int: 40 years.
> Res: 40 years we have nae had a new windeae, understand what I mean? And the draughts! ... They [Glasgow Housing Association] sent this man up, what is it they call him ... he put stuff all round it, that white stuff, to keep out the draught ... but, thay windaes are terrible, terrible, really.'

The inability to up-grade Red Road's windows has, in the light of recent Scottish Government policy, intensified the "problem" of Red Road as a satisfactory housing solution. In 2002 the Scottish Executive issued its *Sustainable Scotland* statement which set down specific goals for windows including the requirement that they be double-glazed with low-emissivity glass. Framing for windows should be uPVC, timber or aluminium, and that all fittings allow for ease of operation and safety. Shortly after, the Scottish Executive published its *Scottish Housing Quality Standard*, which farther expanded the regulatory framework for windows. The *Standard* required that all properties be free from serious disrepair, energy efficient, and healthy, safe and secure. All social landlords were required to produce housing quality delivery plans that outlined a program of action ensuring the standards were met by 2015. Attached as it is to this body of regulation, the Red Road window is increasingly experienced as a "failure" not only by those close to hand (the residents) but also by more distant government agencies which are obliged to comply with these standards. Put another way, this single-glazed, steel-frame, asbestos-encased window "fails" in and of itself, and because it is out of alignment with a re-specified national standard for window performance.

## Conclusion

This chapter has looked into the relationship between window technologies, viewing and "the view" in the context of modernist residential high-rise housing. The story we have constructed is part cultural history, part technology in action. This more performative grasp of the view is forwarded as a way of animating and complicating the existing cultural history of the modern window, such that we can see in it more than a view. We have revisited the scholarship on the ways in which modernism established a specific role for the window, one that restructured the boundary between inside and outside in the home and conceived of the window as a viewing technology. We have followed the bureaucratized middling modernist translation of this "ideal" window into the context of Britain's post-war housing program and Red Road specifically. Our close observation of the Red Road window was not simply bent upon verifying that the Red Road window is a failed translation of a modernist ideal. Rather, we have tried to show how views and viewing proceed in a highly contingent (and variously successful) way. Finally, we have tried to show something of the wider agency of the window technology. At Red Road it is not simply a transparent technology of vision, but also a poorly performing device for creating comfort. And the role of the Red Road window in compromising comfort was ultimately so powerful that it was itself responsible for bringing the high-rise view of Red Road to an end.

## Acknowledgments

This chapter is the result of research supported by the UK's Arts and Humanities Research Council. We would like to thank residents and workers of Red Road and Stephen Daniels for his helpful comments on an earlier draft of this chapter.

## Notes

1   D. Cosgrove, *Geography & Vision: Seeing, Imagining and Representing the World*, New York: I.B. Taurus, 2008, p. 5.
2   D. Cosgrove, "Windows on the City," *Urban Studies* 33(8), 1996, 1495.
3   D. Cosgrove, *Geographical Imagination and the Authority of Images: Hettner Lectures*, Stuttgart: Franz Steiner Verlag, 2006, p. 29.
4   For detailed histories of Red Road, see: M. Horsey, "The Story of Red Road Flats," *Town and Country Planning* July/August, 1982; J.M. Jacobs, S. Cairns and I. Strebel, "A Tall Storey ... but a Fact Just the Same: The Red Road High-rise as a Black Box," *Urban Studies* 44(3), 2007.
5   B. Latour, "On Recalling ANT," in J. Law and J. Hassard (eds), *Actor Network Theory and After*, Oxford: Blackwell Publishers, 1999, pp. 15–25.
6   Ibid., p. 17.
7   J. Law, "After ANT: Complexity, Naming, Topology," in Law and Hassard (eds), *Actor Network Theory and After*, p. 4.
8   Ibid., 4.
9   A. Pickering, *The Mangle of Practice: Time, Agency, and Science*, Chicago: Chicago University Press, 1995.
10  M. Hand and E. Shove, "Condensing Practices: Ways of Living with a Freezer," *Journal of Consumer Culture* 7(1), 2007, 79–104.
11  P. Dunleavy, *The Politics of Mass Housing in Britain, 1945–1975*, Oxford: Clarendon Press, 1981.
12  Le Corbusier, *The Radiant City*, London: Faber and Faber, 1967 (1933).
13  B. Colomina, *Privacy and Publicity: Modern Architecture as Mass Media*, Cambridge, MA: The MIT Press, 1994. See also, A. Friedberg, *The Virtual Window: From Alberti to Microsoft*, Cambridge, MA: MIT Press, 2006, p. 123.
14  P. Ariès and G. Duby, et al., *A History of Private Life, Volume 4: From the Fires of Revolution to the Great War*, Cambridge, MA: Belknap Press, 1990.
15  Colomina, *Privacy and Publicity*, p. 306.
16  W. Benjamin, *The Arcades Project*, Cambridge, Massachusetts: Belknap Press, 1999, p. 221 and 216.
17  Colomina, *Privacy and Publicity*, p. 283.
18  A. Friedberg, *The Virtual Window: From Alberti to Microsoft*, Cambridge, MA: MIT Press, 2006, p. 126.

19 Colomina, *Privacy and Publicity*, pp. 128–30. See also B. Reichlin, "The Pros and Cons of the Horizontal Window," *Daidalo* 13, 1984, 64–78.
20 Colomina, *Privacy and Publicity*, p. 133.
21 Friedberg, *The Virtual Window*, p.128.
22 S. Isenstadt, "Four Views, Three of Them through Glass," in Dianne Harris and D. Fairchild Ruggles (eds), *Sites Unseen: Landscape and Vision*, Pittsburgh, PA: University of Pittsburgh Press, 2007, 213–40.
23 Friedberg, *The Virtual Window*, p. 113.
24 E. Shove, *Comfort, Cleanliness and Convenience: The Social Organization of Normality*, Oxford and New York: Berg, 2003.
25 J.M. Jacobs, S. Cairns and I. Strebel, "Windows: Re-viewing Red Road," *Scottish Geographical Journal* 124 (2–3), 2008, 165–84.
26 J.M. Jacobs, S. Cairns and I. Strebel, "'A Tall Storey …'," 609–29.
27 For examples of scholarship on failed technologies see: B. Latour writing as J. Johnson, "Mixing Humans and Non-Humans Together: The Sociology of a Door-Closer," *Social Problems* 35(3), 1988, 298–310; J. Law, *Ladbroke Grove, Or How to Think about Failing Systems*, Lancaster University: Centre for Science Studies, 2003, available at http://www.comp.lancs.ac.uk/sociology/papers/Law-Ladbroke-Grove-Failing-Systems.pdf

# 25

# TECHNICIAN OF LIGHT

## Patrick Geddes and the optic of geography

*Fraser MacDonald*

> The entire history of our philosophy is a photology, the name given to a history of, or treatise on, light.
>
> *Jacques Derrida*[1]

> See for ourselves!
>
> *Patrick Geddes*

Patrick Geddes' students quickly learnt of their teacher's fondness for drama. Walking up Edinburgh's Royal Mile toward the Castle, the Professor would, without warning, bundle them through a narrow doorway and into a dark staircase. He raced them up cold stone staircases in the pitch black.[2] After passing a dozen landings, the breathless students were pulled through another doorway back into blinding daylight. They would emerge, blinking and disorientated, on an open roof terrace from which cityscape and countryside extend in every direction. To the north, the genteel lines of the New Town are set against the Firth of Forth and the distant Kingdom of Fife. To the south lies the valley of the Esk. The Royal Mile, the backbone of mediaeval Edinburgh, falls away to the east. The Castle Rock squats in the west. A clear day might reveal the distant outline of the Highlands, but then not every day on the Scottish east coast is clear. "Perhaps you're wondering why I rushed you up here?" Geddes would ask. "Well, because the exertion of climbing made your blood circulate more rapidly, thus clearing the fog out of your brain and preparing you physiologically for the mental thrill of these outlooks."[3] For Geddes "fog," either inside or outside the body, was the enemy of insight.

The Outlook Tower is well known as the conceptual space of Patrick Geddes (1854–1932), the renowned if rather quirky Scottish botanist, geographer, sociologist, planner and polymath (Figure 25.1). It was not, however, originally of his design. In the 1850s, Maria Short, the daughter of an astronomer and optician,

**FIGURE 25.1** The Outlook Tower, Castlehill, Edinburgh, from a sketch by Eric Crosbee, 1957.

purchased an old tenement at the top of the Royal Mile to create an observatory with some of her father's equipment. Additional floors and the balcony were later added to accommodate a camera obscura, which became the center-piece of a commercial exhibition.[4] At the end of the nineteenth century, when both Short's Observatory and the Old Town were at their least fashionable, Geddes secured the building with, as it was then, a rather vague intention of turning a popular visitor attraction into a new "Civic Observatory and Laboratory," or what was eventually called the "Outlook Tower."[5] Despite the perpetually provisional character of Geddes' new property – it was never quite "finished" – it still captured his particular way of doing geography, whereby a proper synthesis and analysis could be obtained only through disciplined forms of looking and feeling. The Outlook Tower stands as the embodiment of a strand of geographical thought that persisted long into the twentieth century in which "regional survey" and direct observation in the field gave a conceptual consanguinity not only to the natural and the social sciences, but also to the humanities. Sight was the key to this very civic geography or, in contemporary parlance, "participatory" geography. "The first contribution of this Tower towards understanding life is purely visual," wrote Geddes, "for from here everyone can make a start toward seeing completely that portion of the world he can survey."[6]

To use a favorite Geddes' term, the tower offered a "synoptic" view and one that cannot be easily annexed for an exclusive disciplinary history. While Charles Zueblin, a professor of sociology at Chicago who visited the Outlook Tower in the 1890s, described it as "the world's first sociological laboratory,"[7] historians of geography have also given it a prominent place in their own accounts. For David Stoddart it symbolized "a form of observation and a definition of subject matter which has governed field studies since that time."[8] David Matless and Charles Withers have described the internal layout of the tower with a series of exhibition rooms on consecutive floors, from the particularity of the "Edinburgh room" at the top, moving down (and scaling up) through "Scotland," "Language" (the English-speaking world), "Europe" and "World."[9] More recently, the sociologists Thomas Osborne and Nikolas Rose have portrayed Geddes as a "technician of space" for whom the tower was a means of intervening in the civic consciousness of citizens.[10] My approach is complementary to this body of scholarship but offers a different emphasis. I am concerned with Geddes as a technician of light rather than of space. And I want to use Geddes' interest in the uniqueness of sight to complicate some now familiar discussions about the relationship between visuality and geographical knowledge, a debate that sits at the contemporary interface between geography and the humanities.[11]

In one sense, this chapter is really about contrast. It deals with a series of transitions and translations – from daylight to darkness and back again – that were central to the operation of the Outlook Tower as an experiment in the teaching of geography. I am especially interested in the bodily and epistemological adjustments from light to dark and from sight to blindness. Moreover, this vacillation between light and darkness, between rendering the world picturable and refusing all imagery, is itself a suitable analogue of geography's schizoid relationship with visual culture, caught as it still is between idolatry and iconoclasm, between the primacy of the visual and its disavowal. If such a tension is part and parcel of visuality in the modern era, it is particularly true of disciplinary geography. The well-worn critique of geography's alleged "ocularcentrism" would be an obvious basis for my argument but it is precisely this glib association that I want to put in question. In my view, the abstract and muscular vocabulary of "scopic regimes" and "ocularcentrism" – terms that suggest that vision can be neatly differentiated from a wider bodily sensorium[12] – has inhibited rather than fostered a more careful study of senses and sense-making in geography.

In making this rather bold claim I am, to some extent, both challenging and affirming the work of two outstanding contributors to this literature: Gillian Rose and the late Denis Cosgrove. Though it was not his style to do so, Cosgrove could legitimately have claimed to inaugurate this entire field. His work on the politics of

landscape combined a careful art historical scholarship with the insights of a theoretically invigorated human geography, opening up the way for subsequent explorations of sight, space and reason. I need not rehearse here the important ways in
which Rose, informed by feminism and psychoanalysis, has subsequently reframed
the disciplinary politics of looking.[13] There are two things in particular that I want
to take from these authors. First is the archival and exegetical specificity of Cosgrove's
work across a range of contexts from architecture to imperialism, painting, photography and cartography. No one could accuse him of conceptualizing vision as an
abstract and free-floating discourse;[14] it is always grounded in a specific context.
Secondly, one might argue that the greatest advance in the discussion of visuality
and geographical knowledge has actually been a measured form of retreat. This is
expressed in Gillian Rose's demand to know "how, *exactly*, is geography 'visual'?"[15]
This elementary question has still not, in my opinion, been given a satisfactory
answer; perhaps the question is, in any case, more important. So I make no claim
that this chapter is an adequate response to Rose, only that it has been inspired in
part by her inquiry. I also take up Cosgrove's latter acknowledgment of the varied
modalities of looking and of the indivisibility of sight from other sensory registers,
a theme particularly prominent in the work of the anthropologist Tim Ingold.[16] It
is in this context that I examine the observant practices of Patrick Geddes, acknowledging, of course, that the one thing one cannot see is sight itself.

By all accounts Geddes was an inspiring teacher. His lectures were for the most part
extemporary performances leaving relatively little behind by way of notes. Among
those that have survived, one topic starts with the simple injunction: "See for
ourselves! Hill, tower," before going on to promote a "synthetic geography"
achieved by "developing senses" and an "emphasis on sight, emotion, experience."[17]
At the heart of his pedagogy was a belief that he needed to prepare the bodies of
his students. He would rush them up the stone staircase because he favored a particular way of seeing in which the eye invigorated by physical exertion was better able
to perceive the tableau of Edinburgh than the more detached vision afforded by the
human metabolism at rest. Certain bodily adjustments and attunements were
thought to be conducive to a better understanding of geographical relationships.
Charles Zueblin noted that any student of Geddes, "though first of all *freshened as
an observer*, is regarded not as a receptacle of information, but as a possible producer
of independent thought."[18] Up on the balcony – what Geddes called the "Prospect
Roof" – each of the parapets was devoted to one or other disciplinary perspective.[19] A devoted student, Philip Boardman, remembered that "in quick and brain-
teasing succession Geddes would show how meteorologist, geologist, geographer,
zoologist, botanist, and so on, look at the region visible from the Tower."[20] This was
the synoptic view – literally, a seeing together: "Aristotle … urged that our view be
truly synoptic," he wrote, "a word which had not then become abstract, but was
vividly concrete … a seeing of the city and this as a whole."[21] "Of the educative

value of this synoptic vision," he felt, "every visitor [could] thus ... [have] a fresh experience."[22] Part of this "freshening" was about the observer coming to terms with sight itself. And in order to do this Geddes started with the synoptic view, with its emphasis on wholeness and daylight, before advancing on to the analytic view afforded by the dark, dissecting operations of the camera obscura.

And so his students found themselves once again despatched into darkness. On being ushered into the hollow dome of the turret, Geddes closed the only small window. It became pitch black. "This is the camera obscura, or dark room," he told them, "where for all practical purposes you are inside the bellows of a huge optical photographic apparatus."[23] Only then did Geddes open the aperture: a small disk of light in the surrounding darkness; it had an almost lunar quality. Philip Boardman recalls:

> The professor manipulates a lever and pulls some other cords which are also hanging from the dome, and suddenly there flashes on the white surface of the table a colored moving picture of Edinburgh Castle. People are walking along the Esplanade, trees are blowing in the wind, but, even more striking than these movements, are the colours we see. [...] "Oh it's very simple," says Geddes [...] "Your eyes are first of all made sensitive to light by the darkness of the room. Then the mirror picks up only those light rays that are directly from the objects on which it is directly focused, stopping off all the cross reflections which confuse our vision in broad daylight. The result is just that here you see everything in its true colours with eyes that have been made especially sensitive."[24]

In this way, Geddes used the camera obscura less to examine particular objects of study but more to teach his students *how* to see. It is a form of sight constantly recalibrated by the contrasts of the street, the stairs, the roof and the cupola of the camera obscura. If the student is to think independently, he or she must be able to draw upon this personal experience of seeing things anew, a matter of affect as much as cognition. To this end, Geddes thought that the camera obscura

> ...harmonises the striking landscape, near and far, and this with no small element of the characteristic qualities of the best modern painting ... [it] has therefore been retained; alike for its own sake and as evidence of what is so often missed by scientific and philosophic minds, that the synthetic vision to which they aspire may be reached more simply from the aesthetic and the emotional side, and thus be visual and concrete. In short, here, as elsewhere, children and artists may see more than the wise. For as there can be no nature study, no geography worth the name apart from the love and the beauty of Nature, so it is with the study of the city.[25]

The Outlook Tower was original but not entirely without precedent. Charles Withers situates it within the earlier context of using dioramas and panoramas to give visual form to geographical knowledge; he also reveals it to be an important precursor to a plan for a National Institute of Geography that Geddes was never able to realize.[26] For Withers, Geddes conceived of the camera obscura as "a key instrument in promoting the powers of observation ... and for extending his vision for geography and regional survey."[27] There remains a basic question, however, as to why Geddes was so enthusiastic about an optical instrument that had very much outlived its time, having come to prominence in the middle of the seventeenth century and enjoyed its heyday in the eighteenth century. At its peak, it served as a dominant model for the workings of the human eye;[28] by radically decoupling vision from the body of the observer – using a mechanical apparatus to relocate sensory experience from inside to outside the body – it came to represent an "objective ground of visual truth."[29] By the nineteenth century, however, all of this was to change.

In his seminal book, *Techniques of the Observer*, Jonathan Crary argues that between 1810 and 1840, "the stable and fixed relations incarnated in the camera obscura" gave way to a new valuation of visual experience through science and philosophy that was "abstracted from any founding site or referent."[30] He suggests that this change was not brought about by the invention of photography but rather by a set of intellectual, artistic and sensory changes that made its audacious entrance possible. The appearance of the camera cemented these relations, bestowing an unprecedented mobility to the image that had now been cut adrift from the fixed perspective of appending the human eye to an optical instrument. But in a typically back-to-front fashion, Geddes actually uses photography as the key conceptual referent in understanding the camera obscura, reminding the students that they were "inside the bellows of a hugh photographic apparatus."[31] So by the time that Geddes takes up the instrument as the means by which he can train his students to see geographically, it is, for any purpose other than the quaint amusement of tourists, a rather redundant optical device. It certainly no longer holds any currency as a model of vision. What is significant for our discussion here is that Geddes re-encodes the camera obscura for his own peculiar geographical purposes.

Through the camera obscura, the observer is permitted a return to inhabiting the instrument – something denied by conventional photography – and in so doing withdraws from the familiar visual world. For Geddes, this transition from light to darkness was critically important. Such a withdrawal represents for Jonathan Crary "a kind of *askesis*" or philosophical discipline through which the observer can "regulate and purify one's relations to the manifold contents of the now 'exterior' world."[32] The camera obscura is thus "inseparable from a certain metaphysic of interiority."[33] Something of this older meaning of the instrument was clearly embraced by Geddes even if there was little adherence to the notion that the camera obscura was a benchmark of objective visual truth. But from the intimate distance afforded in the darkened chamber, Geddes and his students were better able to see Edinburgh in its regional context. Their insight came from a newly reconfigured vision founded upon a certain sort of temporary blindness which he hoped would inspire a reflexivity about the role of the senses in everyday encounters with the world. His model of vision is not that of the disembodied eye: for him, looking properly requires a preparation of the entire body. It is significant that the climactic flickering image in the darkness of the cupola was preceded only by the tactile sensation of the round table. Unbeknown to most of his students, this moment in which blindness gives way to luminous knowledge was itself a performative re-enactment of Geddes' own intellectual biography, culminating in his belief that touch was in fact another form of sight.

In November 1879, Patrick Geddes had been botanizing in Mexico when a personal crisis befell him. Exploring outdoors in brilliant sunlight and poring over the microscope from one end of the day to the other put an increasing strain on Geddes' eyesight, which had also been weakened earlier in life. His mother had already been stricken with blindness in middle age;[34] and Geddes would likely have been familiar with the fact that numerous scientists including Joseph Plateau, Gustav Fechner and Scotland's own David Brewster[35] had damaged their sight in the course of their optical research.[36] With the threat of lifelong blindness hanging over him, Geddes agreed to drastic remedial measures: living for weeks in a darkened, shuttered room with a blindfold over his eyes. The prospect of blindness was terrifying. Years after his death his son Arthur wrote about what this might have meant to an "essentially eye-minded [person] ... a visual not an audative." "Once or twice he spoke of this with full meaning, and a haunted expression," he recalled.[37] In this benighted world, Geddes "thought furiously" and explored what other senses were still available to him.[38] And herein lay his conceptual breakthrough. While exploring the walls of the darkened room, his hands came across the shuttered window panes:

> ...his searching fingers used to come to the dark window and feel along the co-ordinates of its frame and explore its rectangular panes that could let through no light to his impassioned observation. It was at that time that this worker in the living and the concrete first acquired the habit of intense abstract thought, for if seeing and acting were to be closed for life, thought must compensate. The diagrams of geometrical shape must be the windows of his thought.[39]

The relief of the wooden window frames stood out in his mind as compartments of a diagram, with each pane or square labelled to represent a concept. Using the idea of the window frame, Geddes tried the same process with pieces of folded paper, "envisioning in his mind's eye a different label for each rectangle on the paper. For instance, a sheet of paper folded lengthwise, then folded into three equal parts, became a chart with six compartments."[40]

Geddes was cheered by this simple invention with which he could maintain a synoptic vision of the unity of knowledge. For instance, he was able to map Frédéric le Play's basic triad – Place, Work, Folk – in a way that represented the constitutive

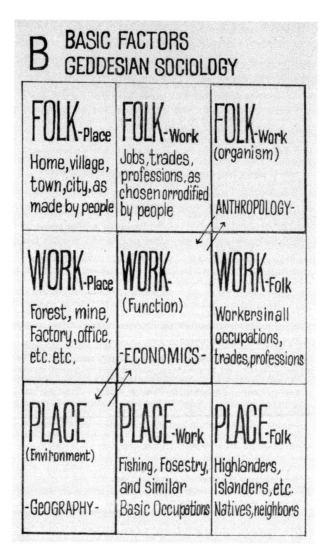

**FIGURE 25.2**   *Geddesian Sociology*, Le Play's "place–work–folk" triad in a Geddesian "thinking machine," from Arthur Geddes, *Patrick Geddes as Sociologist*, 1976.

relations of geography, economics and anthropology (Figure 25.2). "He called this compact and ingenious method for recording ideas and facilitating the understanding of concepts, a 'thinking machine,'" wrote Abbie Ziffren. "By the time he was able to use his eyes again ... he was convinced that in some ways these folded papers were superior to the printed word."[41] In fact he developed such disdain for written analysis that Lewis Mumford – the historian of the city and a wayward disciple of Geddes – disparagingly claimed that the thinking machine was "a contraceptive device that kept him from gestating a commissioned book."[42] Folding paper became an enduring daily ritual for Geddes, a form of "private intellectual gymnastics" that Mumford regarded as pathological.[43] A century later, the French philosopher Gilles Deleuze would popularize a rather different kind of origami metaphysics in which the human subject is figured as the external world "folded" inwards; folding, for Deleuze, is a constitutive process of being human.[44] But Geddes was not given to such abstractions, at least not of this sort. Rather, it was as if the repetition of folding could itself mechanically settle a certain understanding. It was, literally, a guide for the blind, where blindness could be turned back on itself to cognitive advantage. "The diagrams of geometrical shape must be the windows of his thought," wrote Arthur Geddes of his father. "Thus began the process in his way of thinking and expressing himself through diagrammatic notation, permitting ready visualisation and easy comparisons of the facts of science and of the data of sociology in particular."[45] The thinking machines were to prove particularly useful for graphically representing the relationship between science, social science and the humanities. Indeed, Lewis Mumford inaugurated a new Humanities course at Stanford by showing the class one such thinking machine in the opening lecture.[46]

On completing his demonstration of the camera obscura, Geddes and his students poured out of the cupola back into daylight on the Outlook Tower's Prospect Roof. From here they were expected to make one final transition, albeit in their own time, to a more secluded and private place which he called the "Inlook Tower." This was a small dark room furnished with a single chair in which the student was expected to quietly meditate on the emotional experience of all they had seen.[47] These "thinking cells" were recurring motifs in Geddesian architecture, the metaphysical complement to communal spaces of learning.[48] But the location of the Inlook Tower – at the top of the building, neither higher nor lower than the camera obscura – is significant. Volker Welter detects the influence of mountain top temples;[49] it was certainly a place for insight and inner visions. One wonders, however, if the Inlook Tower was as much for recovery as for reflection. The obvious eccentricity of the teacher and the marked grandeur of the view would be stimulating enough, quite apart from the strange, slow, strobing effect of the Outlook Tower with its journey from light (street) to dark (stairs) to light (roof) to dark (cupola) to light (roof) to dark (Inlook Tower). It was a visual experience, to be sure; but it involved such a variety of observant practices, each with its own particular kind of looking and not-looking. In the odd experiment of the Outlook Tower, these varied modalities of the visual are never isolated from other facets of bodily experience. From the breathless disorientation of the Prospect Roof to the haptic regime of the "thinking machine," the familiar language of contemporary visual studies – ocular-this; scopic-that; the bludgeoning rhetoric of "the gaze" – is ill-suited to an account of the sense-making practices of Patrick Geddes.

In saying this, of course, it would be disingenuous to entirely downplay the visual element in Geddesian thinking. The art historian Murdo Macdonald has argued that not only is Geddes a "visual thinker" but he is also a standard bearer for an older tradition of visual intellectualism in Scotland. This is manifest, he believes, in the nineteenth-century triumph of geometry over algebra and in the achievements of Thomas Telford, James Watt, Robert Adam and James Clerk Maxwell.[50] But there is a danger, I think, of being a little too determinate with some of these sensory associations. The challenge for historians of ideas is not to assign biographies or concepts to this sensory category or that (one also wonders what sort of knowledge could be *extra*-sensory?). Rather, we need to explore how particular branches of knowledge are distinctively correspondent to different configurations and imbrications of the senses, remembering too that such knowledge in turn re-writes our perceptive faculties. In other words, the observant practices of geography are simply different after Patrick Geddes. So in thinking about the relationship between visuality and geographical knowledge we must reach beyond the rhetoric of the visual to attend to the multisensory business of looking. In doing this, Patrick Geddes can still instruct us, no longer in how to see but, perhaps, in what it means to see.

## Notes

1 J. Derrida, *Writing and Difference*, London: Routledge and Kegan Paul, 1978, p. 27.
2 A. Ziffren, "Biography of Patrick Geddes," in M. Stalley (ed.), *Patrick Geddes: Spokesman for Man and Environment*, New Brunswick: Rutgers University Press, 1972, pp. 3–91.
3 P. Boardman, *Patrick Geddes: Maker of the Future*, Chapel Hill: University of North Carolina Press, 1944, p. 178.
4 V.M. Welter, *Biopolis: Patrick Geddes and the City of Life,* Cambridge, MA: MIT Press, 2002, p. 78.
5 P. Geddes, *Cities in Evolution: An Introduction to the Town Planning Movement and to the Study of Civics*, London: Routledge, 1998 [facsimile of first 1915 edition], p. 321.
6 Quoted in Boardman, *Patrick Geddes*, p. 179.
7 C. Zueblin, "The World's First Sociological Laboratory," *The American Journal of Sociology* IV.5 (1899), pp. 577–92 (p. 582).
8 D. Stoddart, *On Geography and its History* Blackwell, Oxford, 1986, p. 144.
9 D. Matless, "Regional Surveys and Local Knowledges: The Geographical Imagination in Britain, 1918–39," *Transactions of the Institute of British Geographers* 17, 1992, 464–80; C.W.J. Withers, *Geography, Science and National Identity: Scotland since 1520*, Cambridge: Cambridge University Press, 2001, p 225f.
10 T. Osborne and N. Rose, "Spatial Phenomenotechnics: Making Space with Charles Booth and Patrick Geddes," *Environment and Planning D: Society and Space* 22, 209–28.
11 G. Rose, *Feminism and Geography: The Limits of Geographical Knowledge*, Cambridge: Polity Press, 1993; G. Rose, "On the Need to Ask How, Exactly, is Geography 'Visual'?" *Antipode* 35.2, 2003, 212–21; D. Matless, "Gestures around the Visual," *Antipode* 35.2, 2003, 205–11; J.R. Ryan, "Who's Afraid of Visual Culture?" *Antipode* 35.2, 2003, 232–7.
12 F. MacDonald, "Social & Cultural Geography: Visuality," in N. Thrift and R. Kitchin (eds), *International Encyclopaedia of Human Geography*, 2010.
13 Rose, *Feminism and Geography*.
14 By contrast, compare with D.Z. Sui, "Visuality, Aurality, and Shifting Metaphors of Geographical Thought in the Late Twentieth Century," *Annals of the Association of American Geographers* 90.2, 2000, 322–43.
15 G. Rose, "On the Need to Ask How, Exactly, is Geography 'Visual'?" *Antipode* 35.2, 2003.

16  D. Cosgrove, *Geography and Vision: Seeing, Imagining and Representing the World*, IB Tauris: London, 2007; T. Ingold, *The Perception of the Environment: Essays on Livelihood, Dwelling and Skill*, London: Routledge, 2000.
17  W.I. Stevenson, *Patrick Geddes and Geography: A Biobibliographical study* (occasional paper no. 27), University College, London: Department of Geography, March 1975, p. 13.
18  Charles Zueblin, "The World's First Sociological Laboratory," p. 582; my emphasis.
19  Boardman, *Patrick Geddes*, p. 179.
20  Boardman, "Patrick Geddes and his Many Worlds," p. 5.
21  Geddes, *Cities in Evolution*, p. 13.
22  Geddes, *Cities in Evolution*, p. 321.
23  Boardman, *Patrick Geddes*, p. 179.
24  Ibid., p. 179.
25  Geddes, *Cities in Evolution*, p. 321.
26  Withers, *Geography, Science and National Identity*, p 225f; see also V.M. Welter, *Biopolis*, p. 175f.
27  Ibid., p. 226.
28  S. Clark, *Vanities of the Eye: Vision in Early Modern European Culture*, Oxford: Oxford University Press, 2007.
29  J. Crary, *Techniques of the Observer: On Vision and Modernity in the Nineteenth Century*, Cambridge, Mass: MIT Press, 1990, p. 39.
30  Crary, *Techniques of the Observer*, p. 14.
31  Boardman, *Patrick Geddes*, p. 179.
32  Crary, *Techniques of the Observer*, p. 39.
33  Crary, *Techniques of the Observer*, p. 39; for contemporary resonances of this idea see E. Gilbert, "Eye to Eye: Biometrics, the Observer, the Observed and the Body Politic," in F. MacDonald, R. Hughes and K. Dodds (eds), *Observant States: Geopolitics and Visual Culture*, IB Tauris: London, 2010.
34  Ziffren, "Biography of Patrick Geddes," p. 6.
35  D. Brewster (1781–1868) was the inventor of the kaleidoscope and the lenticular stereoscope, and a key theorist of the properties of light.
36  Crary, *Techniques of the Observer*, p. 141.
37  A. Geddes, "Patrick Geddes as Sociologist," in J.V. Ferreira and S.S. Jha (eds), *The Outlook Tower: Essays on Urbanization in Memory of Patrick Geddes*, Bombay: Popular Prakashan Ltd, 1976, pp. 14–19.
38  Ibid., p. 19.
39  Ibid., p. 19.
40  Ziffren, "Biography of Patrick Geddes," p. 11.
41  Ibid., p. 11.
42  L. Mumford, "The Geddesian Gambit," in F.G. Novak (ed.), *Lewis Mumford and Patrick Geddes: The Correspondence*, London: Routledge, 1995, p. 353–72.
43  Mumford, "The Geddesian Gambit," p. 355.
44  G. Deleuze, *The Fold: Leibniz and the Baroque,* trans. Tom Conley, University of Minnesota Press, 1992.
45  Geddes, "Patrick Geddes as Sociologist,", p. 19.
46  Mumford, "The Geddesian Gambit," p. 356.
47  Boardman, *Patrick Geddes*, p. 184.
48  Welter, *Biopolis*, p. 220.
49  Ibid.
50  M. Macdonald, "Patrick Geddes: Visual Thinker," in G. Prince (ed.), *A Window on Europe: The Lothian European Lectures 1992*, Edinburgh: Canongate, 1993, pp. 229–55 (p. 230); see also G.E. Davie, *The Democratic Intellect: Scotland and her Universities in the Nineteenth Century*, Edinburgh: Edinburgh University Press, 1961.

# 26

# DESERTED PLACES, REMOTE VOICES

## Performing landscape

*Mike Pearson*

All set? Then let's begin. By imagining this…

By imagining yourself jumping, jumping and falling, free-falling in the 15,000 feet of empty airspace above you now. Adjusting pitch and roll and yaw with outstretched arms and legs, for sixty seconds, face down. An aerial view of this place, bird's-eye – a patchwork of wheat, sugar beet, barley, rape at their various stages, dark soils in their different hues…

*Carrlands: Hibaldstow – Air (2007)*

It was in a review, now misplaced, of William Least Heat-Moon's *PrairyErth*[1] – an account of Chase County in the American mid-West – that I first encountered the notion of a deep map: *PrairyErth* as a conflation of "oral testimony, history, topographic details, local folklore, travel anthology, geography, journalism, memoir, natural history, autobiography and everything you need to know about Kansas." A journalistic conceit perhaps but redolent of recent attempts to record and represent the substance, grain and patina of a particular place through juxtapositions and interpenetrations of the historical and the contemporary, the political and the cultural, the factual and the fictional, the academic and the discursive: depth not as profundity but as density of both topics and modes of address, in full cognizance of local and personal knowledge. In this it is akin to Clifford Geertz's *thick description*, the detailed and contextual description of cultural phenomena, to discern the complexities behind the action and from which the observer is not removed; Geertz's *blurred genre* – as Michael Shanks and I appropriated it, "a mixture of narration and scientific practices, an integrated, interdisciplinary, intertextual and creative approach to recording, writing and illustrating the material past;"[2] and Gregory Ulmer's *mystory*, the infusion of personal, popular and expert discourses.[3]

My late colleague Clifford McLucas made one, a deep map, at Stanford University. It is a computer-generated installation work of the San Andreas fault, thirteen metres in length. Writing has provided me another facility: performance writing – the elision and articulation of texts theoretical, descriptive and creative, in forms of inscription that foreground its performative nature. This involves the application of procedures of dramaturgy – of sequencing (figures of *parataxis* – "this and this and this;" *hypotaxis* – "this behind this, adjacent to this;" and *katachresis* – "this, this, this") and of dynamic (accenting; rhythm; timing) – to accretions of texts of differing orders.

This I've tried *in* performance – in *Bubbling Tom* (2000), a guided tour of the landscape I knew at the age of seven, in Hibaldstow, north Lincolnshire. Over a period of two hours we visited ten locations in the village: school, church, stream and others less notable, unmarked; places where nonetheless significant things, memorable events, happened to me: landmarks biographic and personal, though where "you can't tell by looking." A solo voice here is the most flexible medium for conjoining and differentiating varieties of material through rapid shifts in register whilst attending to, and accounting for, the nature of places, with my physical presence shaping the details of the writing. The compositional practices drew upon *gossip*, which invariably includes wide-ranging subjects and in different tones both spoken and unspoken, and upon Walter Benjamin's quotation of the German saying, "When someone goes on a trip he has something to tell about."[4] There are of course ethical responsibilities in employing autobiographical/biographical material – what can, cannot, should not be spoken publicly. And one has to resist becoming a bar-room bore, by alternating allusions to the immediately recognizable and the infinitely strange, and accompanying the textual component with gestural rhetoric that emphasizes, locates, shapes and acts as a mnemonic for the details of the writing. So, I mimicked and gesticulated and posed, dramatizing a familiar past: revisiting and animating a familiar landscape, albeit at a different scale, and frequently to comic effect.

The concept was inspired by the Welsh notion of *y filltir sqwar*, the "square mile," the intimate landscape of our earliest years, that patch we know in close-up, in a detail we will never know anywhere again. Welsh author D.J. Williams claims spatial primacy in memory construction: "When the many things I remember actually happened whether early or late in the course of that six years, I haven't much of an idea. But I can locate most of them with a degree of certainty – where such and such a thing happened and where I was standing when I heard what I heard."[5] Informed too by *cynefin*, the lived environment, totality of land and sea, weather and season, birds and animals, human dwelling and land-use: "Land and language are two strands that tie the Welsh-speaker to his cynefin. There are other links, such as remembrance of things past."[6]

*Bubbling Tom* is an example of site-specific performance, the latest occupation of a location where other and former occupations are apparent and cognitively active in the creation of dramatic meaning, its reception and interpretation. Inevitably, audiences for forms of exposition such as *Bubbling Tom* – performance here as an event bound in space and time, involving live-ness and co-presence – are small. Could its aspirations succeed remotely, in my absence? The monograph *In Comes I: Performance, Memory and Landscape*[7] is structured as a number of *excursions* in north Lincolnshire, guiding the reader – be they in an armchair or in the field – through an itinerary of locales, pausing at each for personal and critical reflection on themes related to, or evoked, by that place. Whilst each short text is relatively discrete, their accumulation within particular choreographic journeys enhances apprehension of the region; and disparate times and places are connected through wormholes and discontinuities in the volume. These themes and approaches are further developed in *Carrlands*,[8] a project funded by the UK Arts and Humanities Research Council "Landscape and Environment" programme.[9]

The Ancholme valley lies between the limestone Heights and the chalk Wolds, one of a series of parallel north-south zones that characterize the topography of the region. The new river, a canal running almost straight for nineteen miles though falling only three metres, from a source near Bishopsbridge to its confluence with the Humber at Ferriby Sluice, was completed in the 1820s; the old river, now little more than a ditch, winds along its course, appearing first on one bank, then the other. The soil here is peaty, the subsoil glacial clay, the underlying geology Upper Jurassic clay. In places there are substantial quantities of bog oak, like a wooden floor in some fields, and occasionally spectacular Bronze Age relics emerge – boats, rafts, jetties. This was long a water-world; this is a shrouded prehistoric landscape.

*Kjarr* is the Danish word for marsh or bog, a surviving echo of immigrant settlement late in the first millennium AD, further evidenced by villages with the suffix *-by* – Snitterby; "carr" came to denote lands close to the river that flooded regularly and were used primarily as seasonal pasture, particularly in the parishes that run down the limestone dip slope to the river. Piecemeal efforts at land drainage by monastic communities – at Cadney for instance, where *-ey* may denote an island in the swamp – were followed by concerted efforts prior to the English Civil War. The valley was eventually successfully drained shortly after Parliamentary Enclosure; in the nineteenth century agriculture became increasing arable, produce shipped out by barge traffic on the new river. This is a man-made landscape, and it is highly administered, maintained through constant, now automated, pumping out and pumping in. This is also the place where I was raised.

But how to share my enthusiasm for an agricultural landscape – simple, flat, often deserted – without moments of conventional scenic heritage: that does not easily reveal itself; that lacks those monumental features which frequently orientate our gaze; where the land seems to draw back from the viewer; where few Xs mark the spot – no blue plaque where my Uncle Fred would jump from Hibaldstow Bridge despite the metal calliper on his leg; no memorial where sky-diver Stephen Hilder cut his own parachute straps, both sets, in an extravagant suicidal display. Through *remote* performance…

*Carrlands* is constituted as a series of three original sound compositions inspired by, and set at, locations in the valley and created in collaboration with composer John Hardy. Each 60-minute audio work – a combination of spoken text, music and effects, with subtle invitations to action and instructions to users – offers directions and recommends walks at a place infrequently visited but that has its own subtle characteristics, qualities and attractions.

The writing process in *Carrlands* involved conventional library and archive research, drawing upon the reports of the Ancholme Inland Drainage Board, Environment Agency flood assessments, M.C. Balfour's late nineteenth-century *Legends of the Lincolnshire Cars* – an extraordinary evocation of the pre-drainage landscape of fogs and fevers, of shattered bog oaks, apparitions, ague and opium eating[10]; and on-site fieldwork to examine and explore details, marks and traces of human activity – the ruins of the farm worked by my father's uncle. It also included interviews, frequently undertaken at site: with disciplinary experts; with regional specialists and enthusiasts – ornithologist, museum curator, water bailiff; with village history groups; and with local inhabitants, family members, and those who have lived and worked in the carrs – those akin to anthropologist Tim Ingold's *dwellers* perhaps,[11] farmers in the main; recordings of voices of different ages, genders and accents; meetings with geographers Stephen Daniels and David Matless and archaeologist John Barrett which resembled old-style fieldtrips. After a day in the field I recorded them, sitting in my car, looking out on Horkstow Carrs, the windscreen a perfect Claude glass, regarding the land, as the English so often do. *Carrlands* is informed by Barrett's perception that as visitors to heritage sites our threshold for data retention is surprisingly low, and that a plethora of historical information can overwhelm us and sully the experience. In developing new practices, he favors equipping the visitor with tools for interpretation, "ways of seeing," rather than lists of dates and periods of building.

The aim is to elicit and draw together insights from archaeology, geography, natural history and folklore, and combine them with the detailed and first-hand experiences, opinions and memories of local people: to illuminate, explicate and problematize the multiplicity of meanings that resonate within and from landscapes; to espouse their imbricated nature. Sometimes interviewees are quoted verbatim. But *Carrlands* is primarily achieved in a single voice: to ensure one level of cohesion, in shifts of material and emphasis – this is not after all a radio documentary; and to lessen technical problems of equalization; a voice that can seduce and reassure, surprise and inform.

The attempt is to demonstrate the intimate connection between personal biography and the biography of landscape, between social identities and a sense of place, through concatenations of local observation and critical academic discourse. One of my principal correspondents was my father's cousin, the only person I met who had lived on the water's edge: she remembers her father pushing blocks of ice away from the house windows with a pole during the snow-melt of 1947; the peaty land on fire, underground, for a year at a time; charred banknotes littering the potato fields after bombing raids on Hull.

The recorded text forms a single layer within a sonic stratigraphy. This mode of characterizing composition is familiar from studio editing within which separate elements are assigned parallel tracks or horizons. The musical components are built up *beneath* the recorded text, emerging more demonstrably in verbal pauses. *Above* the voice there may be more ephemeral features: *effects* – birds calls, aeroplane engines – and *echoes* of the interviewees speaking their words slightly out of phase with me. From moment to moment, the music itself may further evoke or disrupt the nature and atmosphere of these seemingly empty lands. It includes instrumental, vocal and orchestral, and electronic and processed strands; sampling from archival sources – Percy Grainger's folk-song recordings, the voices of Italian prisoners of war; and effects that recall former sound-worlds in these places. And within the matrix, the highly modified, unrecognizable voices of interviewees provide musical textures; or digitally analyzed, the notation for instrumental composition itself. The geological analogy is apposite as the sequence is not necessarily even-bedded, and may exhibit folding, faulting and discontinuity. Efficacy is finally assured through the process of mixing, of relative re-balancing, and even erasure. Become part of a landscape...

The works are available for download as mp3 files from a dedicated website.[12] Each has a distinct tone, character and instrumental voicing and is further divided into four 15-minute movements reflecting a specific aspect of the location – its history, its flora and fauna, agricultural practices, significant events. After download, the listener is free to choose how, where and under what conditions to access the material, at or away from site. If listened to at a distance, a number of photographs are attached in the website gallery to provide visual reference and direction, though the listener is encouraged to imagine the landscape in the mind's eye. On location, the listeners are at liberty to select the time, season, weather, personal mood and social conditions – alone, in a group – of their encounter. Movements may be listened to consecutively, or with pauses between them. Locations are readily accessible and routes for walking are suggested, though in the main these follow paths on the riverbank. None of the text is precise in its address to topography, and hence listeners are free to roam in the surrounding landscape. And none is unremitting: music repeatedly attenuates the narrative and provides interludes. Technology plays a significant and transformative mediating role in the response of art to the environment: a medium that can precipitate and encourage public visitation.

In form, *Carrlands* employs and extends a methodology familiar from art projects such as Janet Cardiff's seminal *The Missing Voice (Case Study B)*[13] in Whitechapel, and contemporary museum trails. Here, the listener is cast not only as an audience of one but also as active participant in the artistic construction: negotiating complex shifts in time and subject matter; bringing their own physical engagement – phenomenological encounter even – to the stories and information embedded in the compositions; working with imagination in a landscape lacking authoritative viewpoints; shifting from optic to haptic apprehension. Betwixt and between ... earth and sky ... land and water ... me and them.

All three vectors in play – place, performance and participant – are generative and in a dynamic relationship. Landscape is *constitutive* rather than scenic backdrop: not only does it counterpoint the text, but it also directly influences the nature of the experience significantly, with a sudden thunderstorm. Performance may be as much a reading onto – a projection of narrative onto the seemingly featureless terrain – as a reading from – the drawing of attention to extant details, animating that which is observable. There may indeed be a creative friction or tension between what is of the place and what is brought *to* the place. The presence of performance might even be inappropriate or anachronistic but in this also revealing: enabling the site to "speak for itself." And in its ambivalence, in its refusal to re-enact all that might have happened here, resistant to closure of interpretation. In full acknowledgment of the interests of various communities, the various social and political constituencies, the various contradictory interests that might lay claim to a location and its past – performance might also challenge pervasive notions of landscape as simply an area of ground or as a purely visual construct. Though of course you may witness, in John Berger's words, "an event (like a dog running or an artichoke flowering) in a field which until then had been awaiting a first event in order to become itself realisable;"[14] a rising green plover, a diving water vole; though if you see someone waving a shovel to attract your attention, be sure to fling a handful of earth in their direction or you too will soon be buried.

*Carrlands* involves an invitation to action. In its fixed texts and walks, it can appear prescriptive and programmatic, orientating and choreographing as it informs. But the participant is free to wander, physically and imaginatively. There are no startling vistas here; only the river bridges offer purchase in the level expanse. This is performance without pointing, either to, or out. The admonitions are not to move or look here or there, but rather – following geographer Mitch Rose's appraisal[15] of anthropologist Kathleen Stewart's work – "Picture this," "Consider this."[16] A landscape perhaps for doing, feeling and contemplating as much as for looking; there is work to be done – walking, looking, musing, imagining.

*Carrlands* is designed to stimulate and extend public appreciation, understanding and enjoyment of landscape through active participation and engagement. It draws attention to and illuminates the historically and culturally diverse ways in which a place is made, used and reused; it proposes performative approaches to acts of interpretation, for anyone who may chose to engage with them. It poses these kinds of questions: How and why does this place come to be as it is? What struggles – natural and human – lie behind its benign façade? Can we regard it as much a network of related stories and experiences, as the outcome of particular processes of human intervention, as a collection of topographic details?

Participant responses fold into the project as feedback, through a questionnaire on the website. But participants are not asked to make a qualitative judgment on the artwork.

Rather, they are invited to contribute their own observations of these places, and how they themselves might describe and interpret them: drawing on knowledge from other academic disciplines, from local understandings, from hearsay; adding expert observation, personal memoir, biographical detail, poetry, fiction, stimulated by the compositions of *Carrlands*: to better elucidate the entangled nature of land, human subject and event, the contested relationships between landscape, experience and identity in this place, and to acknowledge the close link between culture and subjectivity within a given region; to make the map of the Ancholme carrlands deeper.

## Notes

1 W.L. Heat-Moon, *PrairyErth*, London: Andre Deutsch, 1991.
2 M. Pearson and M. Shanks, *Theatre/Archaeology*, New York: Routledge, 2001, p. 131.
3 G. Ulmer, *Teletheory: Grammatology in the Age of Video*, London: Routledge, 1989.
4 W. Benjamin, "The Storyteller," *Illuminations*, London: Pimlico, 1999, p. 84.
5 D.J. Williams, *Hen Dy Ffarm* (The Old Farmhouse), Llandysul: Gomer Press, 2001, p. 6.
6 B. Lewis Jones, "Cynefin – the Word and the Concept," *Nature in Wales*, 1985, pp. 121–2.
7 M. Pearson, *In Comes I: Performance, Memory and Landscape*, Exeter: University of Exeter Press, 2001.
8 M. Pearson and J. Hardy, *Carrlands,* 2006.
9 UK Arts and Humanities Research Council "Landscape and Environment" program [online] available at http://www.landscape.ac.uk, accessed 17 December 2010.
10 M.C. Balfour, "Legends of the Lincolnshire Cars," *Folk-Lore* II, 1891, 145–70; 257–83; 401–18.
11 T. Ingold, *The Perception of the Environment*, London and New York, Routledge, 2000.
12 M. Pearson, 2007, "The Carrlands Project Overview," Carrlands [online] available at http://www.carrlands.org.uk, accessed 6 December 2008.
13 J. Cardiff, *The Missing Voice (Case Study B)* [online] available at http://www.cardiffmiller.com/artworks/walks/missing_voice.html, accessed 6 December 2008.
14 J. Berger, *About Looking*, London: Writers and Readers, 1980.
15 M. Rose, "Gathering Dreams of Presence: A Project for the Cultural Landscape," *Environment and Planning D: Society and Space*, 24 (4), 2006, 537–54.
16 K. Stewart, *A Space on the Side of the Road*, Princeton, NJ: Princeton University Press, 1996.

# 27

# PHOTOGRAPHY AND ITS CIRCULATIONS

*Gillian Rose*

Photography is an imaging technology which has been put to very diverse use, as Patrick Maynard has reminded us.[1] Many scholars in both the humanities and the social sciences are enamoured of some of those uses, and have focused on particular sorts of photographs at some length. Their overwhelming preference is for photographs of people and places, that is, photographs in a figurative, pictorial tradition of imaging. Such photographs have been approached as part of that uniquely human process of making meaning, of recording and interpreting the world by creating images of it. And because photographs always show something that has passed – a pose no longer held, a place no longer looking like that, a person no longer alive – they have persistently been associated not only with what has been, but with death itself. The final book written by Roland Barthes has been hugely influential here. *Camera Lucida*, as is well known, is a sustained meditation on the nature of the photograph, structured by Barthes's search for a photograph that would remind him of his mother after her death.[2] Writing at length on the way a photograph is a "certificate of presence" showing incontrovertibly "what has been," Barthes concluded that "with the Photograph, we enter into *flat Death*."[3] Drawing inspiration from Barthes, not always explicitly, as well as a range of other theorists, there is now a certain elegiac school of writing about photography which approaches photographs through themes of loss, absence, desire and death.[4] All this suggests that photographs are one way of addressing what it is to be human in the sense that, confronted with proof of the inevitable passage of time, of decay and of death, the viewer of a photograph has no choice but to contemplate their own mortality.

Or, of course, that viewer could see a photo as just so cute it simply has to be printed onto a mouse mat, or held on the fridge door with a magnet; or made into a magnet – or a key ring; or turned into a shrine complete with frame, plinth and bronze bootees.[5] A photograph tucked into the frame of a dressing-table mirror can be treasured as a memory of a perfectly happy moment; another might be framed to record a proud one.[6] For as well as those intense meditations provoked in some critics by some photographs are other, very different encounters with photos, which tend not to be accorded quite the same attention or respect by critics – perhaps because they happen not in the archive or the gallery, where detailed and sustained attention and reflection are the expected forms of attention,[7] but in ordinary houses.

Family photography is a hugely popular pastime. Of course its technologies and practices have changed historically, but ever since its invention it seems that those who could afford to do so would own photographs of their loved ones, kept in albums or storage boxes or framed, carried with them in lockets or wallets, sent to others as postcards or tucked in letters or as an email attachment. In the global North now, you'd be hard pressed to find anyone who didn't possess at least a few snaps of some family members, and with the popularity of digital cameras hooked up to home computers it would be possible to find some people with thousands. (In the UK alone in 2005, an estimated 39 million rolls of film were processed, 20 million disposable cameras used, and 2.8 billion digital images taken.[8]) Yet there is little interest in this extensive image-making and image-sharing among the large number of academic critics now writing about contemporary visual culture, and even less in excavating historical examples of everyday encounters with ordinary photos.[9]

The work that has been done on contemporary family photography is generally rather dismissive of it (Geoffrey Batchen and Richard Chalfen are exceptions here).[10] Most discussions of family photographs begin by defining them as photos that show members of a family. It is then demonstrated that family snaps show those family members in particular, limited ways: usually as happy and at leisure.[11] There are no photos of mum doing the ironing, or at work in her office in the family album; there are no photos of teenage tantrums and very few of sick children in the My Pictures folder on the home pc. Instead, members of a family are shown on holiday, or at birthday parties, or in their back gardens, or at a weekend barbecue, or on an outing to the local park. This has led family photographs to be criticized for perpetuating an idyllic image of the nuclear family, cementing only dominant visions of its classed, gendered and racialized identity.[12] Many feminist critics, for example, see family photographs as especially problematic representations of feminine subjectivity. Often citing the work of photographer and writer Jo Spence as inspiration, critics such as Jessica Evans, Deborah Chambers, Valerie Walkerdine and Annette Kuhn all find the images in the family album especially oppressive for women.[13] Family albums, they say, contain distorted and misleading visions of family life in which only happy times and leisure spaces appear. "They will be shared, they will be happy," says Kuhn, "the tone of seduction is quite imperious."[14] The fragility of contemporary family relationships is obscured, they say.[15] Their erasure of domestic labour, and the restricted emotional tones they convey, means that, for these feminist critics, family photo albums are complicit with women's physical and emotional exploitation. Hence as images they are deceptive. When it comes to family photos, warns Simon Watney, "appearances are not to be trusted."[16]

As well as what they show, family photos are also said to be recognizable from how they show it. It is frequently remarked that family photos are not visually innovative. The poses and the events are predictable; the compositions are banal; red-eye and wonky framing are acceptable. So it is that both the conventionality of their subject matter and their unpolished style have contributed to the less-than-positive critical reaction received by family photos. To Susan Stewart, "all family albums are alike,"[17] Jessica Evans claims that it is in family photography that "the most stultified and stereotyped repertoire of composition, subject-matter and style resides,"[18] and even Richard Chalfen has to admit that they have an "overwhelming sense of similarity and redundancy."[19] And if many critics have remarked on the very specific version of family life and domestic space that family snaps help to produce, others have noted the ways in which other places and people have also been represented very selectively in other photographic practices. Several geographers and anthropologists, for example, have looked at archives of photographs, often those brought back to imperial cities by colonial expeditions, and explored the particular ways in which distant places were represented by those images.[20]

And yet family photographs remain extremely popular; so many things are done with them, and done usually by the very women whom they are supposed to be misrepresenting. For many people, putting family photos out on display in a house signifies that the house is now a home, a family home. (Family snaps are often one of the first things that get unpacked after a house move.) They show that this house is a home, to a family perhaps but also to one or more members of other families living elsewhere. Familial relations are also maintained between different family homes in part because copies of photos are sent between family members that trail the connections of family as they travel and come to rest. As many historians of photography have noted, photographs have been made to travel ever since the technology began to develop in England and France in the 1830s,[21] and a large part of what is done with family snaps is to send them to distant family and friends; family photos cross oceans and continents, popped into letters, framed as Christmas cards, attached to emails and up- and downloaded on photo-sharing websites. So grandparents have photos of their grandchildren, children of their parents, and nieces of aunts. In this way the space of one home, one part of a family, infiltrates the walls and shelves of other homes, other families. And in being thus entwined with the familial, these ordinary pictures are also part of the domestic, the intimate and the private.

How should we understand the importance of family snaps to how domestic and familial spaces are made? Some work has begun to consider the effects of photographic images rather differently, thinking somewhat less about what they show and more about what was done with them.[22] How were they made and where were they kept? How and where are they displayed? How are they looked at? And how and where do they travel? And in this work, photographs start to look a little less imperious and a lot more mutable. Many historical accounts of women's photographic practices suggest that photographing family and friends, and doing things with those photos like making albums, far from naively reproducing dominant ideologies of domestic femininity, often negotiates such ideologies with remarkable skill. This body of feminist work has paid most attention to the photographs and albums made by upper-class women in the mid nineteenth century. Some critics, like Julie Lawson, Carol Mavor and Lindsay Smith have examined the photographs made by women in the 1850s and 1860s and argued that they share a distinctive, feminine aesthetic.[23] Others, like Patrizia di Bello and Marina Warner have paid more attention to the albums in which such photographs were displayed.[24] These albums were often heavily worked by their creators, with photographs cut and pasted into watercolor scenes, surrounded by painted flowers, or made part of abstract and surreal geometric schemes. Bello's work is particularly rich, exploring not only the albums themselves but how they would have been looked at in the drawing rooms of these women. She argues that, as Val Williams describes the albums made by Vanessa Bell in the interwar period, these earlier albums were "knowing;"[25] these women were using photographs "to give materiality to their own culturally and socially specific desires and pleasures."[26] An essay by Karina Hof on the current hobby of scrapbooking, unusual in its sympathetic treatment of an aspect of contemporary family photography, suggests that little might have changed. "On a small scale," she says, in a scrapbook, "life can be cropped, embellished and laid out according to available resources, aesthetic preferences and as contemplations on the past and dreams for the future."[27] Drawing on this sort of work, then, family photography might be seen as a more ambivalent and complex field of cultural practice than it has often been given credit for, even by feminists concerned with women's domestic lives.

It is necessary to acknowledge the complexity of family photographs both as images and as visual objects with which a range of things are done, if we are to understand their current mutations. Family photography is going digital, which is enabling some new things, while much is staying the same. Family photography is also going public in ways it has not been before. Increasingly, family snaps are leaving the domestic realm and entering other spaces of circulation and display.

Now, family photographs have always travelled between family members. And, of course, some family snaps have long been visible in more public places. Framed photographs sit on many an office desk; in several European countries, it is taken for granted that a family photograph will embellish a gravestone. In the UK, such practices have been less popular until recently. However, in the past few years it seems that there too, family snaps are entering public spaces of display more and more often. Once mostly restricted to being looked at only by the family and friends of the people pictured, family snaps are now visible more and more often to the gaze of strangers. Increasingly they are appearing on gravestones; they are uploaded onto social networking websites; they are printed onto shopping bags and t-shirts; they are published frequently in the mass media. In this sense, family snaps have gone public.

Why and with what effects? These are questions that it will surely take the skills of disciplines in both the humanities and the social sciences to address. And indeed, scholars from both fields are beginning to offer, if not answers, then at least a sense of an historical change in the constitution of the contemporary public that may provide some context for the public intimacy of family snaps on view to strangers. For it is being argued that qualities associated most with the domestic – the emotional, the familial, the intimate – are more and more at play in the making of the contemporary public sphere of political policy and debate. Lauren Berlant in particular has outlined the historical conditions for this shift toward the affective in public life.[28] While none of these critics would argue that emotions were ever absent from the political public sphere, they do make a persuasive case that contemporary US politics, both domestic and global, depend more and more on emotive rhetoric and less and less on careful analytical debate. The political culture in the UK too is also changing in similar directions, toward what Roger Luckhurst calls a "trauma-culture."[29] "Publics" are now constituted through discussion and debate but also through the collective experience of "feeling," and the self is centered as the conduit of the political through the performance of emotional display. This may be one reason why family photographs are entering public circulation more than they ever have done before. They are objects which induce emotional responses, and they are appearing in public at a time when that public is also getting more emotional.

Berlant argues that such public intimacy by and large only affirms the (geo) political status quo, achieved both by what images and narratives are available but also by how they expect to make us feel. There is evidence that agrees with this claim.[30] However, as this short essay has also suggested, what goes on in private with family photos is far from straightforward; perhaps this suggests that their public manifestations are also more complex than Berlant's arguments imply. This is surely, among the many others suggested by photography's embedding in a hugely diverse range of practices, a topic for further transdisciplinary work.

## Notes

1  P. Maynard, *Engine of Visualization: Thinking Through Photography*, Ithaca, NY: Cornell University Press, 1997.
2  R. Barthes, *Camera Lucida: Reflections on Photography*, trans. R. Howard. London: Vintage, 2000.
3  Barthes, *Camera Lucida*, pp. 87, 85, 92.
4  See, for example, A. Azoulay, *Death's Showcase*, trans. R. Danieli, Cambridge, MA: MIT Press, 2001; U. Baer, *Spectral Evidence: The Photography of Trauma*, London: MIT Press, 2002; J. Hirsch, *Family Photographs: Content, Meaning and Effect*, Oxford: Oxford University Press, 1981; M. Hirsch, *Family Frames: Photography, Memory and Post-Narrative*, Harvard, MA: Harvard University Press, 1997; C. Mavor, *Becoming: The Photographs of Clementina, Lady Hawarden*, Durham NC: Duke University Press, 2000; S. Stewart, *On Longing: Narratives of the Miniature, the Gigantic, the Souvenir, the Collection*, Baltimore: Johns Hopkins University Press, 1984; P. Stokes, "The Family Photograph Album: So Great a Cloud of Witnesses," in G. Clark (ed.), *The Portrait in Photography*, London: Reaktion Books, 1992, pp. 193–205.
5  G. Batchen, "Vernacular Photographies," *History of Photography* 24, no. 2, 2000, 262–71, 266.
6  R. Chalfen, *Snapshot Versions of Life*, Bowling Green, Ohio: Bowling Green State University Popular Press, 1987; G. Rose, "'Everyone's Cuddled Up and it Just Looks Really Nice': The Emotional Geography of Some Mums and their Family Photos," *Social and Cultural Geography* 5, no. 4, 2004, 549–64.

7  G. Rose, "Practicing Photography: An Archive, a Study, Some Photographs and a Researcher," *Journal of Historical Geography* 26, no. 4, 2000, 555–71.
8  PMA Marketing Research, *Consumer Imaging in Great Britain*. USA: PMA, 2006.
9  G. Batchen, "Snapshots: Art History and the Ethnographic Turn," *Photographies* 1, no. 1, 2008, 121–42.
10  Batchen, "Vernacular Photographies," 262–71, 266; Chalfen, *Snapshot Versions of Life*.
11  Chalfen, *Snapshot Versions of Life*; Hirsch, *Family Photographs*; Hirsch 1997; D. Slater, "Domestic Photography and Digital Culture," in M. Lister, *The Photographic Image in Digital Culture*, London: Routledge, 1995, pp. 129–46; J. Spence, and P. Holland (eds), *Family Snaps: The Meanings of Domestic Photography*, London: Virago, 1991; J. Spence, *Beyond the Family Album*, London: Virago, 1986, Stokes, "The family photograph album."
12  P. Bourdieu, et al., *Photography: A Middle-Brow Art*, trans. S. Whiteside, Cambridge: Polity Press 1990; D. Chambers, "Family as Place: Family Photograph Albums and the Domestication of Public and Private Space," in J. Schwartz and J. Ryan (eds), *Picturing Place*, London: IB Tauris, 2002, p. 96–114; Spence, *Beyond the Family Album*.
13  Spence, *Beyond the Family Album*; J. Evans, "Photography," in F. Carson and C. Pajaczkowska (eds), *Feminist Visual Culture*, Edinburgh: Edinburgh University Press, 2000; Chambers, "Family as Place;" V. Walkerdine, *Schoolgirl Fictions*, London: Verso, 1991; A. Kuhn, *Family Secrets: Acts of Memory and Imagination*, London: Verso, 1991.
14  Kuhn, *Family Secrets*, p. 25.
15  Spence, *Beyond the Family Album*.
16  S. Watney, "Ordinary Boys," in Spence and Holland (eds), *Family Snaps* pp. 26–34.
17  Stewart, *On Longing*, p. 49.
18  Evans, "Photography," pp. 105–20.
19  Chalfen, *Snapshot Versions of Life*, 42.
20  J. Ryan, *Picturing Empire: Photography and the Visualization of the British Empire*, London: Reaktion Books, 1997; Schwartz and Ryan (eds), *Picturing Place*; J.M. Schwartz, "The Geography Lesson: Photographs and the Construction of Imaginative Geographies," *Journal of Historical Geography* 22, no. 1, 1996, 16–45.
21  P. Osborne, *Travelling Light: Photography, Travel and Visual Culture*, Manchester: Manchester University Press, 2000.
22  E. Edwards and J. Hart, "Introduction: Photographs as Objects," in Edwards and Hart (eds), *Photographs Objects Histories: On the Materiality of Images*, London: Routledge, 2004, 1–15; D. Poole, *Vision, Race and Modernity: A Visual Economy of the Andean Image World*, Princeton, NJ: Princeton University Press, 1997; J. Larsen, "Practices and Flows of Digital Photography: An Ethnographic Framework," *Mobilities* 3, no. 1, 2008, 141–60.
23  J. Lawson, *Women in White: The Photographs of Clementina Lady Hawarden*, Edinburgh: National Galleries of Scotland, 1997; Mavor, *Becoming*; L. Smith, *The Politics of Focus: Women, Children and Nineteenth-Century Photography*, Manchester: Manchester University Press, 1998.
24  P. di Bello, *Women's Albums and Photography in Victorian England: Ladies, Mothers and Flirts*, Aldershot: Ashgate Press, 2007; M. Warner, "Parlour Made: Victorian Family Albums," *Creative Camera* 315, 1992, 29–32.
25  V. Williams, "Carefully Creating an Idyll: Vanessa Bell and Snapshot Photography 1907–46," in Spence and Holland (eds), *Family Snaps*, pp. 186–98.
26  Bello, *Women's Albums*, p. 5.
27  K. Hof, "'Something You Can Actually Pick Up': Scrapbooking as a Form and Forum of Cultural Citizenship," *European Journal of Cultural Studies* 9, no.3, 2006, 363–84, 381–2.
28  L. Berlant, "Poor Eliza," *American Literature* 70, no. 3, 1998, 635–68.
29  R. Luckhurst, "Traumaculture," *New Formations*, 50, 2003, 28–47.
30  G. Rose, "Who Cares for Which Dead and How: British Newspaper Reporting of the Bombs in London, July 2005," *Geoforum* 40, no. 1, 2009, 46–54.

# 28

# BEYOND THE POWER OF ART TO REPRESENT?[1]

## Narratives and performances of the Arctic in the 1630s

### Julie Sanders

Our journey begins in icy Spitsbergen. In 1631, a gunner's mate called Edward Pelham described the Arctic landscape in which he had recently been residing, and, indeed, trying quite simply to survive, in these terms:

> Greenland is a Country very farre Northward situated in 77. degrees and 40. minutes, that is, within 12. degrees and 20. minutes of the very North Pole it selfe. The Land is wonderfull mountainous, the Mountaines all the year long full of yce and snow: The Plaines in part bare in Summer time, There growes neither tree nor hearbe in it, except *Scurvygrasse* and *Sorrell*.[2]

This description appears as part of a pamphlet entitled "Gods Power and Providence Shewed, in the Miraculous Preservation and Deliverance of 8 Englishmen, left by mischance in *Greenland* Anno 1630, nine moneths and twelve dayes," which was published in London in 1631. If we attempt to think ourselves into the mindset of English and Scottish men in the 1620s and 1630s, those men who formed the bulk of the shipping companies that travelled to Spitsbergen to exploit the possibilities of the newly established whaling trade there (a trade carried out in the face of direct competition with Dutch and Danish companies), it would be hard to underestimate the sheer experiential impact of this encounter with a land of "yce and snow."

Pelham's pamphlet is partly marketed on account of the description it will provide for a 1630s reading public of this alien or hostile landscape, this wilderness – the vicarious pleasure implicit in the reading experience of publications such as this has been discussed by scholars in the fields of both geography and the humanities working more generally on the hermeneutics of narratives of travel and exploration in this period.[3] The title-page of Pelham's pamphlet asserts that it contains "a Description of the chiefe Places and Rarities of that barren and cold Countrey," and one of the more productive sites of encounter between geography, history, and literary criticism as scholarly enterprises has certainly been the return to 'description' as a social art form, an aesthetic that each of these disciplines benefits from reconsidering in its multiple manifestations and formations, not least across time.

But the chief reason that Pelham's pamphlet was being rushed into publication within a few months of his return to England is because of the tale of adventure that it recounts. Because this particular sojourn in Spitsbergen went horribly wrong for Pelham and his seven colleagues, who were there as part of a Muscovy Company enterprise on a ship called the *Salutation*. The men – a mixed grouping of "land-men," whale-cutters, gunners like Pelham himself, and a cooper[4] – were waiting for the wind to turn to enable their ship to head homewards that August after a full season in the whaling township. They were temporarily sent onshore to kill some deer for meat for the return journey when the weather turned and they found themselves trapped in a deep fog unable to locate the ship which they were meant to re-board. When they finally got to Greene-harbour to await the ship there they found it had already called into port, collected the men stationed there, and departed.

A kind of panic set in as it dawned on them that they were stranded for the winter months with no provisions and no precedent for surviving this experience. There were other printed pamphlets available at this time, dealing with similar events, and all sold well in European markets, but inevitably in these versions the stranded sailors perished in the cold: "Well, wee knew that neither *Christian* or *Heathen* people had ever before inhabited those desolate and untemperate *Clymates*" (C1r), says Pelham. What follows, then, is a remarkable account of how these men survive the winter by a combination of tenacity and ingenious labour. They hunt for food and clothing to store up before the hard winter completely closes down the possibility of movement; they develop methods of keeping a fire going almost non-stop in their wooden tented construction, made out of sails and oars left behind on the shore – then as now the detritus of such expeditions littered the supposedly virginal landscapes in which they intervened. The "tent" becomes home for these months; occasionally they set light to sections of it and once let the fire burn out, which could have signalled certain death. The pamphlet is also an astonishing record of these eight men's encounter with the very particular Arctic landscape of ice and snow.

Writing about a much later journey to the Antarctic, the rival expeditions of Roald Amundsen and Robert Falcon Scott in the South Polar region in the early twentieth century, cultural geographer John Wylie has described these particular expeditions as "*enactments*" of landscape, suggesting that we need to understand them in the "concrete, sensuous contexts of practice and performance."[5] If literary criticism as a practice has been deeply shaped and informed in recent decades by what is often called the "spatial turn," which has encouraged sustained attention to topics such as space, place, landscape, and the built environment as constitutive elements in the creation of imaginative literatures – and the attendant appropriation of methods and techniques more familiar to those from the disciplines of geography and archaeology, such as cartography, mapping, and the walking and surveying of sites – Wylie's vocabulary in the above quotations is in turn an indication of geography's own decisive movements toward performance studies in its articulation of intellectual domains of concern.[6]

Wylie is a significant example of a group of cultural geographers and archaeologists who have found particular intellectual purchase in the phenomenological method, which foregrounds ideas of perception and sensory experience.[7] As a result, however, the cited work on Antarctic expeditions aside, Wylie's attentions have tended to remain squarely in the present rather than in the historical past. My interest as a literary historian, and a historian of performance histories at that, is in the possibilities of combining this form of cultural geography with literary hermeneutics and ideas from performance studies, and whether this might provide me with new ways of thinking about historical texts and documents such as Pelham's pamphlet, narrative accounts, as it were, of journeys or experiences of particular spaces, places, and landscapes. This chapter offers up a reading of Pelham's work as a case study of these intellectual aims in practice. In turn, I am aware that my very interest in a piece of writing – a textual *enactment*, to redeploy and redirect Wylie's suggestive terminology – such as a printed pamphlet, published as a "truthful" account of direct experience in the early seventeenth century, is itself exemplary of the impact of the "historicist" turn on literary studies and the attendant expansion of the canon that has taken place. This now finds us regularly juxtaposing poetry, drama, and fiction – the so-called imaginative genres – with documents and artefacts such as pamphlets, letters, and diaries, not as "background context" but on equal terms, as keys to the imaginative and discursive structures of the time.[8]

I am interested, for example, in how a writer such as Pelham in recreating and representing his experience in the land "of yce and snow" brings to bear sensory perception of the landscape and his memory of the same, how either consciously or subconsciously his own cultural upbringing shapes and frames that perception, and also what literary and artistic frameworks he brings to bear both in the writing-up of the experience and in the moment of the experience itself. I am struck reading Pelham's pamphlet by the hybrid set of responses he has to the landscape and environment in which he and his men find themselves marooned for the winter of 1630–31. Describing in one particularly striking passage the men's response when it dawns on them that they may die here and their corpses become hideously disfigured by the polar bears and arctic foxes that are "the chief mammals of this place," he observes:

> All these fearfull examples presenting themselves before our eyes, at this place of *Bottle Cove* aforesaid, made us, like amazed men, to stand looking one upon another, all of us, as it were beholding in the present, the future calamities both of himselfe and of his fellows. And thus, like men already metamorphosed into the yce of the Countrey, and already past both our sense and reason; stood wee with the eyes of pittie beholding one another (C2r).

The recourse to a sub-Ovidian discourse of metamorphosis here is deserving of comment and what a consciously transdisciplinary approach to a text such as this releases is the potential to consider the discursive as well as practical resources that Pelham invokes as part of his survival strategies, and as part of his response after the traumatic experience, alongside each other. The proleptic imagining here in print of a "future" in fact already past is further evidence of the complex reconstructions of memory and experience these texts involve and which an acknowledgment of phenomenology and cognate disciplines such as cognitive poetics can reveal.

The other shaping force in Pelham's expressions other than the literary *per se* is clearly the scriptural and psalmic discourse that would have been comfortingly familiar, part indeed of a coping mechanism for these men in this completely other and horrifying situation. Pelham talks at one point of how thoughts of home also stave off the "thousand sorts of imagination" their minds are subject to – the echoes of Shakespeare's *Hamlet* here are both striking and intriguing (D4r). That *Hamlet* appears to be, albeit allusively, invoked in a passage on grief also suggests an extended understanding of that play's apposite nature in terms of the company of eight men's particular experience. The tracing of literary allusion in this way, within a supposedly "factual" document, one that has obvious material interest for historical geographers, enables us to think about the reading that men such as Pelham undertook both prior to and during their journeys and the ways in which that reading shaped the moments of encounter with the supposedly alien landscape of the Arctic North.[9]

As the title of Pelham's pamphlet suggests, the return of light and the spring after months of almost unimaginable hardships is cast in terms of spiritual regeneration and renewal (the epigraph to the pamphlet is from Corinthians and refers to the dead being brought back to life). In their darkest moments of despair the men imagine they are being punished by God. The sheer poetics of the descriptions of the return of the light and the sun are, then, highly notable: "*Aurora* with her golden face smiled once again upon us, at her rising out of bed" (E2v). The vocabulary here is less scriptural than a newly embodied version of the traditional poetic *aubade*, the same form that John Donne's metaphysical poetry was rethinking and reconfiguring at this time in poems such as "The Sunne Rising" (circulating in manuscript form during Pelham's lifetime). In turn Donne's poetry has been read by many as enlivened by the new geographies of his age.[10] The flow of influence is multidirectional and a transdisciplinary reading of this kind rather than eliding historical and cultural difference can help to render us ever more sensitive to it. As Sara Blair has observed, "literary traditions – of genre, canonization, dissemination – give shape to human interactions with … environment [and] … landscape."[11] A dialogue between geography and literary criticism therefore allows for a fuller reading of a text like Pelham's pamphlet.

Eventually, the ice melts and the ships from Hull return to start the whaling season anew in Spitsbergen only to find Pelham and his half-starved, emaciated company of men still alive in their "tent." There is a remarkable scene where the men invite the returning whalers into their "house" in a strangely spectral performance of the kind of hospitality that was considered a mainstay of social practice and neighborliness in Caroline England: "we shewed them the courtesie of the house, and gave them such victuals as we had; which was Venison roasted foure moneths before and a Cuppe of cold water, which for noveltie sake they kindly accepted of us" (E4v–F1r). In analyzing moments such as these in the printed text, the application of phenomenological geography alone is not enough. Pelham's entire text reconstructs the Arctic North as a site of memory and any sense of direct experience is filtered through and managed by the kinds of discursive and literary constructions of language and descriptive account that I am suggesting a literary historian is particularly equipped to "read." What the invocation of phenomenology and its attention to ideas of sensory perception enables, however, is a far more engaged response to the imaginative force of the experience of a direct encounter with the icy terrains of Greenland that Pelham underwent, and that his pamphlet attempts to convey for an unfamiliar reading audience. These experiences, both as published in their own time and as understood by the modern reader, should not be abstracted from the imaginative, the physical, or the experiential.

If the "spatial turn" has enabled a more informed understanding of the social and material constructs that lie behind imaginative literatures for the literary scholar then for cultural geographers the performative turn has enabled a more dynamic understanding of landscape as both text and embodied space and site. Both these moves have encouraged a sense of landscape and environment as agents in their own right, productive of meaning and practice as well as reflective of it, and for literary criticism an absorption of all these ideas into historicized practices of close reading may in turn allow us, if not to get back to the moment of experience as such, then to respond with new alertness and intimacy to the deep differences across time and space represented by these acts and narratives of encounter.

## Notes

1 The phrase in the title is a reformulation of Captain John Ross's claims that the Arctic wilderness was "beyond the power of art to represent" on his expedition to navigate the Northwest Passage in 1818. In practice, Ross's team engaged in all manner of representation, not least through the acts of diary-writing and the production of watercolors, evidence were it needed of the value of considering these expeditionary journeys as both aesthetic and pragmatic experiences. These details were recounted as part of a BBC4 program in the "Wilderness Explored" series, examining the cultural and imaginary force of the Arctic as a space between the nineteenth century and the present day. The program was first aired in the UK on 8 October 2008.

2 E. Pelham, "Gods Power and Providence Shewed, in the Miraculous Preservation and Deliverance of 8 Englishmen, left by mischance in *Greenland* Anno 1630, nine moneths and twelve dayes," London, 1631, B1r. All references to this pamphlet are henceforth contained within parentheses in the text. My thanks are due to audiences at the Universities of Nottingham, Reading, and Calgary for contributions to paper presentations on Pelham's pamphlet and to novelist Georgina Harding for first alerting me to the existence of this document.

3 See, for example, W.K.D. Davies, *Writing Geographical Exploration: James and the Northwest Passage, 1631–33*, Calgary, Alberta: University of Calgary Press, 2003, p. 258.

4 W. Fakeley, Gunner; J. Wise and R. Goodfellow, Seamen, T. Ayers, whale-cutter; H. Bett, cooper; J. Dawes and R. Kellett, Landmen are the descriptions of the other seven provided in the narrative (A4v).

5 J. Wylie, "'Becoming-Icy': Scott and Amundsen's South Polar Voyages, 1910–1913," *Cultural Geographies* 9, 2002, 249–65, 249.
6 For a useful examination of the effects in both directions, see S. Blair, "Cultural Geography and the Place of the Literary," *American Literary History* 10:3, 1998 544–67. Blair argues that Cultural Geography has "provided powerful new models and vocabularies" for the humanities to revisit the "hotly contested border between literature and culture, the aesthetic and the social" and that in turn "theorists of space and place" have acquired "specific reading practices ... that richly affirm the materiality and texture of spatial experience" (pp. 545–6).
7 For an extended discussion of phenomenology and the study of landscape, see J. Wylie, *Landscape*, London: Routledge, 2007, pp. 139–86. Key figures in the field from archaeology include C. Tilley; see his *A Phenomenology of Landscape: Places, Paths, Monuments*, Oxford: Berg, 1994, and *The Materiality of Stone: Explorations in Landscape Phenomenology*, Oxford: Berg, 2004. A significant example of collaborative work between archaeology and performance studies is M. Shanks and M. Pearson's *Theatre/Archaeology*, London: Routledge, 2001.
8 The key figure in this respect is the pioneer of the New Historicist school or method, S. Greenblatt. Greenblatt himself increasingly turned to the phrase "cultural poetics" in his published writing but his influence on the kinds of texts studied and the ways in which they are studied alongside each other, not least in early modern scholarship, remains seminal and permeates not only literary studies but also historical and geographical practice. For explorations of his "method," see, for example, Greenblatt's "Towards a Poetic of Culture," in H.A. Veeser (ed.), *The New Historicism* London: Routledge, 1989, pp. 1–14. On a related theme, see Paula Yaeger on the "new poetics of geography" in the introduction to her edited volume *The Geography of Identity*, University of Michigan Press, 1996, p. 18.
9 My thinking here is directly informed by G. Beer's pioneering work on Charles Darwin and the Voyage of the Beagle in *Darwin's Plots: Evolutionary Narrative in Darwin, George Eliot, and Nineteenth-Century Fiction*, Ark/Routledge, 1983, which explores the ways in which Darwin's reading, both as a boy and in terms of the books he took with him on the *Beagle* expedition, directly shaped his scientific narrative and method, pp. 7, 31.
10 See, for example, L. Gorton, "John Donne's Use of Space," *EMLS* 42, 1998, 9, pp. 1–27. URL: http://purl.oclc.org/emls/04-2/gortjohn.htm
11 Blair, "Cultural Geography," p. 554.

# 29

# NAVIGATING THE NORTHWEST PASSAGE

*Kathryn Yusoff*

The "opening" of the Northwest Passage, as envisioned by the European Space Agency satellite in 2007, drew an orange loop straight through the sea ice, clearing a passage in the icescape and in the imagination that had been dreamt of, and searched for, over a century ago. The dream of a passage had passed into metaphor, explicating the folly and desire of white man's exploration, set against the backdrop of maritime empires and colonial relations. The passage emerges now as the hot underbelly of that dream of expansion; a line seared through the ice, illuminating global heating. The empire of man, or Anthropocence, expanded to the limits of the atmosphere (Plate XII).

Navigating the Northwest Passage – in the days of exploration and in the contemporary geopolitical contestations of Arctic passage – involves journeys into the ice floes and through the politics of that shifting terrain. In the mythic search for passage, the imaginary of a direct route was always a line of hunger; plotted in advance of travel, for fame, economy and empire. Despite this enflamed arc, the localized reality of navigating this course meant adventures in the labyrinthine passages, between the ice floes and islands, and in the land of the people. If the search for the Northwest Passage is a geographical problem, of territory and movement, of desire and nightmare, of mythic and geopolitical imaginaries, of biophysical and cultural change, how might such a passage be navigated to give account of the different frameworks of thought and material relations, which make such journeys into knowledge and politics possible? And setting out on such a journey, what might we learn from these disjunctures of knowledge that we come upon?

The philosopher Michel Serres likened the search for the Northwest Passage[1] to crossing the path between the humanities and sciences, with all the twisted and disorientating inlets and islands, and shores of abandonment that voyagers washed up on. Serres says;

> This is why I have compared them to the Northwest Passage ... with shores, islands, and fractal ice floes. Between the hard sciences and the so-called human sciences the passage resembles a jagged shore, sprinkled with ice, and variable ... It's more fractal than simple. Less a juncture under control than an adventure to be had.[2]

While Serres' posing of the problem may seem like a metaphorical transposition of a specific geography and set of relations into a philosophical discussion on epistemological boundaries, his understanding of metaphor as means of transport[3] offers a way to see how differing sets of relations move across space and through time. If we take on Serres' assertion that "relations spawn objects, beings and acts, not vice versa,"[4] the possibility of worlds are dependent on the configurations of these relations in the space of knowledge. While these sets of relations are understood as exclusive topographies, be that of science, politics or myth, they remain at an incommunicable distant, like far-off shores. Serres' work suggests how travel between these shores might be possible. If we look at the vectors that are common to all relations, he says, we might find the difficult passages between scientific and humanistic worlds. These passages are difficult for two reasons, according to Josué Harari and David Bell;[5] first, because the divisions are institutionalized, and attempts to cross meet with topographic difficulties; secondly, because the evolution of knowledge calls for more and more specialization and increasing divisions (an ice core scientist and an Inuit hunter may share a relation to climate change, but their worlds of knowledge cannot speak to one another in adequate collaboration). Serres' work emphasizes border crossings, potential "passages" that run between disciplinary and conceptual divisions, making provisional connections between disparate relations, which act as a possible source of intervention.[6] He says: "Relations are, in fact, ways of moving from place to place, or of wanderings;"[7] a geographical practice in thought and deed.

The desire to travel these passages between the sciences and humanities, for Serres, is motivated by an attempt to reinsert the subjective domain into scientific discourse and to challenge the abandonment of the humanities. Serres poses the problem as such:

> recently the main struggle was not between the hard sciences and the social sciences, since both were sciences (real or self-designated) and ignored each other superbly (the one, a world without people; the other, people without a world) ... No, the main struggle was between these two and what they claimed to replace – the humanities ... Just as the hard sciences go their way without man, thereby risking becoming inhumane, just as the social sciences go theirs with neither world nor object, thereby exposing themselves to irresponsibility, likewise, in aggregate and in parallel, in the name of a science that is finally efficient and lucid, the two disciplines together impose the forgetting of the humanities – that continuous cry of suffering, that multiple and universal expression, in every language, of human misfortune. Our short-term powers scorn our long-term frailties.[8]

Serres' explanation for the desertion of the humanities strikes at the intolerance of a practice of knowledge that wants to exclude that which is not immediately useful or accountable. He says:

> They could not abide the accounts and scenes, sung or painted, of the great human passions and sufferings – that immense, continuing clamour, moaning, and lamentation, the psalm of mankind, weeping over the absurd, vain and uselessly mortal drama of its own ineradicable violence – a low and timid lamentation, continuous, barely audible, absolutely beautiful, and the source of all the noise of vengeance ... How could they still tarry over this music, this voice, this moaning preserved by the culture of woe from which we sprang – trans-historic background noise that cannot be attributed to anyone, but springs from the sum of humanity ... Those at the feast said to themselves, "What's the point of preserving what from now on is useless?"[9]

Taking Serres' demand for knowledge that is not divorced from misfortune seriously, at the site in which he articulates this division (the Northwest Passage), how might such passages between the world without people and the people without a world (science – social science) be traversed?

## Geopolitics of cold

Let's look to the geopolitics of cold to see how there is the movement of what Michel Serres terms a "thermal exciter"[10] making passages in the socio-political materialization of cold places, linking, like connective tissue, the technoscience of ice core production to the possibilities of Inuit sovereignty. The change of state in the Arctic that has underpinned the opening of northern passages, and is amplified through all ice, is the transmission of heat: atmospheric, oceanic, terrestrial and solar. Heat is the messenger that runs through all ice-bound relations, opening things up like a hot knife. Russia planted a flag on the seabed of the North Pole, the fever rose, Canada declared the possibility of Arctic military bases, Denmark joined the conversation, asserting its rights via the colonization of Greenland: the talk of stakeholders grew louder. The long ago heat compression that sedimented to form fossil fuels is now a cause for concern in the contemporary energy politics of global heating. The colonizations of rights to these mineral passages are termed the "Scramble for the Arctic." International research programs on the continental shelf and the geology of the seabed, though orchestrating the terms of articulation, alone will not settle this heated confrontation; yet these geosciences are implicated in the articulation of sovereign claims in international governance. Cold War geographies of contested sovereignty return to the Arctic (did they ever go away? historical geographers have repeatedly asked) and the expression of national interest is made through the familiar channels of technoscientific research and the infrastructure of research stations. Elsewhere the heat is activating other kinds of political subjectivities, as Inuit activists protest to the US for their "Right to be Cold," and the

Inuit Circumpolar Council (ICC) call for the recognition of a relational politics of temperature (in the form of protest against UK airport expansions and contestation over global responsibility for maintaining cold environments). Meanwhile, extensive warming in the Arctic and Antarctic Peninsula are causing extreme transformations to cold environments that threaten to overwhelm both polar and temperate communities. As the indigenous peoples group, Many Small Voices suggests, in its alliance between Inuit communities and Pacific Islanders that is based on their mutual displacement by ice, "As we melt, you sink." As heat moves across landscapes it transforms them, and creates new relations, and new passages for politics to emerge between different shores.

## Heat: "theory of transformations"[11]

Articulating a "geopolitics of cold" could take us in many directions, but we might end up with a political terrain that ignores the materiality and effect of this phenomenon of cold in all its manifest qualities – from bio-cultural life force to scientific time tunnel – if we concentrate solely on the topology of politics, science or even the sensation of cold. Common to this geopolitics – from living at the ice-floe edge to ice-core technologies – is the management of heat. This heat management runs across every relation; to cite Michel Serres, temperature "intercepts all the relations between all locations. It captures all the flows"[12] in both space and time. Heat is a vector that works across this cold topology, conjoining them in a way that does not conflate those differences. Cold is a provisional state – fluctuations can cause incremental or abrupt changes in climate; "And as such, it is both atom of a relation and the production of change in this relation."[13] Polar geographies are connected through this hot–cold temperature gradient, and even their status as "places" is afforded only by this gradient of cold; the Polar Regions are not national or international places, they are cold places. So just as we have become more attuned to the material aspects of technoscience, of materializing and dematerializing operations within science, we might look to how materials exchange properties/states in which temperature is a vector. These exchanges in the materialities and states of cold affect geopolitical opportunities and conflict; they, in effect, open more than one passage, and more than one opportunity for journeying. Temperature is a *force*[14] in cold regions geopolitics – glacial retreat seemingly "materializes" potential heating reservoirs while it dematerializes the possibilities for certain kinds of cold weather living practices. According to the Second Law of Thermodynamics, heat is a form of interference in cold geographies, it disrupts and fractures material landscapes. Heat is the entropic force in the equation of cold that both erodes cold places and provides opportunities for the emergence of new places that can be accessed for their oil and gas reserves. New Arctic political configurations (The Declaration of the Rights of Ingenious Peoples 2007, the UNESCO Permanent Forum of Indigenous Peoples 2007, The Circumpolar Inuit Declaration on Arctic Sovereignty, 2008[15]) travel the same passages as this heated petropolitics, but take the politics elsewhere, making exciting connections between people and places that have corresponding shores of climate change. This circle of feedbacks, between the melting of ice and the extraction of new resources and the increase in combustion of fossil fuels that farther boost the carbon economy, and thus melt more ice, recalls Schrödinger's solution for the emergence of new qualities amongst the impending "heat death" (of increasing fossil fuel consumption), as a form of feeding on "negative entropy."[16] In the humanities, it might be called a vicious circle. So what are some of the thermal exciters that quicken these vectors of exchange, and how in their material aspect do they affect each other? (Plate XIII).

Before we arrive at the hot blaze of openings in the Arctic, other journeys and earlier disjunctures in the world of ice need to be put in place; the heat of scientist's ice core work, in Antarctica, Greenland and state laboratories, making biographies of atmospheres and projecting the futures of cold into climate models; and, the blistering cold-heat of ethical/political realization that this cold regions work generated; a realization of responsibility and relationalities to both current and future generations. One part of this equation is the extraction of data from ice cores and their subsequent emergence in climate change models that fed the International Panel on Climate Change's (IPCC) fourth Assessment. The polar landscape and ices cores are at once different and connected. Ice cores have a mimetic relation to ice sheets and shelves, but are subject to some very specific scientific practices with different modes of production and narrative. But this relation between resemblance and difference is not enough; ice and cold relations, in effect, go through a black box – a space of transformation.[17] The black box in this case is the ice core, the vertical drilling into the "events" of temperature change, and the past/future consequences of that spectral haunting – abrupt, massive, fast, changes and "flickering" climates. Now, to the geopolitical relations of cold disclosed through the technoscience of ice coring and how this management of cold is connected to large-scale transformations in the polar environment; the locations of ice core geographies are multiple: field, drill site, freezer, ship, cold storage, laboratory, chemical analysis, General Circulation Models (GCMs), scientific papers, political briefings, policy documents, public spaces. In this sense, ice cores are spacio-temporal envelopes, archiving global atmospheres over nearly one million years, atmospherically constituting many disparate places and events in one. Ice cores have a complex relation to the ice sheets – they are extracted from different geographic locations and from different times (from the International Geophysical Year (IGY) onwards), with different regimes of scientific practice governing data extraction in each location (both site and laboratory). Ice cores may come from one location, but they are often divided up among many countries, different parts of the core going to different labs, with differing practices (there are no standardized methodology, or indices, or governance of data). Storage labs, holdings, facilities and temperature of ice storage vary. There is a geopolitics to the infrastructure; in the field, this relates to who has access to the sites and the size of drilling infrastructures and their potential for being utilized for other geological applications. In the US the ice core laboratory is located in a federal estate with the Department of Homeland Security. Politically, ice core knowledges have been explicitly used to attempt to *force* political ends. For example, The Cold Regions Research and Engineering Laboratory,

**FIGURE 29.1**   National Ice Core Laboratory, Boulder, Colorado 2007 (Photograph by Kathryn Yusoff).

US Army Corps of Engineers uses ice-core and ocean-based sediments for, "research for modelling and predicting impacts of climate change and extreme events on both terrestrial and battle space environments."[18] Potential users are identified as "military strategists and military tactical planners" in the design of "Future Combat Systems."

The cold materials of ice cores are taken through a complex knowledge system in which the material circumstances in which temperature is managed is a critical factor in the governance of the objects and the subsequent knowledge production. At the field site the newly drilled ice must be allowed to "relax" to the surface temperature so that it does not get the equivalent of the bends and the compressed air explodes. In the process of extracting the core, heat is extracted from exposed fleshy bodies – in the laboratory there is a reversal, heat is extracted from the lab. Ice must enter the "black box" to be redistributed, extracted and managed – air entombed in the ice is rendered into data through chemical processes to capture atmospheric circulations. Ice undergoes a form of heat death as information is extracted and the ice destroyed in the process. After air freighting or shipping, the cores are checked into the lab at −36°C. "Aggressive maintenance" is practiced on the heat extractors, then on to processing at −25°C in the examination room. The ice is gently warmed up to reduce the effect of shock when taken out of tubes and sleeves. A horizontal band saw establishes a flat surface, the first cut is saved for measuring stable oxygen/hydrogen isotopes. The planer smoothes the surface of the first cut for electroconductivity measurements – acids, salts, volcanic activity, spectral traces of events in the biosphere; then in visual stratigraphy in the black room textures, times, cracks and breaks are recorded. Then in imaging, the ice is passed under a scanner to make digital photographs of the core. Then, in the division of the core for analysis, ordered samples are identified and labelled and redistributed to different locations in insulated boxes filled with ice packs.

## But it doesn't happen in the labs, it happens outside...

The catalyst to activate (to turn up the heat) on the stream of alarming data – about the relation between carbon dioxide in the atmosphere and abrupt past climate changes – that came from ice cores was the collapse of the Larsen B ice shelf, Antarctica in late 2000. The pitch of the shock wave that travelled around the scientific community was about the speed of collapse – temperature very quickly became time.[19] The collapse of Larsen B made a landscape exhibit of abrupt climate change on a scale of magnitude that science had been unable to achieve with all its operative machines. With the accumulated data and projected scenarios, the collapse could not be predicted, only witnessed. This was because the failure of ice in the models of the IPCC had been excluded on the grounds of its unpredictability, too "hot" to count as acceptable data, too uncertain to be science. But, Larsen B confirmed what was suspected about the power of heat; "Even in the sciences the imagination does the ground breaking."[20] The unplanned test that confirmed the suspicions of scientists released a warm wave of confidence in the reality of global heating. Glacial became a fast word. Ice was moving faster than the models (and continues to do so). The world gained scientific traction as a model of the world. In another part of the equation, witnessing caused a temperature change, in politics and the public. Climate scientists got hot under the collar, and began to respond, and in doing so acknowledged the political responsibilities of their craft. Laboratory work was convulsed into an energetic state. Data flowed. Pixels heated up and graphs exceeded the known. Past data was catapulted into future heat scenarios, creating new senses of time and anxiety and new topologies of unknowing and catastrophe.

There was joy in the ice core, its dizzying, vertiginous history – a new iconic future-orientated object was born – but, what is its place and pace in human existence? Avital Ronell asks, "One has every right, in fact it is a duty, to ask of science if it is capable of devoting itself to securing the conditions for thinking joyousness ... or is science really only able in the end to promote the glacialization?"[21] How do the stories that we tell of ice cores make great "galaxies of joy flare up" and what of its galaxies of time, impermanent, contingent, that make human history seem as a brief coming into being, ready to fall away at the next climate convulsion? Suddenly it seems that our time is but a brief sojourn on a temperate plain of a dynamic biosphere that is ready to flip, and surge, and melt away. For Richard Alley, "interpretation of ice cores, and of many other climate records, has recently revolutionized our view of the earth."[22] Ice core knowledge was an axis, a fulcrum of looking both ways, past–future. The future that this science was portending was opening into a form of questioning, of desire for different kinds of futurity, scenarios, political effectiveness, communication with "publics" and other interdisciplinary excursions. As Serres comments:

Nothing would have happened without this love, this heat, this fire, that comes by and suddenly flares up. Without this light, we would be perhaps have seen nothing. The black box would have stayed black.[23]

There is a feedback to this ice-core knowledge that produced political intervention. The laboratory was opened up (a little) to the public. Without this knowledge, the box as Serres says would have stayed black. Public responsibility and political activism emerged as a consequence of the warming Antarctic Peninsula, and the opening of the Northwest Passage presses home the message, but this time in the land of the people. Ice-core scientists have had to think about what science could mean for human existence – to consider what in this science is vital, vital not just to science but to the possibilities of life – and to consider what in this science is *contra* to this, and thus what must be suppressed in order to secure the political conditions for that which is vital. This scene of repression does "not constitute science's outside but troubles its inner workings,"[24] pointing to disjunctures between the impulses that generated political stances and manifestation of them in political arenas.[25] The heat of realization is tempered out through the spaces of scientific practice, where rational units replace the burst of biosphere energy.[26] There is a dynamic to this knowledge exchange; in drilling down into the ice archive, science must articulate its data in ways that bear on the whole of the ice sheet, making a passage through time, which reconfigures the "times" of climate change. The data orders knowledge itself, creating histories and making scientists see the philosophy inherent in their craft. The circuit of ice-core knowledges burns deeply into the question of what science is and its place in the world. The value of ice and of the Polar Regions more generally also passes through this black box; spaces formerly on the periphery become crucial to the possibilities of knowledge and the future. Heat in this relation can be said to be transcendental; it changes the state of political thinking and political acting (from pedagogy to activism), and it gives rise to a larger geography of engagement for science and scientists. But, where does it take us? According to Serres, "the future will force experts to come quickly to the humanities and to humanity, there to seek a science that is *humane* – since in our language the word signifying our genus also signifies compassion."[27] As we witness together the passages opening in the ice, it is our capacity to situate these movements in the world and to understand and respond to that witnessing with acts of responsibility that burn into the future.

Science isn't the only vector in town, so to speak; as scientists become subjects of the ice, witnesses for global transformations, so do other more intimate ice dwellers. The Inuit witness is another essential witness, alongside science (and the scientist) in the cultures of global heating. Characterized by the media and countless pop climate change books, as a passive bystander "witnessing" the immense changes in the circumpolar north, Inuit activists have not only contested this backgrounding, but exploited it to foreground political concerns, by reconstituting the notion of watching; essentially, witnessing as a political act ... if we watch, you must watch our watching and see with our eyes ... in vision comes responsibility. In witnessing we must embody the political terrain of the landscape that we are inserted into and reconstitute it through our own relational passages.

At Serres' insistence, *knowledge* and *misfortune* cannot be separated,[28] so we must look toward another heat relation; a movement in the opposite direction by oil and gas companies toward the "Last Oil Frontier." According to the *Arctic Oil and Gas 2007* report, more than 5 percent of the world's known oil reserves and over 20 percent of its known gas reserves are in the Arctic. There are estimates that as much as a quarter of the world's undiscovered oil and gas lies in the Arctic.[29] This energetic pathway of fossil fuels links the Arctic to the rest of the planet in the combustion of energy, life and ice. The geographies of mobility – those enabled by sea ice, and those propelled by the economies of oil – are relational geographies in which questions of the centralization of power are crucial to how the inhabitants of the Arctic are *dis*placed, politically, economically and physically (with rising sea levels). "Thinking space relationally,"[30] as the geographer Doreen Massey articulates, is also about thinking about the geographies of responsibility that those relations and non-relations produce, such as the bypassing of oil pipelines that cut through places and bring no benefit to local indigenous peoples. These interlocking geographies require us to reconceive of not just how the inhabitants of the Arctic are situated, but how our practices continue to reinscribe unequal positions in energy politics and the misfortune of those ruptures in the space of knowledge. As Aqqaluk Lygne, poet and President of the ICC, Greenland has stressed for hunters, disappearing ice equates to the disappearance of the usefulness of traditional knowledge. Shelia Watt-Coutier, former chair of the ICC, comments: "And so, nowhere else in the world really does ice and snow represent transportation, mobility and life for a people. And ice and snow, in fact, are our highways that bring us out to the supermarkets, which is the environment, and links us to each other, to other communities." Loss of ice is not just physical loss, but cultural loss, in which valuable knowledge and the connections that make those places what they are become undermined and melt with the ice. The passage from this local reality of ice to the global computations of climate change models is a complex traverse of distant knowledges, where the local is not included in the global, and travel between these two configurations is haphazard.[31] Serres comments: "Henceforth the global does not necessarily produce a local equivalent, and the local itself contains a law that does not always and everywhere reproduce the global."[32] He proposes we see, instead, regional epistemologies; "pockets of local orders in rising entropy."[33] Inuit sovereignty rising in Tar Sands.

## Knowledge and misfortune

[I]n dominating the planet, we become accountable for it.[34]

As hard science lays templates of the future – climate chaos or climate containment – for soft bodies and glacial flows, it is in the humanities that we must ask: What is a proper and humane exercising of this knowledge? How should this knowledge be tied to older knowledges? What are the consequences of what we know *for* life? And how shall others know it? What paths and passages of communication will take this knowledge into the world? How will it infect and reproduce in the knots that already bind us to the world? These are questions for the humanities, for understanding what a humane use of knowledge is, and how responsibilities configure in the making of worlds. The epic planetary scale of science's endeavor is an exercise in measurement, projection and production of new imaginaries, but not new relations *per se*. The present, which is the axis on which climate is flipped forward from using past climate measurements, is a stable blindspot in the equation. Imagining a more *active present*, attuned to the movement of time, and landscapes, and the possibility of politics, might not make such a rigid place from which to cast out scenarios of the future, based on the scope of contemporary configurations, but allow in the possibility of chance, of change, of radical difference. To understand this environment of fluctuating interferences, feedbacks, inscription and change we need a mobile thought that disregards the demands of the all-too-solid structures of academia and the accumulative processes of empiricism. This is part of faithfulness to the state of things, to melting ice and rising shores, to displacement and to the possibilities of sovereignty. As the hot contagion of climate change reworks our landscapes, solidity and fixity are not a possibility of these relations. Different relations are brought into play that suggest the importance of being able to work with flux, chaos and change, and see something other than catastrophe and disaster.

We have systems to understand how heat moves across materials and across landscapes, but no system to understand how such heat knowledge inflames, feeds off the oxygen of sympathetic atmospheres, creates new politics, and finds its reception and feedback in communities of science. It is heat knowledge that has brought science to this place, or it is the burden, the weight of that knowledge as a possible future; a relation loose in the world, unbound from observing machines, wrenched through people's lives and steeped in the sorrow of the violence to come. It is not the science *per se* that does this work – that steers the imperative toward the political – but it is the force of the fracturing power of this knowledge; it is a knowledge that rewrites history and the future. Where this imperative falls short is that, in order to make this arc of past into future, the present must be static, a calcified place on which to plant a leg of the compass, to allow the arc. It is at this false point that the humanities can map out every relation, to reinstate the ground as an active present of growling ice that rubs and burns already in the long history that is already in this place, already active, and already working toward a myriad of futures.

## Acknowledgments

This chapter was first presented as a paper at the "Locating Technoscience" Seminar UCL and at "Polar Productions" BildMuseet, Umeå University, Sweden in September 2007. The ideas discussed here have benefited greatly from the on-going collaboration, "Zero Degrees" with Jennifer Gabrys.

# Notes

1  "MS: I called the book 'The North-West Passage' – you know the passage between the Atlantic Ocean and the Pacific, to the north of Canada, which is very difficult and complicated to negotiate – as an image for the passage between the humanities and the exact sciences. I think the job of philosophy is to open up this passage between the exact sciences and the humanities." M.l. Serres interviewed by R. Mortley, "French Philosophers in Conversation, Chapter III. Michel Serres", 1991, Faculty of Humanities and Social Sciences, Bond University, posted at ePublications@Bond University, http://epublications.bond.edu.au/french philosophers/4. For an extended discussion on critical thresholds see J. Gabrys and K. Yusoff (2011), Arts, Sciences and Climate Change: Practices and Politics at the Threshold", in *Science as Culture*. M. Serres, *Hermès V. Le passage du Nord-Ouest*, Minuit, 1980. English version: M. Serres, *Hermes. Literature, Science and Philosophy*, ed. J.V. Harari and D. Bell, Baltimore, MD: Johns Hopkins University Press, 1982.

2  M. Serres, *Conversations on Science, Culture, and Time*, with B. Latour, trans. R. Lapidus, Ann Arbor, MI: University of Michigan Press, 1995, p. 70.

3  Serres, *Conversations*, "Metaphor, in fact, means "transport," p. 66.

4  Serres, *Conversations*, p.107.

5  Serres, *Hermes*.

6  See S.D. Brown, "M. Serres: Science, Translation and the Logic of the Parasite," *Theory, Culture & Society* 19, no. 3, 2002, 1–27.

7  Serres, *Conversations*, p. 103.

8  Serres, *Conversations*, p.179–80.

9  Ibid.

10  Translator's introduction in Serres, *Parasite*, p. x.

11  M. Serres, *Parasite*, trans. L.R. Schehr, Minneapolis, MN: University of Minnesota Press, 2007, p. 191.

12  M. Serres, *Parasite*, trans. L.R. Schehr, Minneapolis, MN: University of Minnesota Press, 2007, p. 191.

13  For a discussion on the force of materials see J. Bennett, "The Force of Things: Steps Toward an Ecology of Matter," *Political Theory* 32, no. 3, 2004; and E. Grosz, *Time Travels Feminism, Nature, Power*, Durham, NC: Duke University Press, 2005.

14  http://www.itk.ca/circumpolar-inuit-declaration-arctic-sovereignty.

15  See C. Kwa, "Romantic and Baroque Conceptions of Complex Wholes in the Science," in J. Law, and A.-M. Mol (eds), *Complexities: Social Studies of Knowledge Practices*, Durham, NC: Duke University Press, 2002, p. 34.

16  Serres, *Parasite*, p. 7.

17  See www.crrel.usace.army.mil.

18  Steven Connor, discussing Serres, says: "If history is marked by the movements, not from element to element, but between different states of the same element, then time (temps), as Serres often takes pleasure in pointing out, becomes indistinguishable from temperature – or weather (temps)." S. Connor, *Topologies: Michel Serres and the Shapes of Thought*, http://www.bbk.ac.uk/english/skc/topologies/.

19  Serres, *Conversations*, p. 99.

20  A. Ronell, *The Test Drive*, Urbana and Chicago: University of Illinois Press, 2005, p. 15.

21  R. Alley, *The Two Mile Time Machine: Ice Cores, Abrupt Climate Change, and Our Future*, Princeton, NJ: Princeton University Press, 2000, p. 13.

22  Serres, *Parasite*, p. 210.

23  "This scene does not constitute science's outside but troubles its inner workings, pointing at times to a forgotten or displaced germ, a desire, an original need or sense of lack – what both Husserl and Nietzsche would agree belongs to the precincts of an original intensity, circumscribing that which in science is on the side of life. Even so, we are not concerned only uncovering an original vitality but with listening to the future that science portends. In this context, science might be regarded as a kind of questioning, a

structure of exposure that often forgets to turn back on itself in order to interrogate its vital impulses and philosophical point of origin." Ronell, *The Test Drive*, p. 9.

24 Serres: "we are fascinated by the unit; only the unit seems rational to us. We scorn the senses, because their information reaches us in bursts," Serres, *Hermes*, p. xx.

25 See K. Yusoff, "Excess, Catastrophe, and Climate Change," in *Society & Space: Environment and Planning D*, vol. 27, pp. 1010–29.

26 Serres, *Conversations*, p. 182.

27 Serres, *Conversations*, p. 181.

28 Arctic Council's AMAP *Arctic Oil and Gas 2007 Report*, p. 32.

29 D. Massey, "Geographies of Responsibility," 2004, http://en.scientificcommons.org/21338652.

30 Serres, *Hermes*, p. xiii.

31 Serres, *Hermes*, p. 75.

32 Ibid.

33 Serres, *Conversations*, p. 173.

34 Ibid.

# INDEX